ANISOTROPY IN SINGLE-CRYSTAL REFRACTORY COMPOUNDS

Volume 1

ANISOTROPY IN SINGLE-CRYSTAL REFRACTORY COMPOUNDS

Edited by
Fred W. Vahldiek and Stanley A. Mersol

VOLUME 1
- Crystal Structure and Chemistry
- Electronic Structure and Bonding
- Non-Stoichiometry and Phase Relations
- Physical-Chemical Properties

VOLUME 2
- Defect Structure
- Mechanical Properties

ANISOTROPY IN SINGLE-CRYSTAL REFRACTORY COMPOUNDS

Edited by
Fred W. Vahldiek and Stanley A. Mersol
Metals and Ceramics Division
Air Force Materials Laboratory
Wright Patterson Air Force Base, Ohio

Proceedings of an International Symposium on Anisotropy in Single-Crystal Refractory Compounds, held on June 13-15, 1967, in Dayton Ohio. Sponsored by the Ceramics and Graphite Branch of the Air Force Materials Laboratory, United States Air Force.

Volume 1

PLENUM PRESS • NEW YORK • 1968

Library of Congress Catalog Card Number 68-20273

A Division of Plenum Publishing Corporation
227 West 17 Street, New York, N. Y. 10011

Printed in the United States of America

FOREWORD

An International Symposium on Anisotropy in Single-Crystal Refractory Compounds, sponsored by the Ceramics and Graphite Branch, Air Force Materials Laboratory, Wright-Patterson Air Force Base, Ohio, in commemoration of the fiftieth anniversary of materials research in the United States Air Force, was held on 13-15 June 1967 at the Sheraton-Dayton Hotel, Dayton, Ohio. General chairman of the Symposium was Fred W. Vahldiek, assisted by the program chairmen Charles T. Lynch and Stanley A. Mersol, all three being materials research engineers from the Air Force Materials Laboratory. The Symposium coordinator was James M. Miller, University of Dayton Research Institute, who, assisted by his staff, handled the administrative details of the Symposium. Approximately one-hundred and twenty scientists and engineers, representing the United States, Belgium, Canada, France, India, Sweden, West Germany, and Great Britain attended the Symposium. Approximately forty invited papers were presented. An attempt was made to bring together scientists working in the area of anisotropy on single-crystal refractory compounds and in related areas in order to review the state-of-the-art in this important area of research and to insure widespread dissemination of recent findings and ideas.

The welcoming address, given by Alan M. Lovelace, Director of the Air Force Materials Laboratory, emphasized the role of the United States Air Force as the host at this Symposium and set the informal tone for the seven technical sessions of the Symposium. In his opening address, Harris M. Burte, Chief of the Metals and Ceramics Division, Air Force Materials Laboratory, reemphasized the need for proper characterization of materials, especially refractory compounds, and outlined the main reasons for holding this first International Symposium on Anisotropy in Single-Crystal Refractory Compounds. The purpose for holding the Symposium thus firmly crystalized, Fred W. Vahldiek, general chairman, then said a few introductory remarks, and the Symposium was in session.

An attempt was made at this Symposium to cover the principal parameters on anisotropy of refractory compounds relative to their structure, bonding, non-stoichiometry, and thermal and mechanical behavior. Special emphasis was placed on the role of imperfections in crystals, including dislocations, slip, twinning, and stacking faults, and correlating these with the anisotropy in properties observed. Materials primarily discussed were refractory oxides, carbides, borides, silicides, and nitrides of the transition metals. Several contributions necessitated, on fundamental grounds, the discussion of other metallic and nonmetallic materials in order to have a homogeneous understanding of anisotropy. Although many of the papers presented stimulated heated oral discussions, no attempt was made to record these, and thus they are not included in this Proceedings. In general, the questions asked dealt with controversial points, or were the type that required an explanation or further data from the respective speakers, and a relatively select few conferees were involved. It is presumed that many of these points of controversy and/or explanation are being further pursued by the scientists directly concerned.

The overall assessment of the Symposium could be stated as follows:

It was brought out during the course of the Symposium that there is a considerable gap at the moment in relating basic concepts, such as dislocations, electronic crystal structure, etc. to the bulk physical, chemical and mechanical properties of refractory compounds. Whereas considerable progress has been made in the last decade or so on the understanding of these compounds, the state-of-the-art as to the basic properties and behavior of these materials is still lagging far behind metallic systems. The key to many of the inherent problems associated with refractory compounds such as how to make them more ductile, how to increase their high-temperature insulation and thermal shock resistance properties, and how to improve the strength and load-carrying capabilities seems to lie in the controlled changes in the stoichiometry, impurity content, and especially the electronic structure of these materials. Much can be learned about the refractory compounds by referring to the behavior and properties of the respective "host" metals.

It was pointed out that in refractory inorganic nonmetallic or ceramic materials research the problems are often associated with the difficulty in obtaining suitable thin single-crystal specimens for study. Recent developments in transmission electron microscopy, electron microprobe, and microfocus x-ray techniques should allow a careful compositional and structure analysis as well as purity and homogeneity control of refractory compounds, such as are essential if single-crystal and polycrystalline refractories are to be meaningfully compared. Of course, much more work is needed on the physical, chemical, and mechanical properties of the refractory compounds.

Among the principal recommendations are the following: (1) Work in single-crystal refractory compounds such as exemplified by this Symposium is essential, and should be continued at an ever increasing rate if one is ever to understand the behavior and the inherent nature of these materials and how to properly modify them so that they may fulfill a useful function in the presently existent and newly developed aerospace and other systems. In this, fundamental studies, such as for example the understanding and changing of the basic electronic structure, play an increasingly important role. (2) Specialized symposia of this type should be held periodically - perhaps every two or three years - sponsored by the United States Air Force and/or by other interested agencies, in order to insure the best advancement of the state-of-the-art relative to the anisotropy in refractory materials research.

Special thanks are extended to all who substantially contributed to the success of this Symposium: Alan M. Lovelace, Director of the Air Force Materials Laboratory, for his welcoming address; Harris M. Burte, Chief of the Metals and Ceramics Division, Air Force Materials Laboratory, for his opening address; J. Allen Hynek, Director of Northwestern University's Dearborn Observatory and a long-time Air Force consultant on unidentified objects, who was the banquet speaker, for his interesting talk on the UFO's; Mrs. Anna G. Blackwell, Ceramics and Graphite Branch, Air Force Materials Laboratory, for her professional piano recital during the social hour preceding the banquet; James M. Miller, Symposium coordinator, Mrs. Audrey Sachs, Symposium secretary, University of Dayton Research Institute and staff for taking care of the administrative arrangements; Mrs. Jean Gwinn, University of Cincinnati, Symposium secretary and editorial assistant, for helping with setting up the Symposium and editing the manuscripts for the Proceedings, and for taking care of the numerous details associated with a Symposium of this type; and above all, to the conferees themselves who most of all had to do with the success of the Symposium.

The Proceedings of the Symposium are presented in two parts. Part I contains papers of Sessions I-IV, dealing with Crystal Structure and Chemistry, Electronic Structure and Bonding, Non-stoichiometry and Phase Relations, and Physical-Chemical Properties. Part II contains papers of Sessions V-VII, dealing with Defect Structure, and Mechanical Properties.

Dayton, Ohio
October 7, 1967

FRED W. VAHLDIEK
STANLEY A. MERSOL

CONTRIBUTORS

AMELINCKX, S.	Studiecentrum voor Kernenergie, Mol, Belgium
ARONSSON, BERTIL	Swedish Institute for Metal Research, Stockholm, Sweden
AUTIO, G. W.	Cornell University, Ithaca, New York
BERMAN, I.	Air Force Cambridge Research Laboratories, Bedford, Massachusetts
BERNSTEIN, H.	ManLabs Inc., Cambridge, Massachusetts
BLUMENTHAL, R. N.	Marquette University, Milwaukee, Wisconsin
BRITTAIN, J. O.	Northwestern University, Evanston, Illinois
BROOKES, C. A.	University of Bradford, Bradford, Yorkshire, Great Britain
BRUNO, G. W.	Advanced Metals Research Corporation, Burlington, Massachusetts
BUESSEM, W. R.	The Pennsylvania State University, University Park, Pennsylvania
CARTZ, L.	Marquette University, Milwaukee, Wisconsin
CHUNG, D. H.	Massachusetts Institute of Technology, Cambridge, Massachusetts
CONRAD, H.	Franklin Institute, Philadelphia, Pennsylvania
CONSIDINE, D. P.	Air Force Cambridge Research Laboratories, Bedford, Massachusetts
COSTA, P.	O. N. E. R. A., Chatillon-sous-Bagneux (Seine) France
DAMIANO, V.	Franklin Institute, Philadelphia, Pennsylvania

DAY, R.B.	Western Michigan University, Kalamazoo, Michigan
DELAVIGNETTE, P.	Studiecentrum voor Kernenergie, Mol, Belgium
ENGSTRÖM, INGVAR	University of Uppsala, Uppsala, Sweden
FLEISCHER, L.R.	Westinghouse Astronuclear Laboratory, Pittsburgh, Pennsylvania
FREISE, E.J.	Northwestern University, Evanston, Illinois
FRENCH, D.N.	Ingersoll-Rand Research Center, Princeton, New Jersey
GARDNER, W.	Texas Instruments, Dallas, Texas
GIBBONS, D.F.	Case Western Reserve University, Cleveland, Ohio
HARROD, D.L.	Westinghouse Astronuclear Laboratory, Pittsburgh, Pennsylvania
HASSELMAN, D.P.H.	Stanford Research Institute, Menlo Park, California
HAWLEY, J.J.	Air Force Cambridge Research Laboratories, Bedford, Massachusetts
HIRTHE, W.M.	Marquette University, Milwaukee, Wisconsin
HOCH, MICHAEL	University of Cincinnati, Cincinnati, Ohio
HOLLOX, G.E.	RIAS, Division of Martin Company, Baltimore, Maryland
HOSEMANN, R.	Fritz-Haber Institut, Berlin, Germany
HULSE, C.O.	United Aircraft Corporation, East Hartford, Connecticut
HYMAN, A.	University of Maryland, Baltimore, Maryland
JAIN, R.K.	National Physical Laboratory, New Delhi, India
JAIN, S.C.	National Physical Laboratory, New Delhi, India
JORGENSEN, P.J.	General Electric Company, Schenectady, New York

KAUFMAN, L.	ManLabs Inc., Cambridge, Massachusetts
LAUBACH, J.E.	Marquette University, Milwaukee, Wisconsin
LOBODA-ČAČKOVIĆ, J.	Institut "Ruder Bošković", Zagreb, Yugoslavia
LONDON, G.	Franklin Institute, Philadelphia, Pennsylvania
LUNDSTRÖM, TORSTEN	University of Uppsala, Uppsala, Sweden
LYE, R.G.	RIAS, Division of Martin Company, Baltimore, Maryland
LYNCH, C.T.	Air Force Materials Laboratory, Wright-Patterson Air Force Base, Dayton, Ohio
MARSHALL, R.C.	Air Force Cambridge Research Laboratories, Bedford, Massachusetts
MERSOL, S.A.	Air Force Materials Laboratory, Wright-Patterson Air Force Base, Dayton, Ohio
MOLL, SHELDON H.	Advanced Metals Research Corporation, Burlington, Massachusetts
MOORE, J.W.	E.I. Dupont Company, Wilmington, Delaware
MURRAY, M.J.	Cavendish Laboratory, Cambridge, England
NADEAU, J.S.	General Electric Company, Schenectady, New York
NEWEY, C.W.A.	Imperial College, London, England
O'NEILL, J.B.	University of Bradford, Bradford, Yorkshire, Great Britain
PALMOUR, HAYNE III	North Carolina State University, Raleigh, North Carolina
PINZ, B.A.	Marquette University, Milwaukee, Wisconsin
PONS, L.	University of Caen, Caen, France
RADFORD, K.C.	Imperial College, London, England
RYAN, C.E.	Air Force Cambridge Research Laboratories, Bedford, Massachusetts
SATO, Y.	University of Minnesota, Minneapolis, Minnesota

SAUER, R. W.	Northwestern University, Evanston, Illinois
SCALA, E.	Cornell University, Ithaca, New York
SMAKULA, A.	Massachusetts Institute of Technology, Cambridge, Massachusetts
STOKES, R. J.	Honeywell Corporate Research Center, Hopkins, Minnesota
STRAUMANIS, M. E.	University of Missouri, Rolla, Missouri
THOMAS, D. A.	Ingersoll-Rand Research Center, Princeton, New Jersey
TIGHE, N. J.	National Bureau of Standards, Washington, D. C.
TOTH, L. E.	University of Minnesota, Minneapolis, Minnesota
VAHLDIEK, F. W.	Air Force Materials Laboratory, Wright-Patterson Air Force Base, Dayton, Ohio
VAN VLACK, L. H.	The University of Michigan, Ann Arbor, Michigan
VENABLES, J. D.	RIAS, Division of Martin Company, Baltimore, Maryland
WESTBROOK, J. H.	General Electric Company, Schenectady, New York
WILKE, W.	Fritz-Haber Institut, Berlin, Germany
WILLIAMS, WENDELL S.	University of Illinois, Urbana, Illinois
WITTER, D. E.	North Carolina State University, Raleigh, North Carolina
ZBASNIK, J.	University of Minnesota, Minneapolis, Minnesota

CONTENTS
Volume 1

CONTENTS
Volume 2

SESSION VII

Mechanical Properties • PART 2

SESSION I

Crystal Structure and Chemistry

Chairman: M. E. Straumanis
University of Missouri
Rolla, Missouri

Co-Chairman: A. Smakula
Massachusetts Institute of Technology
Cambridge, Massachusetts

SOME ASPECTS OF THE CRYSTAL CHEMISTRY OF BORIDES, BORO-CARBIDES AND SILICIDES OF THE TRANSITION METALS

Bertil Aronsson

Swedish Institute for Metal Research
Stockholm, Sweden

and Torsten Lundström and Ingvar Engström

Institute of Chemistry
University of Uppsala
Uppsala, Sweden

Abstract

New results on Me_2B borides with the $CuAl_2$ structure and on boro-carbides with the $Cr_{23}C_6$ structure are presented. The results are discussed with particular emphasis on the occurrence of B-C substitution.

Detailed structural data on NbB_2, CrB_6, MoB_4, WB_2 and Ru_2B_3 are reported. Some general comments are given on the crystal chemistry and stoichiometry of borides with compositions in the range MeB-MeB_4. Their relationship to the refractory carbides is particularly considered.

The crystal structures of Rh_4Si_5 and Rh_3Si_4 are presented. They are closely related to the MnP structure.

* * *

It is the main purpose of this paper to present some new results on borides of transition metals. In particular we shall discuss features of the crystal chemistry with an emphasis on structural relationships between borides and carbides.

The first section is devoted to metal-rich borides and carbides in which the non-metal atoms have a square antiprismatic environment of metal atoms. In the second section borides of the IVth to VIIIth group transiton metals with compositions in the range MeB-MeB_4 will be discussed. The crystal chemistry will be presented and possible explanations for deviation from stoichiometry will be particularly considered. Summarizing comments on boron-carbon substitution are given in the third section. The fourth section, finally, deals with some new silicides which are structurally related to MnP.

Much information on borides will be found in a recent monograph[1] and when appropriate we shall refer to this source rather than to original publications.

1. Borides and boro-carbides crystallizing with the $CuAl_2$ and $Cr_{23}C_6$ structures

In most metallic borides the boron atoms are surrounded by a triangular prism of metal atoms with additional metal or non-metal neighbours outside the square faces of the prism. The nine neighbours are at the corners of a tetrakaidekahedron (fig.1). Among metal-rich phases the only borides with a different co-

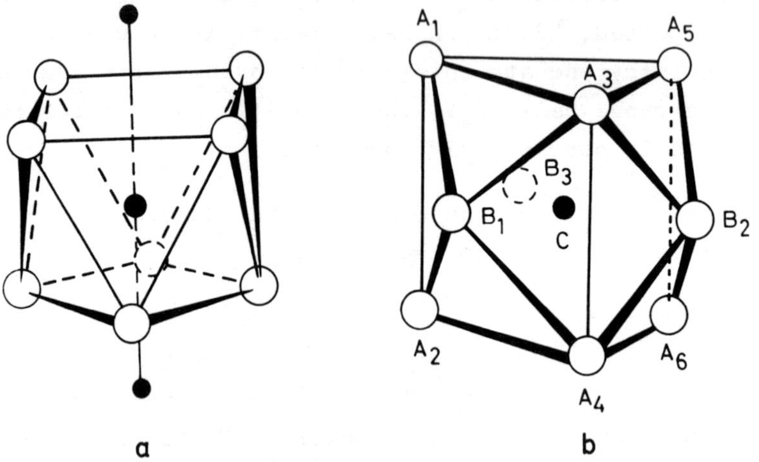

Fig. 1. Usual arrangements of neighbours around boron atoms. (a) A square (Archimedian) antiprism and (b) a tetrakaidekahedron. The B atoms may be metal or boron atoms.

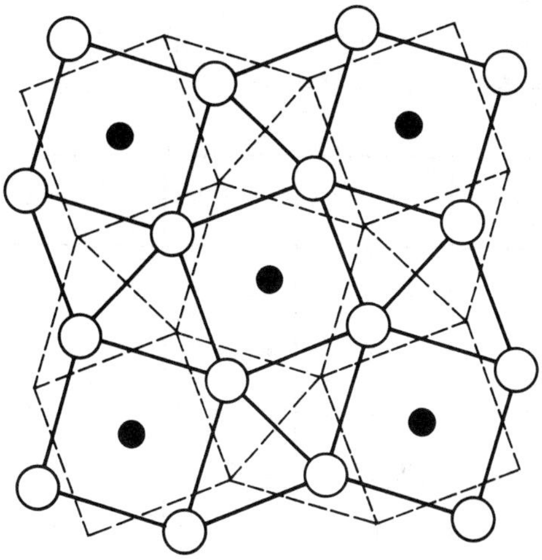

Fig. 2. A projection of the $CuAl_2$ structure along the tetragonal axis.

ordination around the boron atoms are those crystallizing with the $CuAl_2$, Mn_4B, Cr_5B_3 or $Cr_{23}C_6$ structure types. In these the close metal neighbours of the boron atom are situated at the corners of a square (Archimedian) antiprism (fig.1). The structures of Mn_4B and Cr_5B_3 are not very common and are closely related to $CuAl_2$; therefore they will not be further considered here. The $CuAl_2$ and $Cr_{23}C_6$ type structures will be described and recent results on borides and boro-carbides with these structures will be presented.

Me_2B phases crystallizing with the $CuAl_2$ structure

A Me_2B phase possessing the $CuAl_2$ structure is conveniently described as being built up of plane metal layers perpendicular to the tetragonal axis as shown in fig.2. There are two orientations (A and B) of the layers and the sequence is A-B-A-B... There is only one crystallographic position for the metal atoms. It has 15 neighbours in an arrangement often found in complex alloy phases: three close (metal) contacts have six neighbours in common with the central atom; the twelve more distant contacts have five neighbours in common with the central atom. The boron atoms are located at the centre of the square antiprismatic holes between the metal layers. The structure is characterized by a high symmetry, two parameters c/a and x_{Me} being sufficient for defining the ratio between any pair of interatomic distances in the structure.

TABLE 1

Crystallographic data on Me_2B borides

	W_2B	Mn_2B	Fe_2B	Co_2B	Ni_2B	Notes
a, Å	5.567	5.148	5.109	5.016	4.989	
c, Å	4.744	4.208	4.249	4.220	4.246	
c/a	0.8522	0.8174	0.8317	0.8265	0.8511	
V, Å^3	147.0	111.5	110.9	106.2	105.7	
x_{Me}	0.1693	0.1619	0.1661	0.1663	0.1684	Stand.dev. ±0.0002
R, %	7.0	5.4	6.3	5.0	4.8	Reliability index
B, Å^2	(-0.05)	0.21	0.23	0.22	0.27	Average isotropic temp. factor
			Interatomic distances (Å)			
Me-1Me(1)	2.666	2.357	2.400	2.360	2.377	
Me-2Me(2)	2.717	2.464	2.446	2.421	2.415	
Me-4Me(3)	2.980	2.729	2.694	2.645	2.624	
Me-4Me(4)	3.026	2.684	2.719	2.690	2.708	
Me-11Me(5)	2.920	2.631	2.631	2.595	2.594	
B-8Me	2.418	2.198	2.181	2.147	2.138	
B-2B	2.372	2.104	2.125	2.110	2.123	

(1) Atoms in the same z-level, having 6 metal neighbours in common

(2) Atoms at different z-levels having 4 metal and 2 boron atoms as common neighbours

(3) Atoms in the same z-level

(4) Atoms in different z-levels

(5) Average Me-Me distance

Crystallographic data for a number of Me_2B borides with the $CuAl_2$ structure are collected in table 1. The lattice parameters have been obtained with an accuracy better than 0.05% from powder photographs using silicon ($\underline{a}$=5.43054Å) as an internal standard; they do not deviate significantly from values reported previously. The determination of the atomic parameter was based on single crystal investigations using Weissenberg cameras and Mo$\underline{K}\alpha$ radiation. Chemical analyses of fairly pure samples (>99%) prepared by arc-melting showed that all the phases displayed in table 1 have narrow ranges of homogeneity and a composition closely corresponding to the formula Me_2B.

When this investigation was started it was hoped to find some systematic trend in the interatomic distances. However, one has to consider very subtle details in order to observe any such trends in the values shown in the table. In particular one does not observe any tendency towards shorter Me-B distances with increasing group number of the metal, as has been found[1] among other borides. Thus our results support earlier statements[2,3] that the $CuAl_2$ structure is essentially a size-factor structure.

The Me-B distances are fairly insensitive to changes in $\underline{x}_{Me}$ and $\underline{c}/\underline{a}$; the large shifts in these parameters necessary for any substantial changes in the Me-B distances to occur would imply the formation of very unlikely Me-Me contacts. Probably the parameters $\underline{c}/\underline{a}$ and $\underline{x}_{Me}$ are mainly determined by the requirements of the short Me-Me contacts. It is perhaps not unreasonable to say that the stability of the Me_2B borides mainly depends on a favourable arrangement of the metal atoms and that they would not exist if the Me-B interactions were very critical. The antiprismatic holes in the metal skeleton have a size permitting relatively favourable Me-B interactions, but the structure is not adjusted so as to optimize these interactions.

In view of the inflexibility of the $CuAl_2$ structure it is not surprising that no carbides with this structure exist. It is not possible to adjust the parameters so as to make the antiprismatic hole sufficiently small for favourable C-Me contacts to be formed without making some intermetallic contacts unlikely short. According to available results on Me-C-B systems the solubility of carbon in the Me_2B phases is negligible. In a recent study[4] of the Mo-C-B system, for instance, the lattice parameters of Mo_2B were the same in all samples containing this phase.

The Me_2B phases just described may serve as a good example of borides structurally closely related to silicides and complex alloy phases, amongst which the $CuAl_2$ structure is quite common. Borides of this type show little resemblance to the carbides; nor has boron-carbon substitution been observed among them.

$Me_{23}X_6$ (X=B,C) phases crystallizing with the $Cr_{23}C_6$ structure

The $Cr_{23}C_6$ structure is more complicated than that of $CuAl_2$. As shown by fig. 3, taken from Westgren's original article,[5] it contains no less than four crystallographically different metal atoms. The Cr_{III} atoms are at the corners of a cubo-octahedron, in the center of which a Cr_I atom is located. The Cr_{IV} atoms form cubes which are bonded together by Cr_{II} atoms situated at the extension of the long diagonals of the Cr_{IV} cubes. This description is of course merely one of convenience and does not imply that bonding within any of these structural units is different from that between atoms in adjacent units. Similar to the boron atom in the Me_2B phases recently described, the carbon atoms are located in square antiprismatic holes of the metal skeleton. In addition, there are relatively large holes in the Cr_{IV} cubes. The centre of these holes corresponds to a 4(b)-position with coordinates of the type $\frac{1}{2}$,0,0. In phases containing more than stoichiometric amounts of X atoms these holes may be partially occupied. (In passing it may be noted that the determination of the $Cr_{23}C_6$ structure dates from 1933. In view of the practical importance of phases crystallizing with this structure a more accurate redetermination appears overdue.)

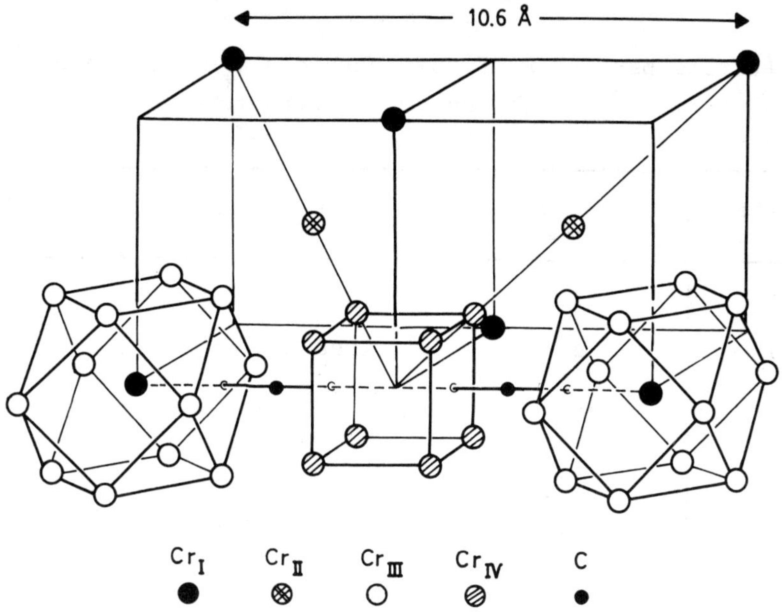

Fig. 3. The $Cr_{23}C_6$ structure. A description of the structure is best achieved by dividing the unit cell into eight small cubes (two of which are shown). The Cr_I atoms and the surrounding cubo-octahedron of Cr_{III} atoms are situated at every second corner of the small cubes, while the Cr_{IV} atoms are arranged around the other corners. The Cr_{II} atoms are at the centre of the small cubes.

Of the different metal atoms, Cr_{II} has 16 neighbours in an arrangement common in complex alloy structures (e.g. the Laves-phases). The surroundings of the other metal atoms rather resemble those found in the metallic elements. The large number of metal positions makes the $Cr_{23}C_6$ structure adaptable to a variety of compositions, and it occurs in a large number of Me'-Me''-C and Me'-Me''-B ternary systems.[6-11]

As already shown by Westgren[5] $Fe_{21}W_2C_6$ crystallizes with the $Cr_{23}C_6$ structure. A number of other $(Me'Me'')_{23}X_6$ (X=C,B) phases have recently been reported by Stadelmaier and collaborators,[6,7] by Nowotny and his school,[8] and also by Kus'ma and collaborators.[9-11] The composition often corresponds closely to the formula $Me'_{21}Me_2''X_6$ with Me' mostly Co and Ni (occasionally Fe) and Me'' a transition metal from the IVth to VIth group or a non-transition metal. It is most probable that the transition Me'' atoms mainly occupy the Me_{II} position, since

TABLE 2

Lattice parameter of some carbides and boro-carbides with the $Cr_{23}C_6$ structure

Approximate composition	Lattice parameter (Å)	Heat treatment	Ref.
$Cr_{23}C_6$	10.60	1000°C/7 days	12
$Cr_{23}(C,B)_6$	10.67	1000°C/7 days	12
$Mn_{23}C_6$	10.59	950°C/5 days	12
$Mn_{23}C_{\sim4}B_{\sim2}$	10.65	950°C/5 days	12
$Fe_{23}C_3B_3$	{10.59 10.63	(carbon-rich) (boron-rich)	6 6
$Fe_{21}Mo_2C_6$	10.63	700°C/10 days	4
$Fe_{22}Mo(C,B)_6$	10.57	700°C/10 days	4

they are commonly located in positions with a similar surrounding in complex alloy phases. Predictions about the location of non-transition Me'' metals are more uncertain. Sometimes very large homogeneity ranges are found. Thus, in the Re-Fe-B system[8] the composition varies from $(Re_{0.9}Fe_{0.1})_{23}B_6$ to $(Re_{0.4}Fe_{0.6})_{23}B_6$.

Information concerning $Me_{23}(C,B)_6$ phases is more scanty. The reported phases are assembled in table 2. The boro-carbide $Fe_{23}B_3C_3$ only exists at ternary compositions,[6] but in the Mn-C-B and Cr-C-B systems[12] the homogeneity range of the isostructural phase includes the binary composition $Me_{23}C_6$. In all these phases a considerable B-C substitution may take place.

The abundant occurrence of carbides and boro-carbides possessing the $Cr_{23}C_6$ structure is associated with the anti-prismatic 'hole' being considerably smaller than in the $CuAl_2$ structure. However, the number of independent parameters is very limited also in $Cr_{23}C_6$ and a large variation in the size of the antiprism is not possible without the formation of very unlikely intermetallic contacts. The fact that there are four different metal positions opens many possibilities for obtaining favourable combinations of interatomic interactions by changing the composition.

There is considerable evidence to show that the electron concentration plays an important role in the stability of the $Cr_{23}C_6$ structure. This has been particularly pointed out by Stadelmaier[6,7] and may be exemplified by the following observations: $Fe_{21}W_2C_6$ exists but not $Fe_{21}W_2B_6$, while increasing the number of electrons by replacing Fe by Co or W by Re results in the formation of a stable $Me_{23}B_6$ phase. A further example[4] is offered by the $Me_{23}X_6$ phase occurring in quenched and annealed molybdenum steels. Adding boron to these results in a change of the composition from $Fe_{21}Mo_2C_6$ to approximately $Fe_{22}Mo(C,B)_6$. (These compositions were obtained by electron probe analyses of precipitates in a Fe-2%Mo-0.2%C steel.) In the last example the critical electron concentration is maintained by an accompaning change in metal composition on substitution of boron for carbon so as to compensate for the decrease in the number of valence electrons.

In addition to the $Me_{23}X_6$ phases only a few examples of B-C substitution have been reported. In cementite, Fe_3C, a large fraction of carbon can be replaced by boron. Similarly in Co_3B, isostructural with Fe_3C, carbon can replace up to 20% of the boron atoms.[13] However, the isomorphous phase Ni_3B does not dissolve carbon. In most metallic phases replacement of boron by carbon and vice verse only occurs to a very limited extent. This was exemplified in a recent study[4] of the Mo-B-C system where no significant B-C substitution was observed in any of the metal-rich phases.

2. Borides with compositions in the range MeB-MeB_6

Some new experimental results

The composition and the lattice parameters of the phases investigated are assembled in table 3. They were prepared by

TABLE 3

Lattice parameters of the phases NbB_2, CrB_6, MoB_4, WB_2 and Ru_2B_3

Phase	Structure type	Lattice parameters (Å)	Annealing treatment
NbB_2	AlB_2	{ a=3.115$_0$, c=3.264$_5$ +) a=3.085$_3$, c=3.306$_2$ ++)	2000°C/1 day 2000°C/1 day
CrB_6	-	a=5.478, b=4.744, c=2.865	1350°C/3days
MoB_4	(WB_4)	a=6.345, c=5.203	1400°C/3 days
WB_2	W_2B_5	a=2.985, c=13.878	(arc-melted)
Ru_2B_3	W_2B_5	a=2.904, c=12.812	(arc-melted)

+) boron-poor (about 67 at% boron)

++) boron-rich (about 72 at% boron)

arc-melting mixtures of the elements. Subsequent annealing was carried out in a tantalum-tube vacuum furnace. Lattice parameters were calculated from powder photographs obtained in cameras of the Guinier-Hägg and Debye-Scherrer types using CuKα- and CrKα radiation. Single crystals of WB_2 and Ru_2B_3 were studied with MoKα and AgKα radiation in a goniostath. The results are summarized in table 4. More complete information will be found in ref. 14.

It has been corroborated that NbB_2 has an extensive range of homogeneity. Within the accuracy of measurements the lattice parameters of this boride vary in a linear way with composition, the a-axis decreasing and the c-axis increasing with increasing boron content. As yet, no results from accurate chemical analyses are available, but it appears that the homogeneity range lies mainly on the boron-rich side of NbB_2 - from NbB_2 (67 at% boron) to $NbB_{2.5}$ (72 at% boron) at 2000°C. At the metal-rich compositions large single crystals were easy to obtain; at the other end of the homogeneity range the crystals were invariably very small (~5 μm) and seemed much more brittle.

The chromium boride CrB_6 could only be prepared by prolonged annealing in the temperature range 1350-1400°C. The $\underline{d}$-values of the strong powder lines showed that the phase prepared by us is identical with the earlier reported phase with the same composition (see ref. 1). However, complete indexing of the powder photograph showed that the symmetry is

TABLE 4

Crystallographic data for Ru_2B_3 and WB_2

(W_2B_5-structure, space group $\underline{P}6_3/\underline{mmc}$, No 194)

	Position	Ru_2B_3	WB_2
$\underline{x}_{Me}$	4($\underline{f}$)	0.1397_4 ++)	0.1375_7
$\underline{B}_{Me}$+), Å^2		0.33	0.09
$\underline{x}_{B(1)}$	4($\underline{f}$)	0.0311	0.0235
$\underline{B}_{B(1)}$		0.76	0.21
$\underline{B}_{B(2)}$	2($\underline{d}$)	0.31	0.40
$\underline{B}_{B(3)}$	2($\underline{b}$)	-	0.78
No of reflexions		83	278
$\underline{R}$, %		2.8	5.9

+) Individual isotropic temperature factor

++) Standard deviation of coordinates of metal atom ±0.0001
" " " " " B(1) " ±0.001

Note: More complete information will be found in ref. 14.

orthorhombic and not tetragonal as claimed earlier. Small variations of the lattice parameters indicate the existence of a range of homogeneity. The exact composition is still unknown.

The boride MoB_4 was present in arc-melted samples. Results of preliminary X-ray studies strongly indicate that it is isomorphous with WB_4. As with this latter phase[15] a narrow range of homogeneity has been observed which probably extends to the boron-rich side of the composition MeB_4.

Accurate X-ray investigations of WB_2 ("W_2B_5") and Ru_2B_3 ("$RuB_{1.5}$") were undertaken in order to solve the longstanding question concerning the distribution of boron in these structures. The results (assembled in table 4) show beyond doubt that there are no boron atoms in the 2 (a) positions as suggested previously (see ref.1) for the structure of "W_2B_5". In Ru_2B_3 the 2 (b) positions are also vacant. Thus, the structural composition of the phases is WB_2 and Ru_2B_3 and these formulae will be used in the following. WB_2 has a narrow range of homogeneity whereas accurate chemical analyses showed the composition of Ru_2B_3 to be $RuB_{1.50 \pm 0.02}$ without noticeable variations in the composition.

Crystal chemistry and stoichiometry of borides in the composition range MeB-MeB_4

The crystal structures of many borides in the composition range 50 to 80 atomic % boron are conveniently described by using the structural elements shown in fig.4: closepacked layers of metal atoms (A,B,C), plane (H,H,') or puckered (K,K') layers of boron atoms. In the AlB_2 structure the stacking sequence is AHAH... . The accurate structure determination of WB_2 and Ru_2B_3 reported in the previous paragraph has confirmed that the stacking sequence in these phases is AHAK'BHBK'... and AH'AK'BH'BK', respectively. Thus, no K-layers occur in these borides, as previously suggested for WB_2 and also ReB_2 and Mo_2B_5. The structures of AlB_2, Ru_2B_3 and WB_2 are illustrated in fig. 5.

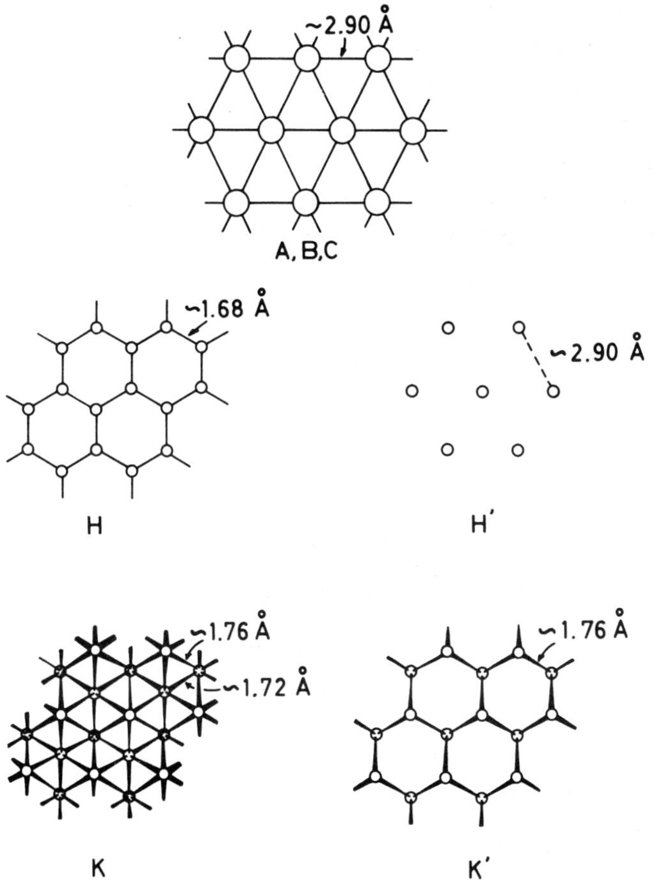

Fig. 4. Structural elements found in boron-rich borides.

The structure proposed for WB_4 by Romans and Krug[15] is also easily visualized by using the structural elements of fig.4. If, in the AlB_2 structure, one third of the metal atoms is replaced by a pair of boron atoms, the structure proposed for WB_4 is obtained. (The line connecting the boron atoms is parallel to the hexagonal axis. The mid-point of the line is at the previous metal position.) In this way adjacent H sheets of boron atoms are connected by a B_2 pair, which is situated above the centres of the open hexagons of the H-layers. Romans and Krug[15] have suggested atomic parameters so as to make the distance within a B-pair 1.73Å and that between a "pair-atom" and the six boron neighbours in the nearest H sheet 1.88Å. For reasons given below we consider these distances quite reasonable.

The structure proposed for WB_4 suggests a possible explanation for the extended homogeneity ranges exhibited by NbB_2 and TaB_2: metal atoms (up to 10%) are replaced by pairs of boron atoms. This is consistent with the observations that $\underline{c}/\underline{a}$ increases with increasing boron concentration and that the homogeneity range extends chiefly on the boron-rich side of the composition MeB_2. A similar mechanism might explain the composition found for Mo_2B_5. This phase has the same metal frame-work as WB_2 but is characterized by a higher boron concentration.

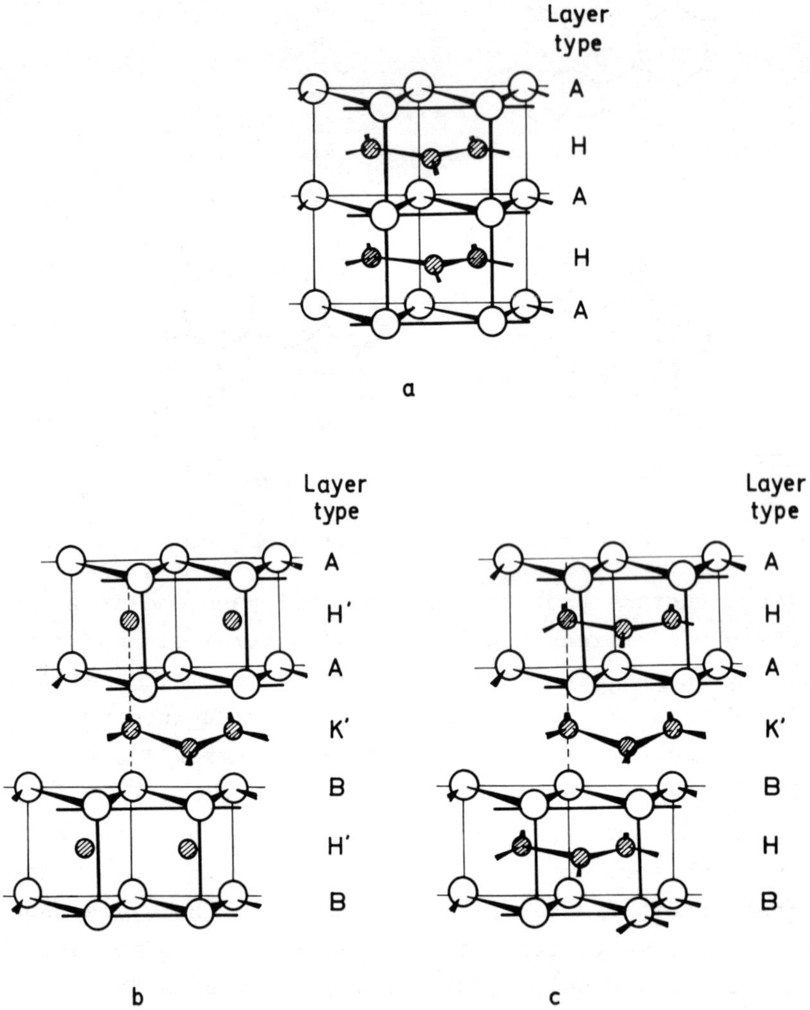

Fig. 5. The crystal structures of (a) AlB_2, (b) Ru_2B_3, and (c) WB_2.

When discussing the crystal chemistry and stoichiometry of borides two general observations are useful.

(1) Each boron atom has only five boron neighbours within a distance of 1.7-1.8Å. Further boron-boron contacts are longer than 1.9-2.0 Å. (The only exception is the recently determined AlB_{10} structure.[16])

(2) With increasing group number of the metal, Me-B distances tend to become shorter, whereas Me-Me and B-B distances become longer.

The structure proposed for WB_4 as well as the explanations of the extended homogeneity ranges of NbB_2, WB_2 and Mo_2B_5 are consistent with the first observation. Accordingly, they appear more likely than earlier suggestions. In particular, it does not seem probable that K-layers of boron exist at all. For WB_2 this has been shown by the experimental results quoted above; the high boron content of Mo_2B_5 may depend on a Mo-B_2 "substitution" of the type envisaged for NbB_2 rather than on the presence of K-layers as claimed earlier.

Observation (2) rationalizes a number of results concerning the composition of borides. A comparison of WB_2 and Ru_2B_3 affords a good example. In both phases the metal frame-work is the same. As expected, the Me-B contacts are considerably shorter in Ru_2B_3 (and the isomorphous boride Os_2B_3) than in WB_2. However, this shortening is not possible in the WB_2 structure without causing some B-B contacts to become very close. By replacing the boron H sheets in WB_2 with H' sheets in Ru_2B_3 the short B-B contacts are avoided. Analoguously MoB_2 (AlB_2-structure) contains H layers but RuB (with the same metal arrangement) contains H' layers.

Some comparisons with carbides

There are few similarities between carbides and borides in the composition range under discussion. In the metallic carbides no short C-C contacts are present, whereas short B-B distances are a common feature of boride structures. There are numerous examples of the reluctance of carbon atoms to form close bonds with other non-metal atoms in metallic phases. The MeC phases crystallize with the NaCl or WC structures and not with any of the MeB or the anti-NiAs structures in which short X-X contacts are present. The metal frame-work of many Me_2C phases is the same as that of "anti-NiAs" but the non-metal content is sufficiently low for close C-C contacts to be avoided.

Two phases in which B-C substitution might be expected are WC and RuB which both prossess the WC structure (stacking sequence AH'AH'.. with the notations of fig.3). In these the environment of the non-metal is triangular prismatic (as in most borides) and no short X-X contacts are present. The experimental results on Me-C-B (Me=Mo,W,Ru,Os) are scanty. An investigation[17] on the quasi-binary system W_2B_5-WC did not indicate any B-C substitution but further studies appear necessary before any definite conclusions should be drawn.

3. Concluding remarks on B-C substitution in metallic phases

As evident from the preceeding sections substitution of boron for carbon and vice versa is not as common in metallic phases as one might expect in view of the small difference in atomic size. Apparently, the individual electronic properties of the elements impose stricter conditions for substitution to occur than is generally the case for metallic elements.

According to available results boron and carbon replace each other to a large extent only in $Me_{23}X_6$ phases in which Me is mainly Cr,Mn,Fe,Co or Ni, and in cementite-type phases Me_3X with Me=Fe or Co. We have no explanation for these observations. They may be associated with the trend (mentioned earlier) towards shorter Me-B contacts with increasing group number of the metal. Because of this, boron has a size quite close to that of carbon in phases of the iron group metals. However, as already indicated electronic factors certainly have a decisive influence.

With the above exceptions, B-C substitutions is quite restricted in metallic phases. It is of particular interest to consider the refractory borides and carbides of the IVth to VIth group transition metals. The solubility of carbon in MeB_2 phases is negligible; this is consistent with the general observation that carbon does not form close contacts with non-metal atoms in the phases under discussion. One expects a fortiori that carbon does not dissolve to a significant extent in the boron-rich phases described in the previous section. Boron, on the other hand, can replace a fraction of the carbon atoms in the refractory MeC carbides, although the octahedral environment of the non-metal atoms in MeC is not exhibited by borides (the earlier reported NaCl-type phases TiB, ZrB and HfB do not exist).

Thus, there appear to be considerable possibilities to modify the properties of MeX carbides (and also nitrides) by alloying with boron. This element can be in solid solution (at high temperatures) as well as precipitated as borides. In this context the interesting paper[18] on precipitation of TiB_2 in TiC should be mentioned. Additions of carbon to the MeB_2 borides, on the other hand, are not likely to cause similar changes in microstructure and properties. The refractory diborides are stable in the presence of graphite - a property that may be useful in many applications.

4. The crystal structure of Rh_4Si_5 and Rh_3Si_4 and their relation to the MnP structure family (by Ingvar Engström)

During the study of the phase relationships in the Rh-Si system, single crystals of the compositions Rh_4Si_5 and Rh_3Si_4 have been synthesized. The crystal structures of these compounds have been determined and the present communication gives an account of the structures and their relation to structure types already known.

Results of the structure determinations

Rh_4Si_5. The crystals are monoclinic with a = 12.335Å, b = 3.508Å, c = 5.924Å, β = 100°.181 and Z = 2. The space-

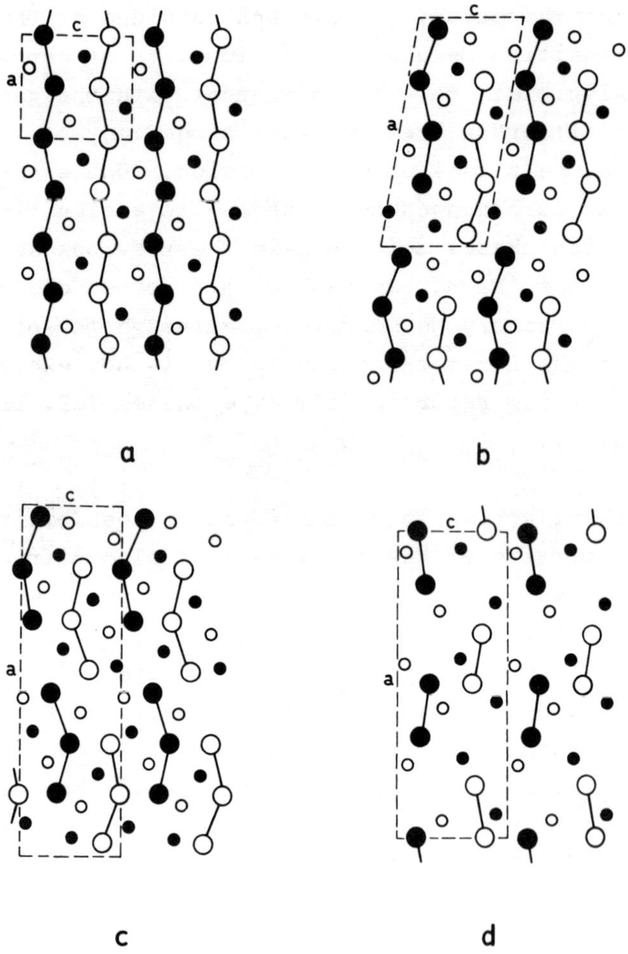

Fig. 6. Projections of the crystal structures of (a) MnP, (b) Rh_4Si_5, (c) Rh_3Si_4, and (d) Pt_2Ge_3.

group is $P\ 2_1/m$ with all the atoms in the twofold position 2(c). The structure was determined and refined using 396 independent h0l-reflexions. In fig. 6b the structure is projected on the ac-plane.

Rh_3Si_4. The crystals are orthorhombic with a = 18.810Å, b = 3.614Å, c = 5.813Å and Z = 4. The space-group is Pnma with all the atoms in the fourfold position 4(c). The structure was determined and refined using 332 independent h0l-reflexions. In fig. 6c the structure is projected on the ac-plane.

Description and discussion of the structures

The structures are most simply described by comparison with the MnP structure (fig.6a). Metal atoms in the MnP-type structure form infinite parallel zig-zag chains, and if these are systematically broken down to shorter chains a structural series is formed where this chain length is different for different members of the series. Disruption of the chains can be either considered as due to the systematic creation of vacancies by removal of metal atoms, or as the result of a systematic replacement of metal atoms by non-metal atoms. This leads to an ordered structure, although in both cases the change results in a deformation of the original MnP structure.

In Rh_4Si_5 the metal atoms are arranged in four-membered chains. The structure contains layers having the MnP structure and separated from each other by silicon-rich layers. This deformation results in a displacement of adjacent MnP layers so that the structure formed is monoclinic.

In Rh_3Si_4 the length of the metal chains is reduced to three atoms. This results in thinner layers of MnP-structure type than in Rh_4Si_5, whilst the silicon-rich layer has the same appearance in both structures. Despite these deformations, the symmetry of the MnP-structure is retained.

A further step in the degradation of the chains results in a structure with only pairs of atoms. Pt_2Ge_3 is related to the MnP-structure type, as has been pointed out by Bhan and Schubert.[19] As evident from fig. 6d this structure can be considered as belonging to the same series as Rh_3Si_4 and Rh_4Si_5. In Pt_2Ge_3 the sheets with the MnP-arrangement have completely disappeared, and the structural arrangement is very reminiscent of that found in the silicon-rich layers of Rh_3Si_4 and Rh_4Si_5. Pt_2Ge_3 has the same symmetry as MnP.

Acknowledgments

The authors wish to thank Professors Gunnar Hägg and Roland Kiessling for their kind interest. Many thanks are due to Mrs. Gerd Bergman and Miss Sonja Asplund for invaluable aid in the preparation of the manuscript, and to Mr. Sven Andersson for permission to publish result on CrB_6.

References

1. B.Aronsson, T.Lundström and S.Rundqvist, "Borides, silicides and phosphides". Methuen & Co Ltd, 1965
2. F.Laves in "Theory of alloy phases" page 124. ASM Cleveland 1956
3. H.H.Stadelmaier and J.G.Avery, Z.Metallk. 56(1965)508
4. M.Aldén (University of Uppsala, Sweden). Private communication
5. A.Westgren, Jernkont.Ann. 88(1933)501
6. H.H.Stadelmaier and R.A.Gregg, Metall 17(1963)412
7. J.D.Schöbel and H.H.Stadelmaier, Metall 19(1965)715, 20(1966) 31
8. E.Ganglberger, H.Nowotny and F.Benesovsky, Monatsh.96(1965) 1144, 97(1966)101, 97(1966)494
9. Yu.B.Kuz'ma, M.V.Chepiga and A.M.Plakhina, Izv.Akad.Nauk SSSR Neorg.Mat. 2(1966)1218
10. Yu.B.Kuz'ma, O.V.Pich and R.V.Skolozdra, Izv.Akad.Nauk SSSR Neorg Mat.2(1966)1975
11. Yu.B.Kuz'ma, V.I.Lakh, Yu.V.Voroshilov and B.I.Stadnik, Dop.Akad.Nauk, Ukr.RSR, 1966 (nr 6)772
12. A.Hede (Swedish Institute for Metal Research) Private communication
13. S.Rundqvist, Acta Chem.Scand. 12(1958)658
14. T.Lundström, Acta Chem.Scand. To be published.
15. P.A.Romans and M.P.Krug, Acta Cryst. 20(1966)631
16. G.Will, Nature 212(1966)175
17. G.V.Samsonov, Zhur.Tekh.Fiz. 26(1956)716
18. W.S.Williams, Trans.AIME 236(1966)211
19. S.Bhan and K.Schubert, Z.Metallk. 51(1960)328

APPLICATION OF THE ELECTRON MICROPROBE TO THE STUDY OF REFRACTORY METAL CARBIDES, OXIDES, AND SILICIDES

Sheldon H. Moll and George W. Bruno

Advanced Metals Research Corporation
Burlington, Massachusetts

Abstract

The electron microprobe has been successfully applied to the study of single crystals of the refractory metal carbides, oxides and silicides. The compositional homogeneity, impurity distribution and stoichiometric range have been examined in both as-grown and thermally treated materials.

The necessary equipment and techniques for the point by point analysis of both the interstitial element (C, O...) and the heavy element (Mo, W, Ti....) will be described.

Particular applications to be discussed include the study of second phases in as-grown Mo_2C and WSi_2 single crystals and the analysis of decomposition and impurity reaction products in heat-treated crystals. Stoichiometric variations were also determined in TiC crystals by direct analysis of the local carbon content.

* * *

This research was supported by the AF Materials Laboratory under contracts AF 33(657)-10390 and AF 33(615)-3786.

Application of the electron microprobe to practical materials studies has become commonplace since the introduction of the first working instrument by R. Castaing in 1950. The electron probe is rapidly being recognized as an analytical tool matching in importance the optical and electron microscopes.

As a result of its ability to perform a point by point chemical analysis of a sample volume as small as 1-2 cubic microns, the microprobe has been useful in solving materials problems associated with compositional inhomogeneity, the presence of foreign inclusions or reacted surface layers, corrosion products, etc. A wide variety of materials such as metals, ceramics, glass, semi-conductors, organics, etc., have been examined.

The purpose of this paper is to review some typical applications of the electron microprobe to the study of the micro-

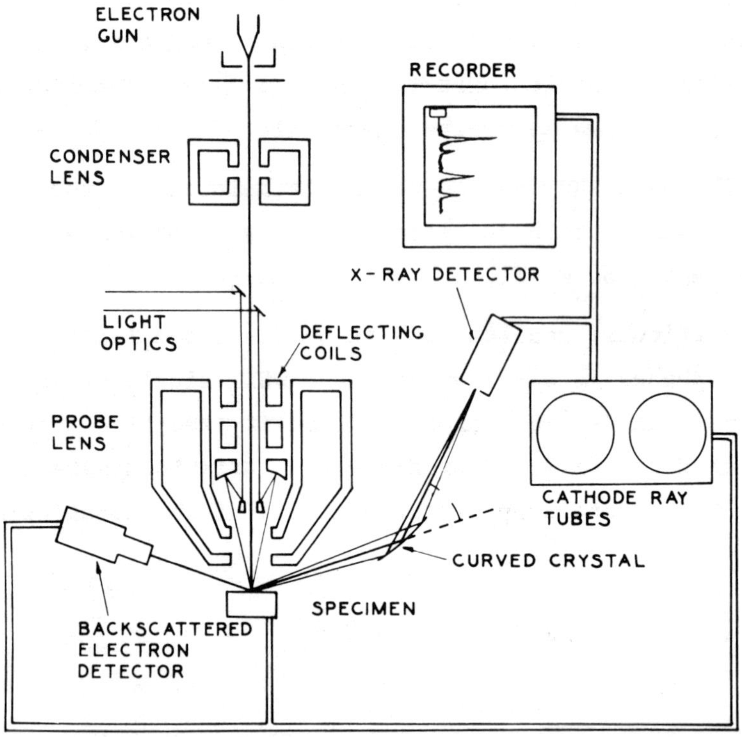

Fig. 1. Schematic drawing of the electron microprobe.

chemistry of single crystal materials. In particular, single crystals of Mo_2C, TiC and WSi_2 were analyzed.

All of the analyses to be discussed were performed with one of three AMR electron beam microanalyzers (Figure 1). A beam of electrons is accelerated with a high potential and focused by means of electro-magnetic lenses to a diameter of about 1 micron at the surface of the specimen to be investigated. As in an x-ray tube the specimen will act as a primary source of x-rays. A continuous or white spectrum is produced as well as fluorescent x-rays characteristic of the elements excited by the electron beam. A chemical analysis of the excited area is afforded by analysis of the characteristic x-ray lines by means of a single crystal x-ray spectrometer.

The original development of the electron microprobe allowed the analysis of all elements from Na to U in the periodic table. In the last year or two this analytical range has been extended to include the light elements B, C, N, O, F and Ne. Analysis of the light elements is an extremely important additional capability of the microprobe and it would be well to review briefly the analytical components required.

The K emission x-ray wavelengths available from the light elements range from 10 to 100A°. The two most important characteristics of these wavelengths pertaining to their detection in the analytical procedure are their high absorption in matter and the large atomic spacing required for crystal monochromators capable of diffracting them.

All of the light element emissions are readily absorbed in the detector windows, air or gas paths and detector gases. As a result, the x-ray spectrometer must employ a vacuum path and any windows placed in the x-ray path must be thin and low in average atomic number.

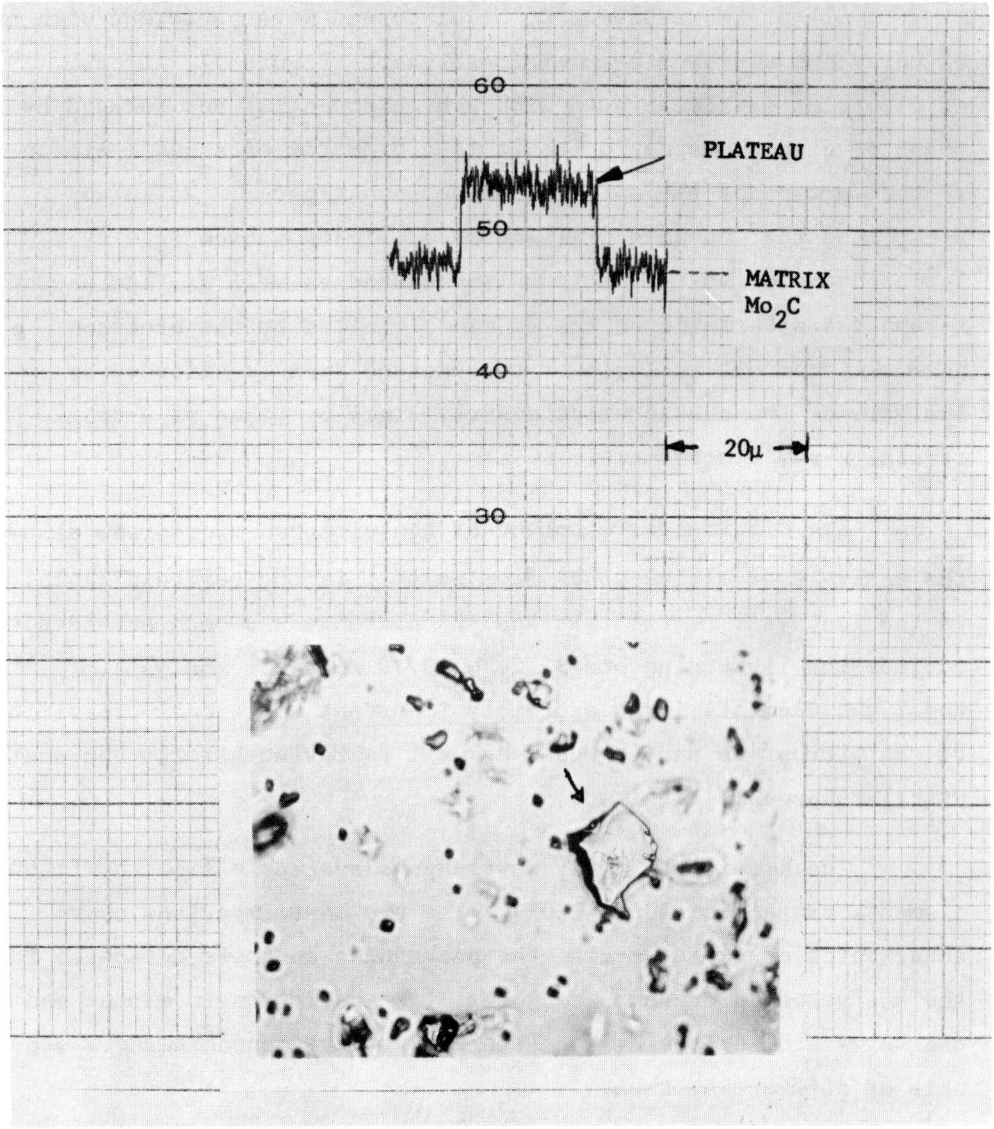

Fig. 2. Mo distribution, single crystal Mo_2C, C-1, plateau.

The monochromator crystal now utilized is an artificially formed lead stearate decanoate overlay formed on a suitable substrate.

The x-ray detector is a traditional gas flow proportional counter utilizing a thin formvar or collodion window to separate the detector gas from the vacuum x-ray path. Gas mixtures used include various methane/argon mixtures at atmospheric or less than atmospheric pressure.

Primary or secondary electrons back-scattered from the specimen may also follow the x-ray path and enter the detector producing a high background counting rate. This problem is eliminated by trapping the electrons in a magnetic field most easily generated with small permanent magnets.

Analysis of Mo_2C Single Crystals

The electron microprobe was first applied to the study of unusual features which appeared on the surface of Verneuil grown crystals. Specimens had been metallographically polished and lightly etched prior to vacuum annealing at 1800°C for two hours. The surface then exhibited small plateaus (Fig. 2) which appeared to have grown out of the surface. A Mo concentration distribution traverse revealed that the plateau was essentially pure Mo.

Many as-grown and annealed crystals exhibited an included microconstituent which was quite evident on metallographically prepared surfaces examined under polarized light (Fig. 3). As the Mo traverse reveals the phase is identical to the Mo_2C matrix in Mo content. No enrichment or depletion of C was observed. The "phase" is probably a small sub-grain with a high angle boundary at the interface.

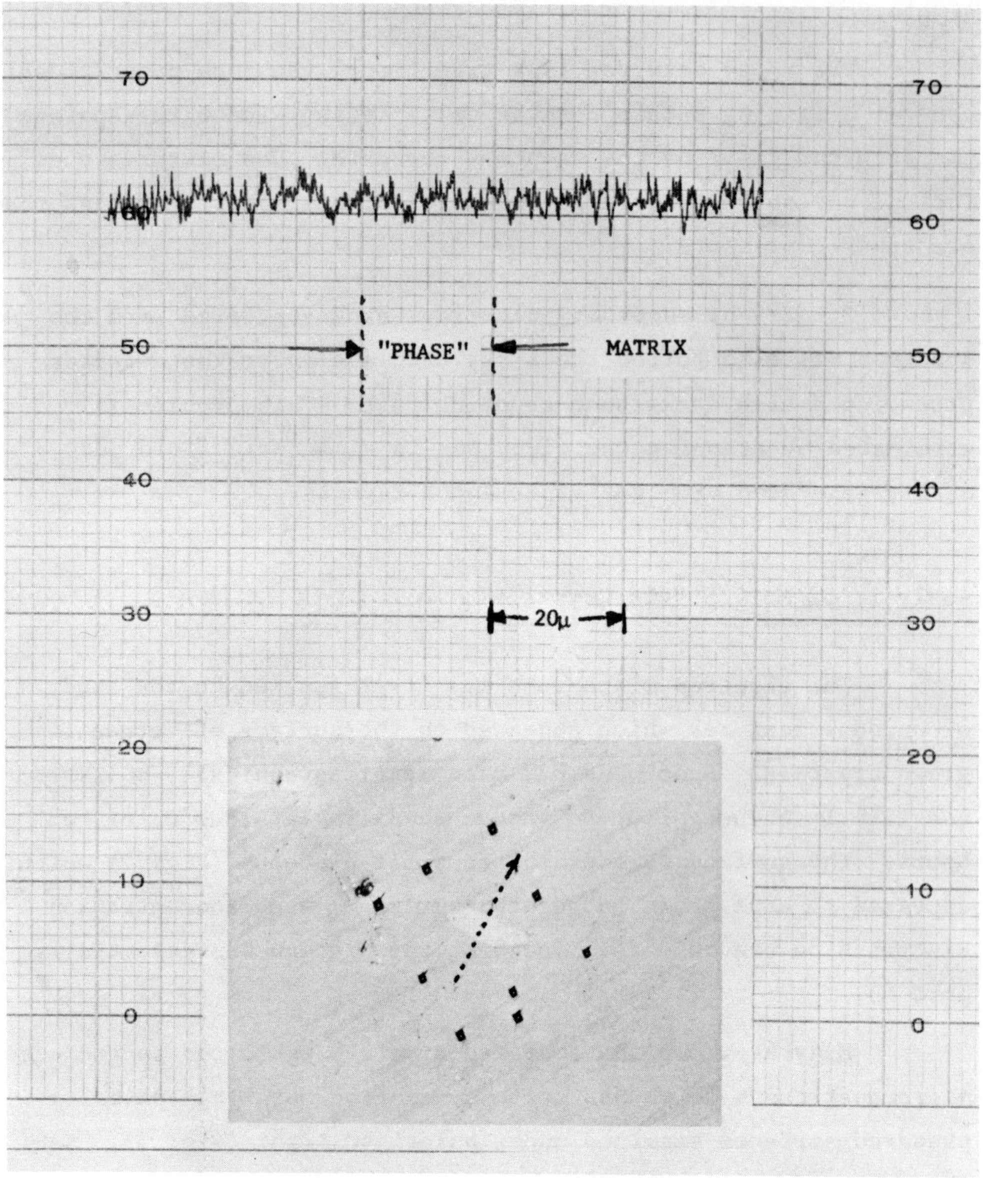

Fig. 3. Mo distribution, single crystal Mo_2C, C-2b, included "phase."

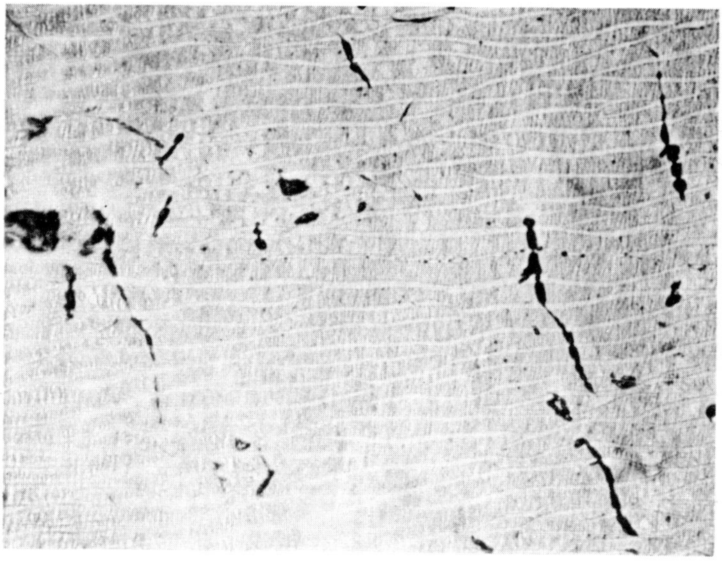

Fig. 4. Mo_2C crystal (C-9), longitudinal surface, 606 ×.

All as-grown and low temperature annealed crystals exhibited a curious veining structure typified in Figures 4, 5 and 6. The veins appear to orient themselves in the crystal growth direction (longitudinal). A series of discrete points along a

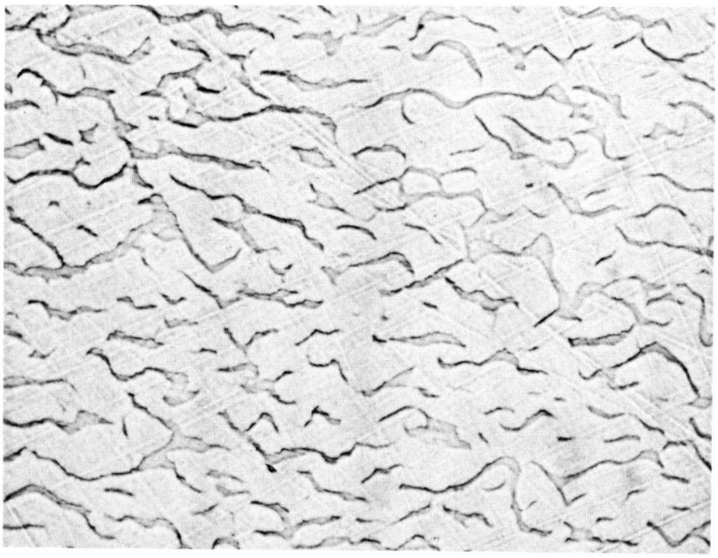

Fig. 5. Mo_2C crystal (C-9), transverse surface, 606 ×.

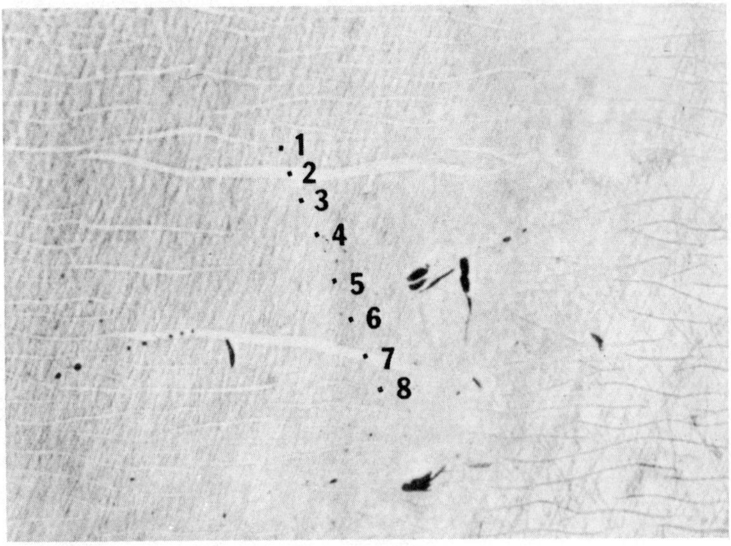

Fig. 6. Mo_2C crystal (C-9), longitudinal surface, 606 ×.

Table I

Vacuum Annealed Mo_2C Crystal
Concentration Distribution

Position	Mo	C (w/o)
1	94.0	6.1
2 (vein)	90.9	9.0
3	94.1	6.0
4	94.2	6.0
5	94.0	5.9
6	93.8	6.0
7 (vein)	90.8	8.9
8	94.5	5.9

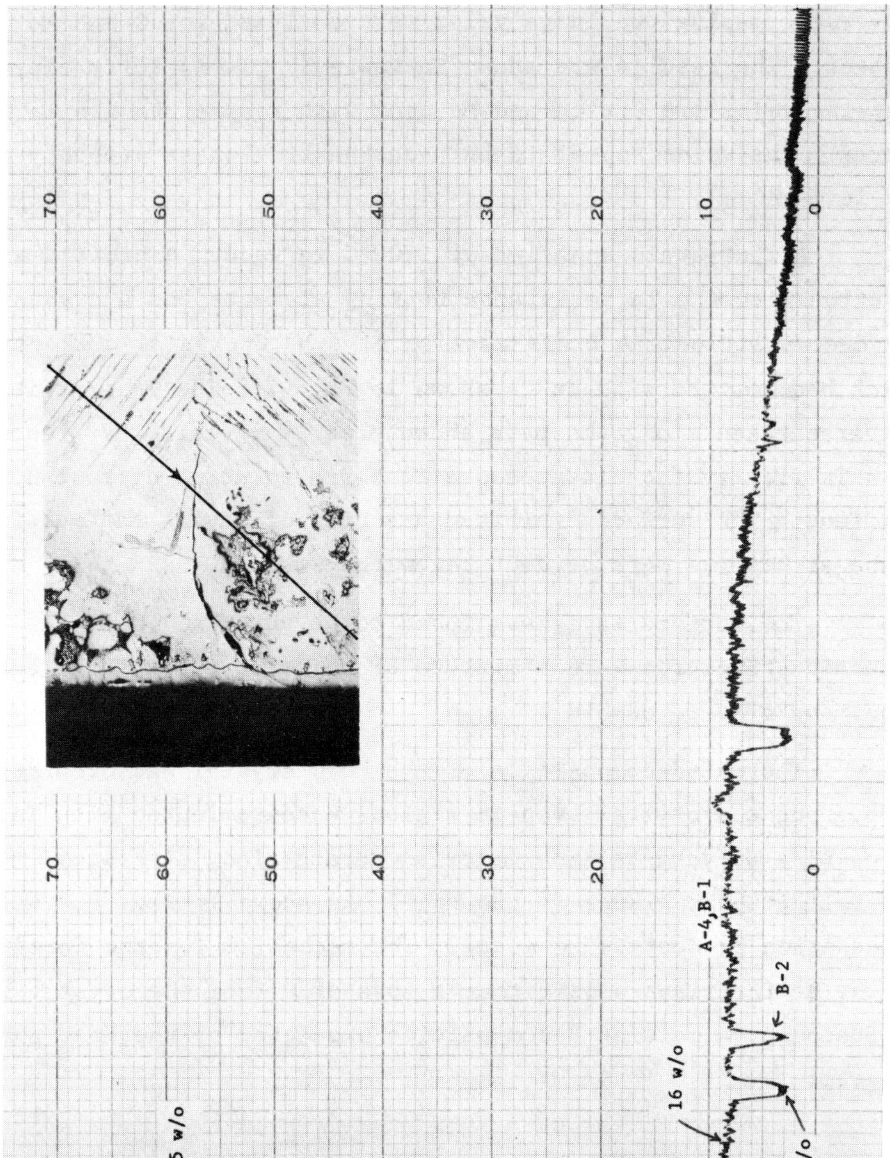

Fig. 7. Ta distribution, single crystal Mo_2C, C-8 (CS), edge.

traverse crossing two large veins were analyzed for C and Mo content. The results are shown in Table I. It is quite clear that the veins are C rich and Mo poor with respect to the Mo_2C matrix. The veins appear to be a carbon rich phase probably of the form MoC_{1-x}.

All crystals annealed at 2200°C in vacuum exhibited a reaction with the Ta resistance heating elements and a crystalline decomposition. A cross-section of a Mo_2C crystal surface which had reacted with Ta is shown in Fig. 7. The Ta distribution traverse taken along the path shown clearly reveals the presence of a Ta rich surface layer and a Ta solid solution diffusion gradient. The surface exhibited a series of layers and multiphase structures were evident below the surface.

The surface layer was found to be pure TaC while the eutectic structures were a mixture of Mo metal (with Ta in solution) and a Mo-Ta carbide.

Thermal decomposition of the Mo_2C crystal had occurred throughout the crystal even in regions uncontaminated with Ta. Veins of a polycrystalline material formed along discrete crystallographic planes in the Mo_2C. As shown in Fig. 8, the decomposition product is Mo metal. The mechanism for the formation of the Mo interlayers is either a loss of C from the crystal by diffusion or a melting decomposition predicted by the Mo-C phase diagram.

Analysis of TiC Single Crystals

A large number of as-grown and also annealed TiC single crystals were examined for evidence of compositional inhomogeneity, impurities, surface deposits, etc. In general, the single crystals were found to be uniform in chemistry with no detectable impurities (< 0.05 w/o for most elements).

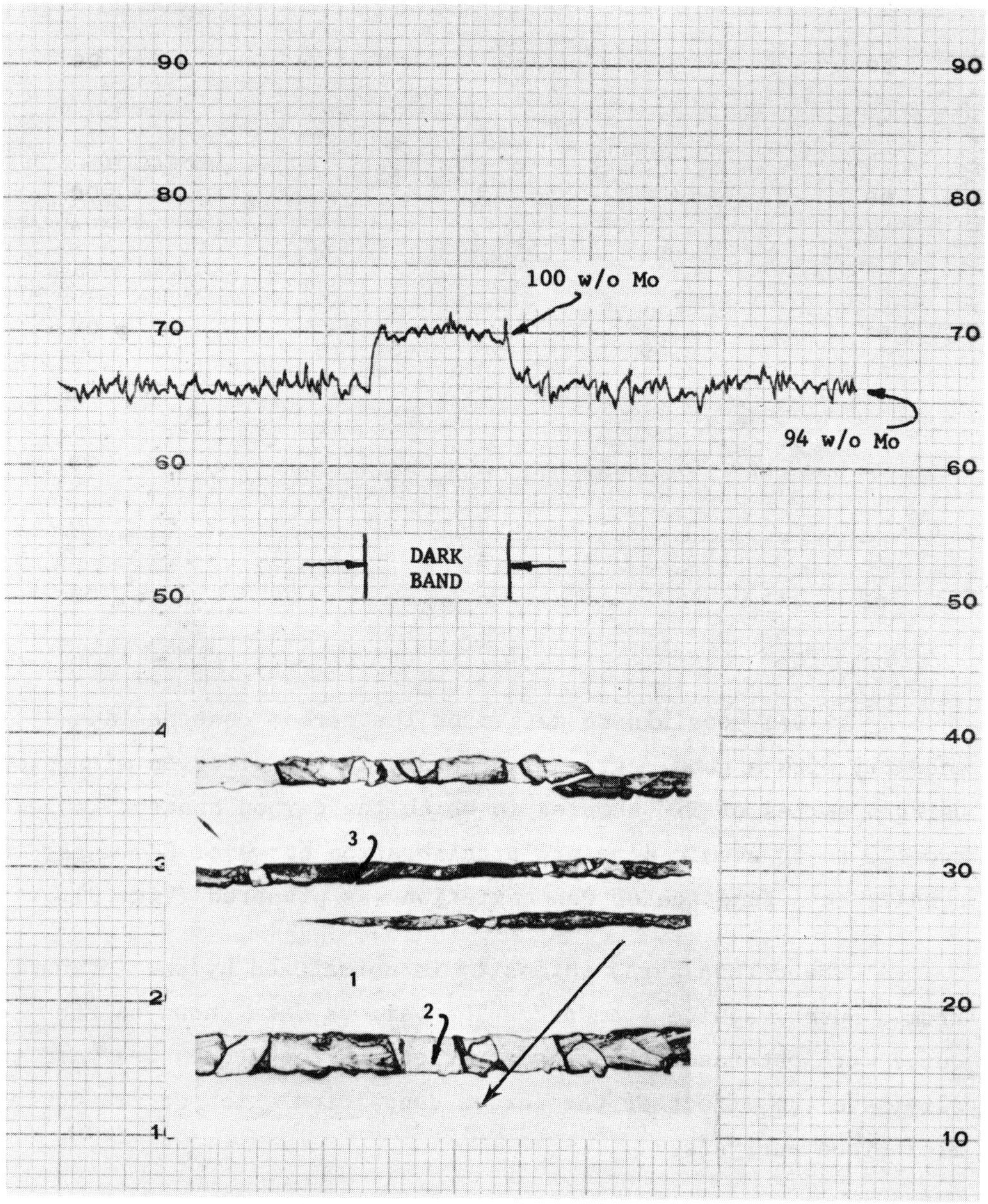

Fig. 8. Mo distribution, single crystal Mo_2C, C-4, (diagonal) vein.

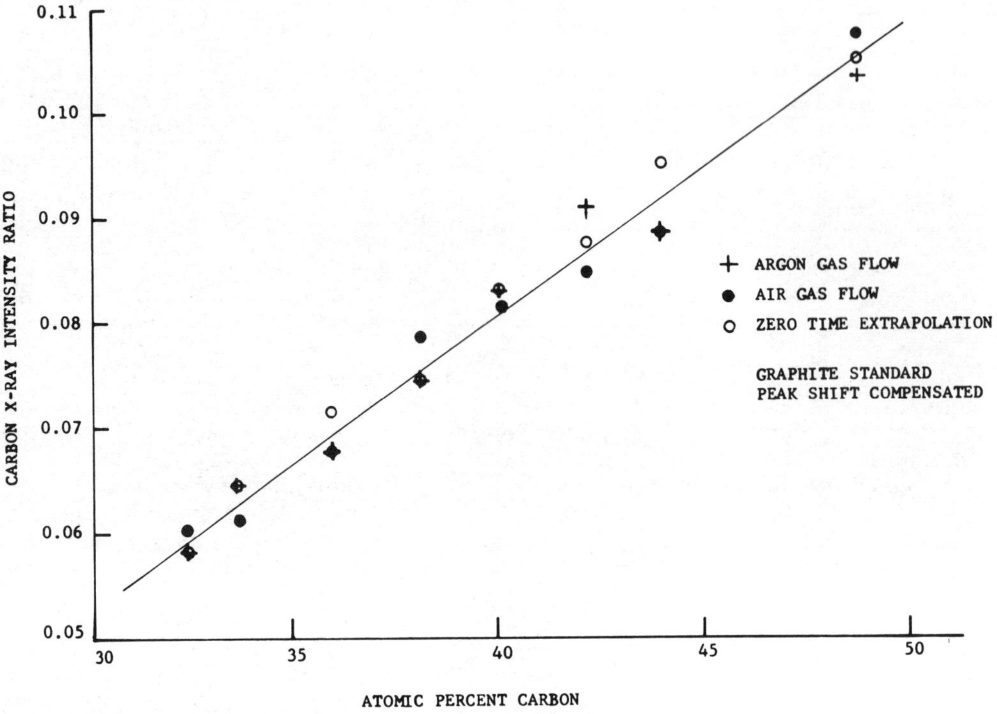

Fig. 9. Carbon X-ray intensity calibration curve for non-stoichiometric titanium carbide (30 kV).

It was possible to determine the carbon content independently with a good degree of accuracy in the electron microprobe. Using a series of TiC samples in which the carbon content ranged from 30 to 50 atomic percent, a calibration curve of CKα x-ray intensity as a function of concentration was prepared (Fig. 9).

The carbon x-ray intensity is normalized by the intensity from a pure graphite standard. The various data shown on the curve were obtained utilizing a variety of techniques designed to eliminate the effect of the carbon deposition which occurs during microprobe analysis.

The analytical precision obtained is generally within a few atomic percent. Most of the crystals contained a carbon content which corresponded to $TiC_{0.92 \pm 0.02}$ and the distribution was uniform within the analytical precision.

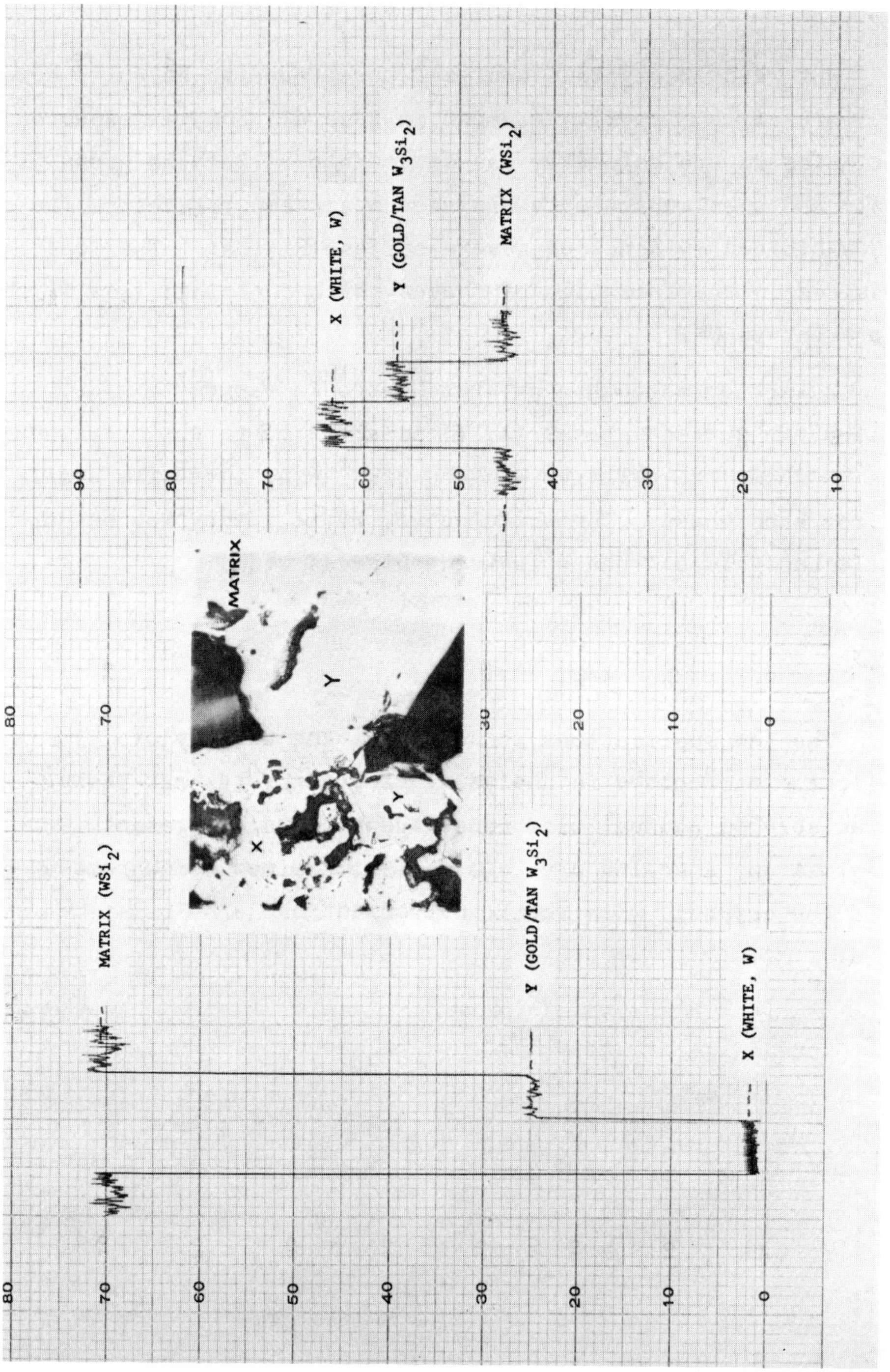

Fig. 10. (left) Si distribution, single crystal WSi_2, W−3g, h, surface phases. (right) W distribution, single crystal WSi_2, WSi_2, W−3g, h, surface phases.

Analysis of WSi_2 Single Crystals

WSi_2 single crystals analyzed were found to be quite pure (< 0.05 w/o of any impurity element in general) and, as expected, no variation in stoichiometry was detectable. Both as-grown crystals and specimens treated in vacuum at temperatures as low as 1600°C exhibited evidence of severe surface Si loss. The loss of Si resulted in the formation of phases which contained less Si than that present in WSi_2.

A section through a crystal heated in vacuum to 1700°C exhibited two surface phases (X, Y) in addition to the matrix WSi_2. Figure 10 shows the relative $WL\alpha$ and $SiK\alpha$ x-ray intensity available from each phase. The white phase, W, is identified as pure W, while the gold/tan phase is identified as W_3Si_2.

Summary

The few applications shown reveal the utility of the electron microprobe in analyzing single crystal materials. The chemistry of unusual microconstituent features present in as-grown and annealed crystals may be examined. The reaction of the crystals with their environment can also be studied.

EQUILIBRIUM PARTICLE SIZE AND PARACRYSTALLINE DISTORTIONS

R. Hosemann and J. Loboda-Čačković x)
and W. Wilke
Fritz-Haber-Institut der Max-Planck-Gesellschaft
Berlin, Germany

1. Introduction

It is well known that the size of crystals depends on conditions of crystallisation. There is, in principle, no limit of their sizes. Starting from a solution, for instance, the lower the rate of cooling the larger and more perfect will be the crystals formed. Imperfections, for example contaminations, grain boundaries as conglomerates of step- and screw-dislocations are removed step by step by careful heat-treatment. The undistorted, ideal periodic crystalline lattice domains within a real crystal (so called "mosaic blocks") can reach any size desired.

In nature, on the other hand, we often meet with substances, which are noncrystalline or "amorphous". They cannot form single crystals in a macroscopic scale. What is the atomistic structure of such substances as glasses, mica, rubber, plastics, fibers or liquids? "Noncrystalline" is a negative definition, "amorphous" a concept of the macroscopic world. We must be very careful, since many shapeless substances exist, which nevertheless consist of an assembly of unoriented microcrystallites. Annealing them and using the zone melting, one can transform them into big single crystals, which now have the typical shape of a crystal. So the word "amorphous" (from the Greek

---

x) Institut "Ruder Bošković", Zagreb.

αμορφοσ = shapeless) gives no information of the atomistic structure of an amorphous substance. It is an empty concept for atom physics.

2. Fourier analysis of amorphous matter

A polycrystalline powder gives rise to an interference function I ($\underline{b}$), which in the reciprocal $\underline{b}$-space is spherical symmetric.

$$I(\underline{b}) \rightarrow I(b); \qquad |b| = \frac{2 \sin \zeta}{\lambda} \tag{1}$$

$$\underline{b} = \frac{\underline{S}-\underline{S}_o}{\lambda}$$

λ wavelength of the radiation, 2ζ scattering angle, $\underline{S}$ resp. $\underline{S}_o$ unit vectors in the direction of the diffracted resp. primary beam (so-called Ewald-construction). The Fourier inverse transform of I ($\underline{b}$), the so-called Q-function, is given by

$$Q(\underline{x}) = \widehat{\rho(\underline{x})\,\rho}(-\underline{x}) = \int \rho(\underline{y})\, \rho(\underline{y}-\underline{x})\, dv_y \tag{2}$$

The vector $\underline{x}$ expands the three dimensional physical space, the symbol $\frown$ is well-known as convolution (or folding) integral of the electron density distribution $\rho(\underline{x})$. For an unbounded ideal crystal Q ($\underline{x}$) degenerates to the Patterson function of crystallography (cf Hosemann, Bagchi (1962)). The Fourier inverse transform is defined by

$$F^{-1} = \int e^{2\pi i(\underline{bx})}\, dv_b \tag{3}$$

dv_b is a volume element of the three dimensional (reciprocal) Fourier space, dv_x resp. dv_y a volume element of the physical space. Now, in the case of spherical symmetric functions (1) F^{-1} degenerates to

$$Q(x) = F^{-1}\, I(b) = \frac{2}{x} \int_0^{\infty} I(b)\, b \sin 2\pi bx\, db \tag{4}$$

Q(x) is well-known in the theory of liquids as density distribution function of $\rho(x)$.

No crystallographer, analyzing a crystalline powder pattern would start with eq. (4) because of large mathematical difficulties.

He, in the reciprocal $\underline{b}$-space, firstly expands the reciprocal lattice by well-known constructions (indexing of reflections with Miller numbers h_1, h_2, h_3).

Herefrom he gets the crystalline lattice by other well-known operations. Going in the opposite way from Q (x) to I (b), one has to use the Fourier transform (symbol F)

$$I\,(\underline{b}) = F\,(Q\,(\underline{x})); \qquad F = \int e^{-2\pi i\, \underline{bx}}\, dv_x \tag{5}$$

which for spherical symmetric function degenerates to

$$F = \frac{2}{b} \int_0^\infty \sin 2\pi bx \cdot x dx \tag{6}$$

Now starting with a polycrystalline powder with unresolvable small peaks, Q (x) has "pointlike" peaks at $x = |p_1\underline{a}_1 + p_2\underline{a}_2 + p_3\underline{a}_3|$ (7)

where p_1, p_2, p_3 are integers and $\underline{a}_1, \underline{a}_2, \underline{a}_3$ are the vectors of a lattice cell. Especially for an onefold primitive cubic lattice cell with the edgelength a (6) degenerates to

$$I\,(b) = F\,(Q) = \frac{1}{2\pi b} \sum_{n=o} N_n \sin(2\pi \sqrt{n}\; b\, a) \cdot (\sqrt{n}\; a)^{-1} \tag{8}$$

Here $N_o = 1$; $N_1 = 6$; $N_2 = 12$; $N_3 = 8$; $N_4 = 6$, $N_5 = 24$; $N_6 = 24$; $N_7 = 0$; $N_8 = 12$, $N_9 = 30$ a. s. o.

No crystallographer will perform this tedious and cumbersome summation of (8), which for noncubic and nonprimitive lattices is much more intractable. Instead of averaging all lattices to Q (x) and then calculating their Fourier transform he firstly calculates the Fourier transform of one single crystalline lattice and then averages over it in Fourier space to get the powder pattern. The contrary is the case for those, who are concerned with noncrystalline matter: They discuss the "density distribution function" (4) as if Q ($\underline{x}$) itself does not consist of single bounded lattices. In other words: They take not into account that Q ($\underline{x}$) as a convolution square of ρ ($\underline{x}$) averaged over all individual lattice domains in the sample.

This seems to us one of the reasons why till up to date matter is classified in a dualistic way into crystalline and amorphous or noncrystalline atomistic structures.

3. The concept of paracrystals

The I-function of a noncrystalline substance does not contain Bragg-like peaks (Debye-Scherrer lines) as polycrystalline powder, but offers only some more or less diffuse interference-maxima. Nevertheless, there arises no necessity to deny the reality of bounded lattices within the structure. The only necessity exists in generalizing the concept of a lattice.

Starting from the classical structure theory of liquids (Ornstein, Zernike, Prins, Debye), we learn it is convenient to introduce an "a priori statistic" for the distance statistics between the center $\underline{x}_n$ of a particle and the centers $\underline{x}_{n+k}$ of all other particles $k \neq 0$, which is independent of n. Contrary to the classical concept of the Q-function of a liquid, which contains all particles in the sample, we now as a first step only take into account the particles of one of the lattices in accordance with crystallographers. How is it possible to combine the concept of a generalized lattice with that of the same a priori distance statistics?

Let us discuss the one dimensional case: If $H_{10}(x)$ is the density distribution function for the next neighbours at the distance x, than the overnext neighbour has the distance statistics $H_{10}(y-x)$, to find its center at y. The same a priori distance statistics means that there is no statistical correlation of y-x and x. Whatever may be the position x of the nearest neighbour, the overnext neighbour is apart from it by $H_{10}(y-x)$. Hence, to find the probability of two adjacent distances x and y-x, one has to multiply the single probabilities $H_{10}(x)$ and $H_{10}(y-x)$. If we are interested in finding the distance statistics H_{20} of overnext neighbours without being interested in the position x of the next neighbour, whatever it may be, we have to integrate over all x and, hence find

$$H_{20}(y) = \int H_{10}(x) H_{10}(y-x)\, dx \tag{9}$$

This is nothing else but the convolution product (symbol $\frown$) of H_{10} with itself x)

$$H_{20}(y) = \overset{\frown}{H_{10}(y)\ H_{10}(y)} \tag{10}$$

x) see for details Hosemann Bagchi (1962)

The distance statistics to the nth neighbour then is given by

$$H_n(x) = P(x-o) \overbrace{H_{10}(x)\, H_{10}(x) \ldots H_{10}(x)}^{n\text{-times}} \tag{11}$$

P(x-o) is a Dirac delta function at x = o, which practically only at x = o has (large) values, but has the weight one:

$$\int P(x-o)\, dx = 1 \; ; \; P(x) = o, \text{ if } x \neq o \tag{12}$$

For three dimensional paracrystalline lattices one finds for the a priori probability function $H_{pqr}(x)$ of the center of an atom, whose averaged position is given by

$$\underline{x}_{pqr} = p\bar{\underline{a}}_1 + q\bar{\underline{a}}_2 + r\bar{\underline{a}}_3 \tag{13}$$

the expression

$$H_{pqr}(\underline{x}) = P(\underline{x}-o)\, \overbrace{H_1 H_1 H_1 \ldots H_1}^{p\text{-times}}\, \overbrace{H_2 H_2 \ldots H_2}^{q\text{-times}}\, \overbrace{H_3 H_3 \ldots H_3}^{r\text{-times}} \tag{14}$$

Here

$$H_1(\underline{x}) = H_{100}(\underline{x}) \; ; \; H_2(\underline{x}) = H_{010}(\underline{x}) ; \; H_3(\underline{x}) = H_{001}(\underline{x}) \tag{15}$$

is the distance statistics to the next neighbour in the direction (100) resp. (010) resp. (001). The Q-function of a single unbounded paracrystal then is given by a convolution polynom:

$$Z(\underline{x}) = \sum_{p=-\infty}^{+\infty} \sum_{q=-\infty}^{+\infty} \sum_{r=-\infty}^{+\infty} H_{pqr}(\underline{x}) \tag{16}$$

Here, for instance

$$H_{\bar{p}oo}(\underline{x}) = P(\underline{x}-o)\, \overbrace{H_1(-\underline{x})\, H_1(-\underline{x}) \ldots H_1(-\underline{x})}^{p\text{-times}} \; ; \; p > o \tag{17}$$

Then one obtains for the standard deviation of a peak (poo) in the direction of $\bar{\underline{a}}_1$

$$\Delta x_p = \sqrt{p}\, \Delta a_1 \; ; \tag{18}$$

The standard deviation Δa_1 of the paracrystalline lattice cell edge $\underline{a}_1$ in the direction of the averaged lattice cell edge $\bar{\underline{a}}_1$ is defined by

$$\Delta a_1 = \left[\int H_{100}(\underline{x}) \left(\underline{x} - \bar{\underline{a}}_1, \frac{\underline{a}_1}{a_1}\right)^2 dv_x \right]^{1/2} \tag{19}$$

In a crystalline lattice H_{100}, H_{010}, H_{001} degenerate to point functions (see eq. (12)):

$$H_{100}(\underline{x}) \rightarrow P(\underline{x} - \bar{\underline{a}}_1)\ ;\ H_{010}(\underline{x}) \rightarrow P(\underline{x} - \bar{\underline{a}}_2);\ H_{001}(\underline{x}) \rightarrow P(\underline{x} - \bar{\underline{a}}_3) \tag{20}$$

hence all Δa_k are zero. The Fourier transform of (16) is called paracrystalline lattice factor Z ($\underline{b}$) and easily can be calculated from (14):

$$Z(\underline{b}) = F(z) = \frac{1}{v_r} P(\underline{b} - o) + \prod_{k=1}^{3} \operatorname{Re} \frac{1 + F_k}{1 - F_k} \tag{21}$$

F_k ($\underline{b}$) is the Fourier transform of one of the coordination statistics H_k ($\underline{x}$) (k = 1; 2; 3;), defined by (15):

$$F_k(\underline{b}) = F H_k(\underline{x}). \tag{22}$$

4. The Diameters $\bar{D}$ of paracrystals and their g-values

Paracrystalline lattice distortions directly are observable by studying a set of reflections (nh, nk, nl) with n = 1, 2, 3, and constant (hkl)[x)]. The integral width δb_n of the reflection n directly can be calculated from (23) and is given by (cf. Hosemann Bagchi (1962))

$$\delta b_n = \frac{1}{\bar{a}} \left[\frac{1}{\bar{N}^2} + \pi^4 g^4 n^4 \right]^{1/2} \tag{23}$$

$\bar{a}$ is the averaged distance of the net planes, which correspond to the reflection (h, k, l),

$$g = \frac{\Delta a}{\bar{a}} \tag{24}$$

is the relative standard deviation of this distance a and $\bar{N}^2$ is the squared and

---

x) It is absolutely necessary to use diffraction cameras with a high resolution up to 3000 Å to study the line widths and line profiles. We used the AEG-Guinier camera with Johansson quartz monochromator after Jagodzinski and Cu-Kα_1 or Mo Kα_2 radiation.

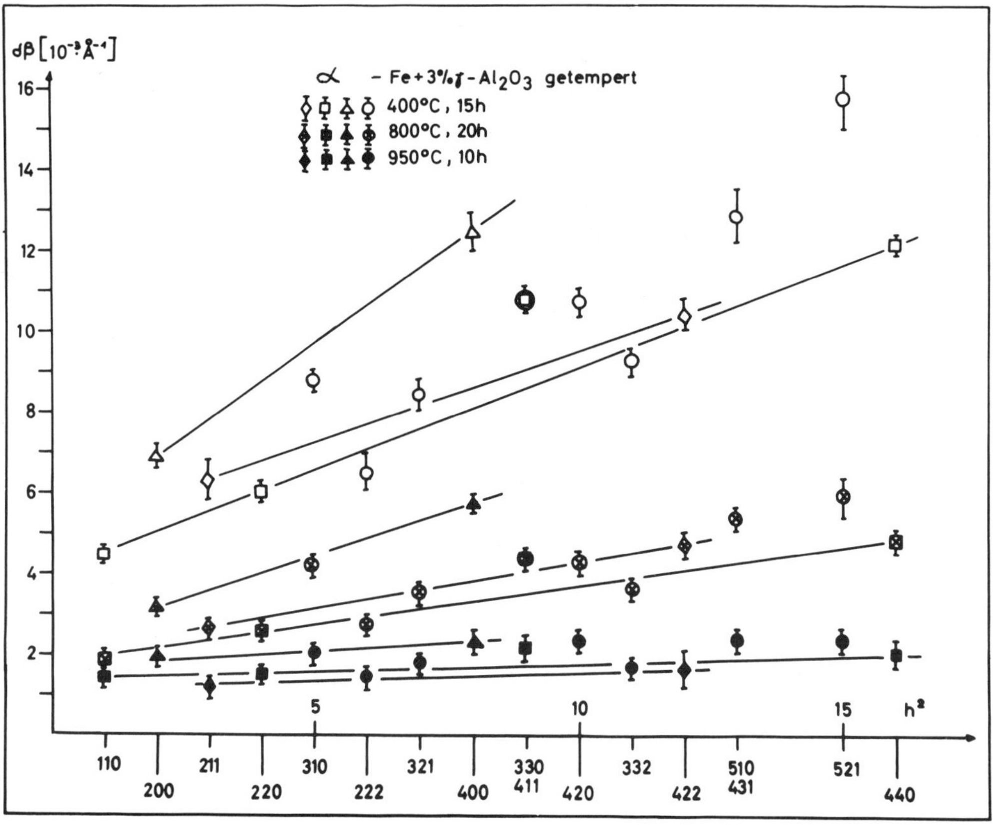

Fig. 1. Integral widths $\Delta\beta$ of α-Fe with 3% Al_2O_3.

averaged number of net planes (see for details Hosemann Bagchi (1962)). If one can observe at least three orders n of one net plane, one can correct, if necessary, for strain broadening and calculate the values of g and $\bar{N}$.

In Fig. 1, as an example, are given the integral widths of Debye-Scherrer lines of an ammonia catalyst which was promoted with 3 weight % of Al_2O_3. In the $\delta b_n - n^2$-plot in accordance with (23) straight lines arise for the single families of net planes. From the slope of these lines one obtains the respective g-values, which are given in Table 1.

From the intersection of these lines with the ordinates n = o, one obtains the respective diameters

$$\bar{D} = \overline{N\,a}$$

Table 1

Paracrystals in Al-promoted ammonia catalysts

Annealing	Lattice planes	$\bar{a}$ (Å)	$\bar{D}$ (Å)	g (%)	α
400°C	h h h	1.24	250±60	1.2±0.07	0.17±0.
	h 0 0	1.43	200±40	1.64±0.08	0.19±0.
15 h	h h 0	2.03	250±30	1.02±0.03	0.11±0.
800°C	h h h	1.24	500	0.76	0.15±0.
	h 0 0	1.43	435	1.1	0.19±0.
20 h	h h 0	2.03	540	0.63	0.10±0.
950°C	h h h	1.24	800	< 0.1	not in equilibrium
	h 0 0	1.43	650	< 0.1	
10 h	h h 0	2.03	710	< 0.1	

of the paracrystalline lattices. After annealing at 400°C, g-values of 1.2% for (hhh) are observed and diameters D of 250 Å. After annealing at 800°C, g drops down to 0.76% and $\bar{D}$ increases to 500 Å. Annealing at 950°C, g becomes immeasurably small and $\bar{D}$ increases to 800 Å.

Obviously there exists a connection between g and $\bar{D}$. The larger g, the smaller $\bar{D}$. Introducing a dimensionless factor α this connection quantitatively is given by

$$\bar{D} = \bar{a} \left(\frac{\alpha}{g} \right)^2 \quad (25)$$

The factor α, on the other hand, can be calculated by the observable quantities D, $\bar{a}$, g:

$$\alpha = g \sqrt{\frac{D}{\bar{a}}} = g \sqrt{N} \quad (26)$$

Multiplying this with the mean net plane distance $\bar{a}$ one obtains from (24) and (18)

$$\Delta a \sqrt{N} = \alpha \bar{a} = \Delta a_N \quad (27)$$

This equation demonstrates, that the standard deviation Δa_N of the most distant net planes in a paracrystalline lattice is given by distance $\bar{a}$ of adjacent net planes, multiplied with α. According to Table 1, α of the Al-promoted α-Fe-catalysts is given by

$$\alpha = 0.16 \pm 0.03 \tag{28}$$

for the (hhh) net planes. In the body centered α-Fe-lattice the g-value of these net planes is given by nearest neighbours, whose distance statistics are called "coordination statistics". They build up the convolution polynom (16). By means of this one can calculate the g-values of the (h00) and (hh0) net planes (Hosemann - Preisinger - Vogel (1966)).

From (23) we learn that in a crystalline lattice, whose g= o, crystals with unbounded large diameters $\bar{D}$ can be obtained in principle. In a paracrystalline lattice on the other hand the diameter $\bar{D}$ is limited. This is of importance to understand the physicochemical behaviour of catalysts, colloids, high polymers, plastics a. s. o.

5. Some examples of paracrystalline distortions

One of the most important examples, why crystalline lattices cannot be built up, is given by mixed crystals, whose different sorts of atoms in the solid solution state are randomly distributed to the sites of the lattice and have different diameters.

A simple two dimensional example is given in Fig. 2: Steelballs of 4 mm diameter build up a two dimensional crystalline lattice. In the moment, however, where some of them are replaced by 4.5 mm balls, after a short time of shaking, an equilibrium state is reached where straight rows of balls are replaced by statistically bent files and ranks of balls with g-values unlike zero. An example was found in metallic Fe-Al compounds. The metallic Al-atoms have a diameter R_1 5% larger than R_2 of the Fe-atoms. If β atom percent of Al statistically are inserted into the Fe-lattice, g-values occur and were observed with a value not too far away from the theoretical:

$$g = \left(\frac{R_1}{R_2} - 1 \right) \frac{\sqrt{\beta\,(1-\beta)}}{1 + \beta\left(\frac{R_1}{R_2} - 1\right)} \tag{29}$$

Fig. 2a. Model of a two-dimensional crystal made of steel balls. In the central region some of them are replaced by greater ones.

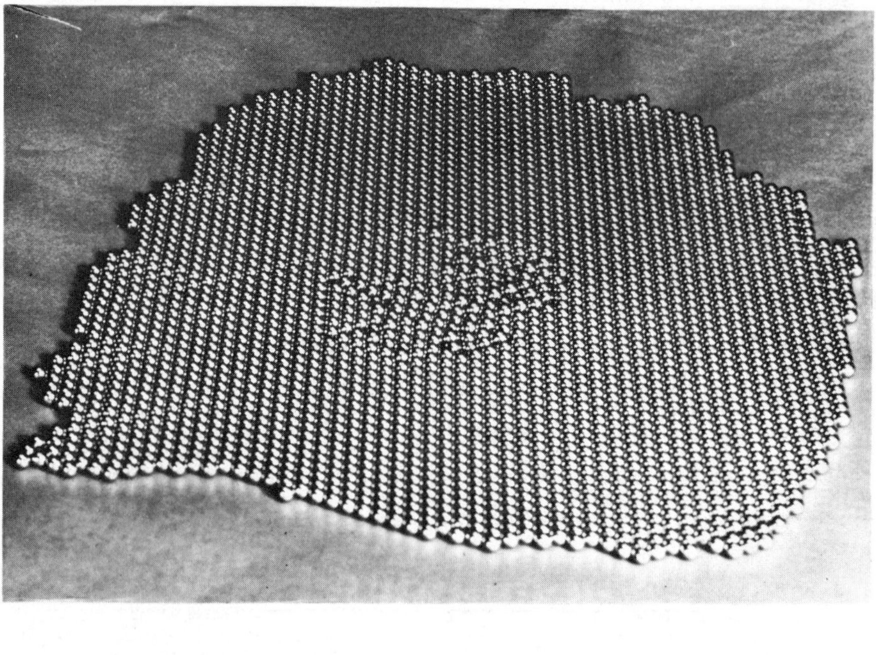

Fig. 2b. The same model after a longer time of oscillation. The central region has become paracrystalline.

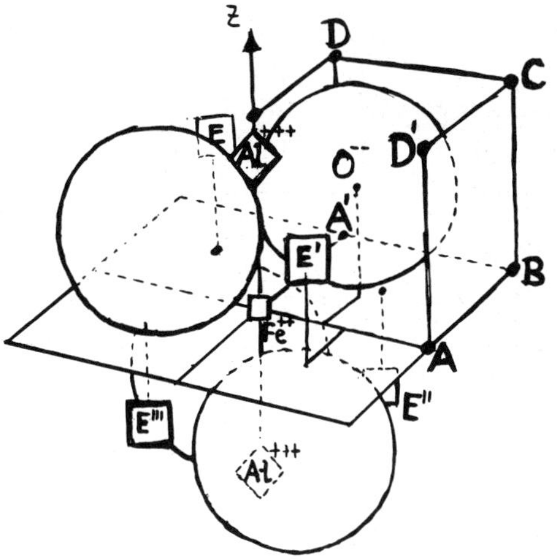

Fig. 3. An $FeAl_2O_4$ molecule, built into the α-Fe-lattice "endotaxy"

See for further details Hosemann, Bialas, Schonfeld, Wilke, Weick (1966).

The above mentioned Al-promoted ammonia catalyst is a mixed crystal of metallic α-Fe-atoms, where statistically for every seven Fe-atoms, one ionic molecule $FeAl_2O_4$ is built in. These seven ions all have need for the same space as the removed seven α-Fe-atoms. Hence the lattice constant of

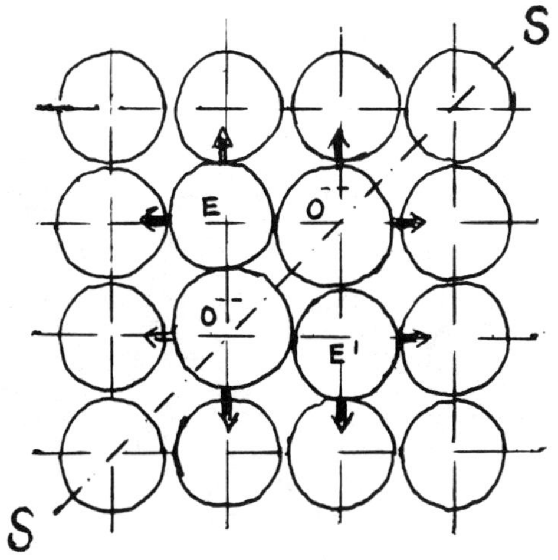

Fig. 4. An 100 netplane of α-Fe. The $FeAl_2O_4$ molecule is too thick.

Table 2

Paracrystals in $Mn_x Fe_{3-x}O_4$ spinels

x	lattice plane	$\bar{a}$ (Å)	$\bar{D}$ (Å)	g (%)	α
≦ 1.67	h h 0	3.01	> 2000	0.1	
1.88	h h 0	3.01	1540	0.85	0.19

the promoted catalyst is the same as that of the unpromoted one.

The shape of this molecule, however, is not the same as that of the seven Fe-atoms, since the O^{--}-ions have too large diameters, the Fe^{++} - and Al^{+++}-ions too small ones. This kind of filling in bricks ($FeAl_2O_4$) into a lattice is called "endotaxy".

In Fig. 3, one $FeAl_2O_4$ molecule within the α-Fe-lattice schematically is drawn. Figure 4 shows a net plane (100) in which two of the oxygen ions are imbedded. Arrows denote the regions of expansion. In Fig. 5 the correspondent (110) net plane is drawn. Arrows indicate the regions of free volume. Another example of a mixed crystal is given in Table 2.

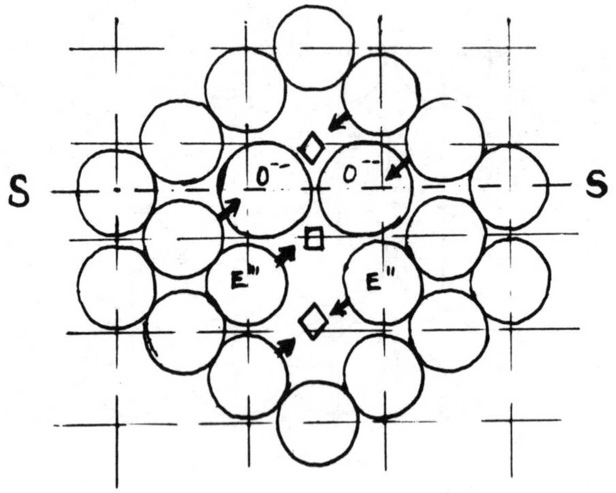

Fig. 5. An 110 netplane of α-Fe. Here the α-Fe lattice can be compressed.

Here paracrystalline g-values of 0.85% were observed for higher Mn-concentrations with x = 1.88. The paracrystalline lattices still remained (pseudo)-cubic if the sample was cooled down from 900°C to room temperature not slow enough. By cooling down with 7°C per minute, the tetragonal structure appears and the paracrystalline distortions still remain, but have only a value two-thirds of the former. The Debye-Scherrer lines in the first case have typical feet demonstrating that in microdomains Mn-octahedra are oriented parallel with their tetrahedral axes. The quotient c/a of the tetrahedric subdomains varies statistically and reaches the equilibrium only for some domains. In the slow cooled sample, all domains have the exact c/a-value of equilibrium, the Bragg reflection, hence, now are split. Nevertheless, paracrystalline distortions still remain since the subdomains have different shapes.

A fourth sample was detected in linear polyethylene (Table 3). Hotstretched bulk material contains simultaneously microfibrils with diameters of 91, 186, 278 Å (cf. Bonart Hosemann (1962) and Hosemann Balta - Calleja, Wilke (1965)).

Table 3

Paracrystallites in linear polyethylene

Sample	net planes	D (Å)	$\bar{a}$ (Å)	g (%)	α
hotstretched 1 : 60	110	91	4.1	3.15	0.15
	110	186	4.1	2.2	0.15
Single crystals 24 h crystallized nonisothermally	110	332	4.1	2.14	0.19
Single crystals crystallized at 85°C 24 hr	110	530	4.1	1.42	0.16
	200	337	3.7	1.59	0.15

Again we find established relation (23) and (24) with a factor α identical with (26) within the errors of experiment. Results from single P. E. crystals are given in Table 3 too. Here paracrystalline mosaic blocks exist, whose D-values can be found too (cf. Hosemann, Wilke, Balta Calleja (1966)). The g-values are lower as for the hotstretched material. Especially the isothermally grown crystals have extremely small values, for instance, g = 1.42% for the (hh0) net planes.

It could be proved that these paracrystalline distortions again correspond to a thermodynamic equilibrium. Annealing the single crystals, for instance, at 125°C, the g-value did not change, but the D-value slowly grew up until the values given in Table 3. It seems to us very interesting that again relations (23) and (24) hold with the same α-value of eq. (26).

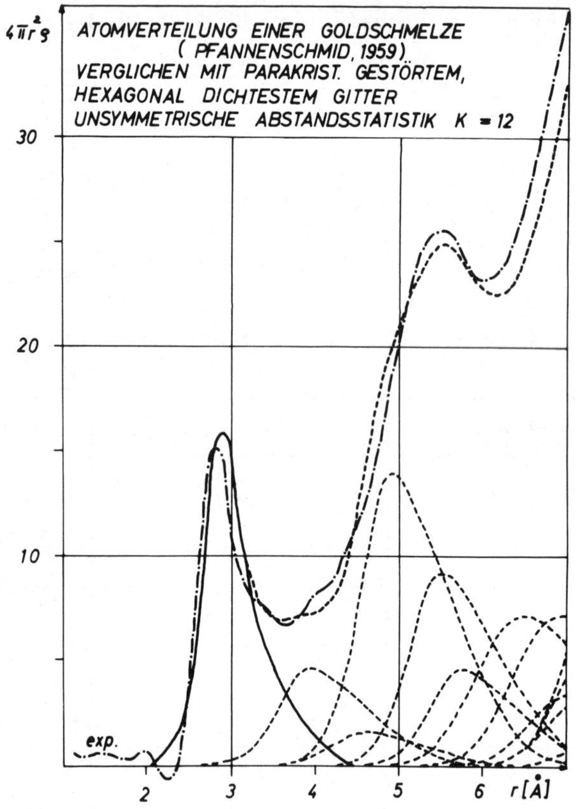

Fig. 6. Atomic density distribution of liquid gold: from experiment (—·—·—), from convolution polynom (— — —), of the coordination statistic (———).

Another example of polycrystalline structures we found in molten Al, Ag, Au, Pb, Mg, Na, K, Pb (cf. Hosemann-Lemm (1965)). Here near the melting point D-values of the order of 30-50 Å were obtained and g-values in the order of 5%. This corresponds again to a factor α, given by eq. (26).

A special example is given in Fig. 6. The atomic density distribution of liquid gold experimentally was obtained by Pfannenschmidt. The synthetic distribution, calculated with a convolution polynom (14), fits if simultaneously five conditions are satisfied:

1.) lattice type
2.) weight and position of the coordination statistics
3.) integral width of coordination statistics
4.) composition of an symmetric and asymmetric part
5.) amount of paracrystalline distortions

we found in all cases

to 1.) hexagonal dense packing. A certain amount of stacking faults is possible. c. f. c. lattice and all other types must be rejected.

to 2.) in the order of 14 (mystic number)

to 3.) and 4.)

The symmetric part has a width in full agreement with Debye's theory. Its position corresponds exactly to the crystalline lattice constant, its weight is about 10. The asymmetric part has a position 0.8 Å more removed, a weight of about 4.

to 5.) paracrystalline distortions of at least g = 5% are absolutely necessary.

6.) Physical reasons for paracrystalline distortions

The simplest reason for paracrystalline distortions was given in the model of Fig. 2 b: The two kinds of atoms in a mixed crystal have different diameters R_1 and R_2. The g-value is given by eq. 29.

The examples of molten metals demonstrate, that one has to take into account not only the radii of the atoms, but the volumes, occupied by them. They consist of the volume of the atom at rest and the "oscillation volume". If these oscillation volumes statistically fluctuate from brick to brick, paracrystalline distortions occur.

The example of spinels as $Mn_xF_{3-x}O_4$ demonstrates that paracrystalline distortions occur moreover if the volumes of the bricks are identical but their shapes are different. The same is the case for the Al-promoted ammonia catalyst where a $(FeAl_2O_4)$-complex has the same volume as 7 metallic α-Fe-atoms, but another shape.

Finally, in the case of linear polyethylene all monomers $-CH_2-$ have the same shape and volume, but paracrystalline distortions occur, since CH_2 is not a sphere. Along a chain molecule trans- and gis conformations occur. Instead of a crystalline zigzag structure (2 helix) statistic irregularities of the conformation give rise to paracrystalline distortions. Moreover, the oscillation volumes may give a contribution to these distortions. Investigations are in progress.

In biological systems as fiber proteins, moreover, not only the statistically varying conformation but also the statistically varying configuration and chemical constitution of the different monomers give rise to paracrystalline distortions, which were observed in the superlattices of these substances (cf. Hosemann (1956)). Table 4 gives a summary.

Table 4

Paracrystalline structures still observed

	Lattice bricks are different with regard to			
Sample	Volume	Shape	Oscillation Volume	Bricks
mixed crystals (Fe, Al)	+	-	-	atoms
spinels ($Mn_xFe_{3-x}O_4$)	-	+	-	microdomains
reduced spinels ($FeAl_2O_4$, Fe)	-	+	-	ionic complexes
linear high-polymers (P. E.)	-	+	-	monomers
molten metals	-	-	+	atoms
fibroins	+	+	-	units of the polypeptides

Synopsis

Many noncrystalline or amorphous substances give rise to diffraction patterns, which quantitatively can be explained by the concept of a paracrystal. They in subdomains consist of three dimensional lattices, which contrary to crystalline lattices have no long range order. They can quantitatively be described by a convolution polynom.

Careful studies of line profiles of mixed crystals (Fe-Al), ($FeAl_2O_4$-α Fe); $Mn_xFe_{3-x}O_4$;) or of high polymers (polyethylene hotstretched; polyethylene single crystals) and the synthesis of experimentally given atomic density distribution curves of molten metals with convolution polynoms prove the existence of paracrystals with g-values and diameters D, which hold the relation

$$\bar{D} = \bar{a}\left(\frac{\alpha}{g}\right)^2$$

($\bar{D}$ mean diameter of the paracrystallites, $\bar{a}$ lattice constant, g paracrystalline distortion).

In all cases $\alpha \approx 0.16 \pm 0.03$ if families of high densely packed net planes are observed. The different paracrystals present a special type of thermodynamic equilibrium with a finite particle size.

Literature

Bonart, R. und R. Hosemann (1962) Koll.Z. u.Zs.f.Polymere 186, 16

Debye, P. P. (1930) Phys. Zs. 31. 348

Hosemann, R. (1956) Erg.Ex.Nat.Wiss. 24 Springer Verlag "Die Erforschung der Struktur hochmolekularer und kolloider Stoffe mittels Kleinwinkelstreuung"

Hosemann, R. und Bagchi (1962) Direct Analysis of Diffraction by Matter, North Holl.Publ.Comp.Amsterdam

Hosemann, R., A. Preisinger und W. Vogel (1966) Ber.Bunsen Ges. 70. 769

Hosemann, R. , W. Wilke, Balta Calleja, (1966) Acta Cryst.21.118

Hosemann, R., Balta Calleja, W. Wilke (1965) Abhdlg.d.Deutsche Akad. d. Wiss. Bln. Klasse f. Chemie, Geologie, Biol. 3, 79

Hosemann, R., K. Lemm (1965) in "Physics of non crystalline Solids" North-Holl. Publ.Comp. S. 85 edited by J. A. Prins

Hosemann, R., D. Bialas, A Schönfeld, W. Wilke u. D. Weick (1966) Advances in Materials, Pergamon Press

Ornstein, L.S. und F. Zernike (1918) Phys. Zs. 29.134

Pfannenschmidt , O. (1960) Z.f. Nat. Forschg. 15 a (7) 6o3 (196o)

Zernike F. und I.A. Prins (1927) Z. Phys. 41,184

CRYSTALLOGRAPHIC POLARITY OF WC

D. N. French and D. A. Thomas

Ingersoll-Rand Research Center
Princeton, New Jersey

ABSTRACT

The hexagonal crystal structure of WC is non-centrosymmetric, resulting in polar character of crystal faces of the $\{10\bar{1}0\}$ type. One consequence of this is the formation of triangular-prism crystals bounded by $\{10\bar{1}0\}$ faces. A number of simple experimental methods have been investigated for revealing the polar behavior of WC, similar to the extensive studies by others on A_{III}-B_V and A_{II}-B_{VI} compounds. The experiments include K_2CO_3 electrolytic etching, contact angles of liquid drops, and microhardness. These results are discussed in terms of the crystal structure and bonding of WC.

INTRODUCTION

Polarity in the properties of crystals with non-centrosymmetric structures is well known. Extensive recent studies of A_{III}-B_V and A_{II}-B_{VI} compounds with the zinc blende (cubic) and wurtzite (hexagonal) structures have

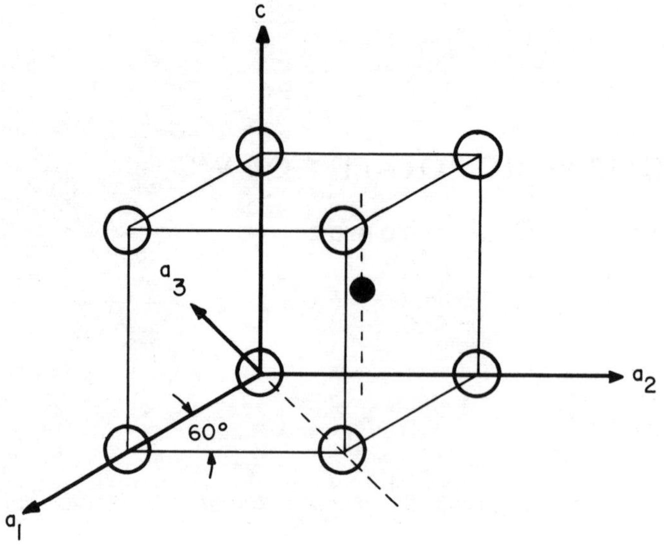

Fig. 1A. The unit cell of WC, with reference to the hexagonal axes a_1, a_2, a_3, and c. Open circles are tungsten atoms; solid circle is carbon.

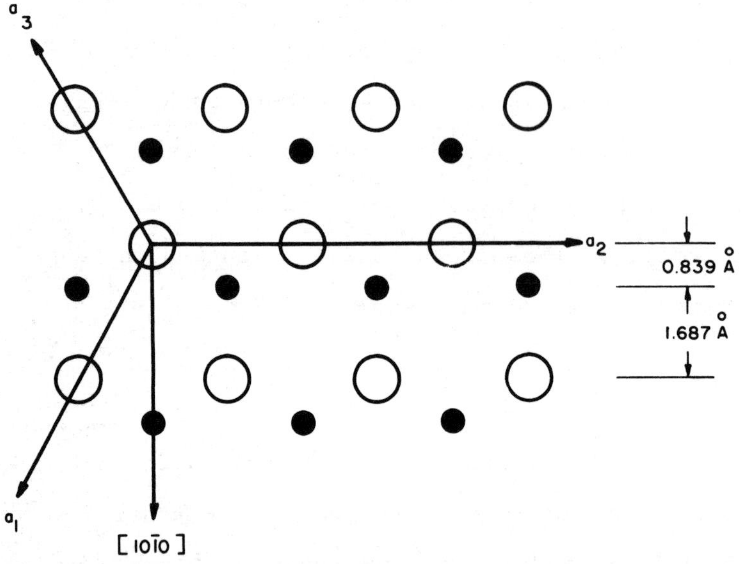

Fig. 1B. The crystal structure of WC viewed along [0001], with carbon atoms above the plane of the drawing. (10$\bar{1}$0) tungsten and carbon planes, which are perpendicular to [10$\bar{1}$0], have unequal spacings shown.

revealed many differences in properties of surfaces that terminate exclusively in A atoms or B atoms.[1-3] X-ray diffraction methods also make possible absolute identification of the A and B surfaces.[4]

The hexagonal crystal structure of WC is also non-centrosymmetric and hence should show polar characteristics. We have used a WC crystal available from previous work[5] to do simple experiments to confirm the expected polarity.

The crystal structure[6] of WC, as shown in Figure 1A, is simple hexagonal with two atoms per unit cell: tungsten at coordinates 0,0,0 and carbon at $\frac{1}{3},\frac{2}{3},\frac{1}{2}$. The lattice constants are a = 2.906 Å, c = 2.837 Å, and c/a = 0.976. Crystal planes of the type $\{10\bar{1}0\}$ are expected to be polar, as shown in Figure 1B. The polarity is apparent from the differences in spacing between tungsten and carbon planes in the $[10\bar{1}0]$ direction. With the asymmetry of the W-C spacing, there are two sets of three equivalent $\{10\bar{1}0\}$ planes, rather than six equivalent $\{10\bar{1}0\}$ planes.

EXPERIMENTAL RESULTS

The WC crystal, which was grown from liquid metal solution, was a right equilateral triangular plate with the sides parallel to $\{10\bar{1}0\}$ planes. Similar shapes were observed by Pfau and Rix,[7] Corteville and Pons,[8] and Takahashi and Freise[9] on crystals grown from liquid cobalt. This shape is a striking consequence of the polarity of $\{10\bar{1}0\}$ surfaces, and the crystals are undoubtedly bounded by one set of equivalent surfaces, either tungsten or carbon. For convenience, planes parallel to the natural growth faces are called $(10\bar{1}0)$ and the opposite planes $(\bar{1}010)$.

It was impossible to use the x-ray techniques reported by others[4] to identify the tungsten and carbon surfaces because the geometric scattering factors for the $(10\bar{1}0)$ and the $(\bar{1}010)$ families of reflections differ by less than 2% on either side of the L absorption edge. The K absorption edge could not be used as it occurs at a 2θ of 4°. Hence the only way to identify tungsten and carbon surfaces positively would be to use neutron diffraction techniques, which we did not attempt.

Etching, sessile-drop wetting, and hardness experiments have all revealed differences between the polar surfaces. The crystal was mounted and metallographically polished on a natural growth face, $(10\bar{1}0)$. The specimen was ground and polished on the back side to obtain the $(\bar{1}010)$ face. The polished faces were within ½° of $(10\bar{1}0)$, as determined by Laue back-reflection x-ray patterns. For comparison, some experiments were repeated on the polar surfaces of a crystal of InSb, of the A_{III}-B_V type.*

Etching. The two polar surfaces were simultaneously electrolytically etched in an aqueous solution of 50 grams K_2CO_3 per liter for 5 seconds at 6 volts. Figure 2 shows the results of this experiment. The $(10\bar{1}0)$ surface shows a rougher, more pock-marked character than does the $(\bar{1}010)$, which appears more uniformly attacked.

*This crystal was kindly supplied by Prof. A. F. Witt of the Department of Metallurgy, Massachusetts Institute of Technology.

$(10\bar{1}0)$ $(\bar{1}010)$

Fig. 2. Polar surfaces of WC electrolytically etched in 50 g/l K_2CO_3 for 5 secs at 6 V. 225 ×.

Sessile-Drop Wetting. Two nonpolar liquids, mercury and purified mineral oil, and two polar liquids, water and a phosphate ester (phosphate ester of alkylphenol ethoxylate, similar viscosity to mineral oil), were used for the sessile-drop experiments. The crystal surfaces were degreased but no exceptional precautions were taken to prevent atmospheric

$(10\bar{1}0)$ $(\bar{1}010)$

Fig. 3. Sessile drops of distilled water on the polar surfaces of WC. Contact angle for $(10\bar{1}0)$ is about 22° and for $(\bar{1}010)$ is about 42°. About 20 ×.

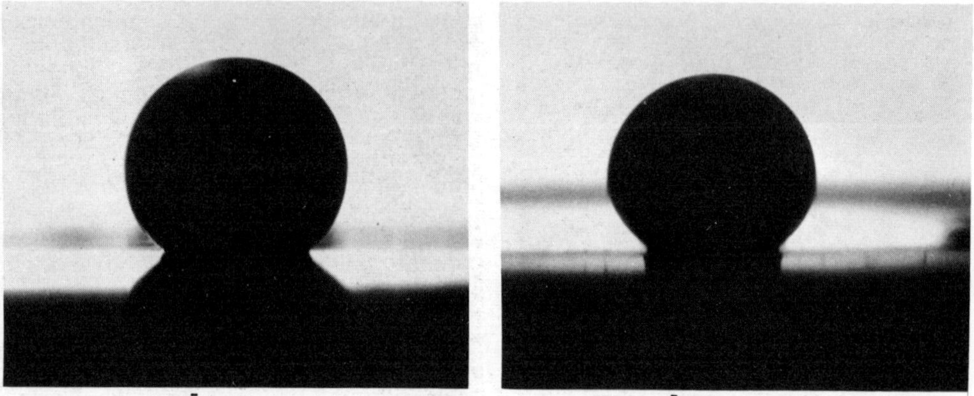

$(10\bar{1}0)$ $(\bar{1}010)$

Fig. 4. Sessile drops of mercury on the polar surfaces of WC. The contact angles for the two surfaces are approximately equal. About 20 ×.

contamination of the surface. A small droplet of the liquid being investigated was placed on the crystal with a hypodermic needle and the shape of the drop photographed.

Figure 3 shows a water drop on the $(10\bar{1}0)$ and $(\bar{1}010)$ crystal faces of WC. Figure 4 shows the same for mercury. Table I summarizes the results. The experiments were repeated in InSb, with only visual estimates of the contact angles made. Table II gives the results. The two polar liquids exhibit different contact angles on the opposite polar surfaces of both WC and InSb crystals. On the other hand, no appreciable differences in contact angles can be detected using the nonpolar liquids.

Hardness. Table III summarizes Knoop microhardness data, at 100 gram load, on the polar surfaces of WC. Two orientations of the long axis of the Knoop indenter were used, parallel and perpendicular to [0001]. Each hardness number is the average of at least 4 measurements. In both directions the hardness on the $(10\bar{1}0)$ surface is greater than on the $(\bar{1}010)$ surface.

TABLE I

Sessile Drop Contact Angles on WC Crystal

Liquid	Contact Angle $(10\bar{1}0)$ Surface	Contact Angle $(\bar{1}010)$ Surface
Hg	138°	134°
Mineral Oil	7°	6°
H_2O	22°	42°
Phosphate Ester	28°	18°

TABLE II

Estimated Sessile Drop Contact Angles on InSb Crystal

Liquid	Contact Angle In Surface	Contact Angle Sb Surface
Hg	135°	135°
Mineral Oil	10°	10°
H_2O	60°	45°
Phosphate Ester	25°	45°

TABLE III

Knoop Hardness of WC Crystal

Surface	Orientation: long axis parallel to [0001]	Orientation: long axis perpendicular to [0001]
$(10\bar{1}0)$	1228 (1189-1254)	2570 (2505-2622)
$(\bar{1}010)$	967 (958-977)	2408 (2361-2467)

Similar measurements of the InSb crystal gave Knoop microhardnesses at 100 gram load of 220 on the In surface and 197 on the Sb surface, confirming other results on $A_{III}-B_V$ materials.[10] An arbitrary orientation was chosen but was similar for both the In and Sb surfaces.

DISCUSSION

The polar character of $\{10\bar{1}0\}$ surfaces in WC is confirmed by the observations of crystal shape and of differences in etching, contact angles, and hardness. For InSb, similar differences are found for contact angles and hardness, in addition to the many other differences previously reported in the literature.

The polarity of WC crystals will be discussed briefly in terms of the crystal structure and atomic bonding of the material.

Crystallography. The crystal structure of WC, Fig. 1, may be viewed as alternate close-packed (0001) layers of tungsten and carbon atoms, such that each small carbon atom is at the center of a trigonal prism formed by six comparatively large tungsten atoms. Figure 5A again shows the arrangements of planes of tungsten and carbon atoms as viewed in the [0001] direction. Separation of a crystal along the dashed line in Fig. 5A is energetically favorable because tungsten atoms on side AA remain bonded to four carbon atoms and separate from only two carbon atoms. Similarly each carbon atom in side BB retains four tungsten bonds and breaks only two. The surfaces generated are shown in Fig. 5B. The AA surface thus consists of tungsten atoms alone and is called the "tungsten surface."

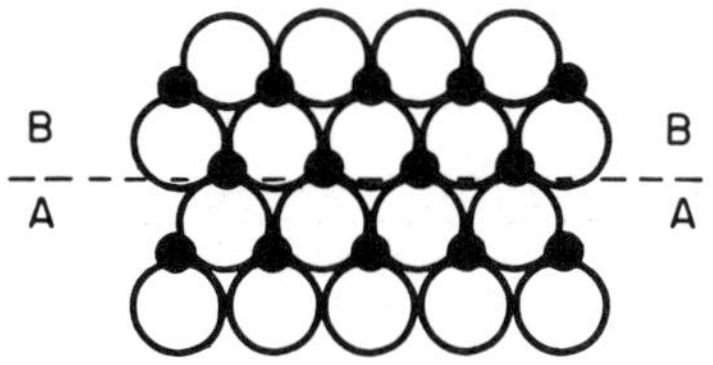

Fig. 5A. Crystal structure of WC viewed along [0001], with carbon atoms above the plane of the drawing. The dashed line indicates the energetically favorable separation plane, as each W atom breaks bonds with only 2 C atoms.

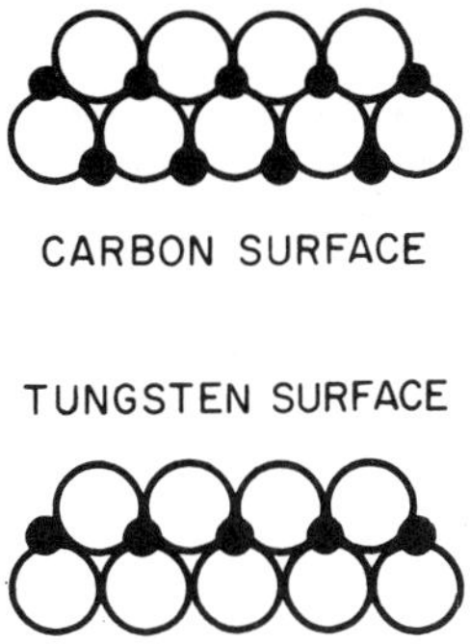

Fig. 5B. Similar view of the crystal structure of WC as in Fig. 5A but separated to show the "carbon" and "tungsten" surfaces.

The BB surface consists of carbon atoms nestled between tungsten atoms and is called the "carbon surface." It is reasonable to expect these different surfaces to exhibit different properties, such as those reported above. Pfau and Rix[7] and Corteville and Pons[8] also clearly recognized the different structure of the two types of prism faces, but they related these only to the observed crystal shape. Other work[5,9,11] has shown that WC crystals also slip on $\{10\bar{1}0\}$, and the plane of slip is also expected to be along the dashed line of Fig. 5A.

<u>Bonding.</u> A simple covalent bond model accounts for many differences between the polar surfaces in A_{III}-B_V compounds, particularly chemical differences.[1] On the surface consisting of group V atoms, such as Sb, 3 of the 5 outer electrons are shared with 3 subsurface In atoms. The fourth In atom, to which Sb would normally be tetrahedrally

TABLE IV

Electrical Resistivity and Young's Modulus of Carbides

Substance	Resistivity (microhm-cm)	Young's Modulus (psi)
TiC	180-250	45×10^6
VC	150	39×10^6
NbC	147	49×10^6
HfC	109	-
TaC	200	41.5×10^6
WC	53	102.5×10^6

coordinated, is missing at the free surface, leaving an unshared pair of electrons on each Sb surface atom. On the In surface, however, each In atom shares its 3 electrons with 3 subsurface Sb atoms, leaving no unshared electrons at the surface. The presence of unshared electrons on one surface and none on the other surface helps account for many property differences of the polar surfaces.

No simple bond model of WC is possible because of the probably complex metallic-covalent bonding. Strong tungsten-carbon bonding is expected, possibly by metallic hybrid orbitals of the d^4sp or d^5p types.[12] However, a carbon atom can complete its stable octet by sharing only two-thirds of an electron from each neighboring tungsten, so tungsten-tungsten bonding is also expected. This is consistent with the fact that tungsten atoms approach almost as closely as the nearest neighbor distance in pure tungsten, 2.735 Å versus 2.837 Å in WC. A strong metallic component of the mixed covalent-metallic bonding accounts for the high electrical conductivity and high modulus of elasticity compared with other carbides, as in Table IV.[13]

On the polar surfaces of Fig. 5B, surface bonding is also expected to differ. On the carbon surface, for example, each surface C atom could bond strongly with the metallic orbitals of each of the four subsurface W atoms. On the tungsten surface, however, each W atom neighbors on four subsurface C atoms and extra bonding electrons are available at the surface, presumably strengthening the bonding between W surface atoms. Without further development, this picture suffices only to rationalize the expected and observed differences in behavior of the polar $\{10\bar{1}0\}$ surfaces in WC.

REFERENCES

1. H. C. Gatos and Mary C. Lavine: Journal of the Electrochemical Society, vol. 107, no. 5, pp. 427-433, 1960.
2. R. E. Hanneman, M. C. Finn, and H. C. Gatos: J. Phys. Chem. Solids, vol. 23, pp. 1553-1556, 1962.
3. H. C. Gatos, P. L. Moody, and Mary C. Lavine: J. Appl. Phys., vol. 31, pp. 212-213, 1960.
4. A. N. Mariano and R. E. Hanneman: J. Appl. Phys., vol. 34, pp. 384-388, 1963.
5. D. N. French and D. A. Thomas: Trans. Met. Soc. AIME, vol. 233, pp. 950-952, 1965.
6. E. Parthé and V. Sadagopan: Monatshefte Für Chemie, vol. 93, pp. 263-270, 1962.
7. H. Pfau and W. Rix: Metallkunde, vol. 45, pp. 116-118, 1954.
8. J. Corteville and L. Pons: C. R. Acad. Sc. Paris, vol. 260, pp. 4477-4480, April 26, 1965.
9. T. Takahaski and E. J. Freise: Phil. Mag., vol. 12, no. 115, pp. 1-8, 1965.

10. R. E. Hanneman: private communication.

11. J. Corteville and L. Pons: C. R. Acad. Sc. Paris, vol. 257, pp. 1915-1918, Sept. 23, 1963.

12. W. Hume-Rothery and G. V. Raynor: The Structure of Metals and Alloys, p. 272, The Institute of Metals, London, 1956.

13. P. Schwarzkopf and R. Kieffer: Refractory Hard Metals, Macmillan Co., New York, 1953.

TABULATION OF INDICES AND INTERPLANAR ANGLES FOR RAPID IDENTIFICATION OF CRYSTALLOGRAPHIC PLANES

David E. Witter and Hayne Palmour III
North Carolina State University
Raleigh, North Carolina

ABSTRACT

A computer program (PIMAX) has been developed to facilitate unambiguous identification of the orientation of single crystals. This program systematically tabulates and indexes all possible triads of interplanar angles between prominent poles occurring in low index zones. Identification is quickly accomplished by selecting three prominent poles on a stereographic projection of diffraction data, measuring and ranking the three interplanar angles in decreasing order, and matching these angles with similarly ranked angles in the table to find the required indices. The identification may be verified by using triads involving additional poles.

PIMAX is not limited to any particular crystal symmetry; it permits computation and ranking of angular spacings and indices specific to the particular crystalline substances under investigation. As input, the program requires axial angles α, β, and γ; unit cell dimensions a_o, b_o, and c_o; and indices of prominent planes, all obtained from standard X-ray data. Tabulations are usually compiled for some characteristic portion of the stereographic projection, whose selection depends upon the symmetry elements involved and the magnitude of the angles to be measured. Applications are illustrated for the cubic, hexagonal, and triclinic crystal systems.

REVIEW OF LITERATURE

Introduction

By definition, all anisotropic materials are characterized by dependence of the magnitude of various properties on the crystallographic direction of measurement. It is interesting to note (13) that "the centre of gravity in the crystal kingdom" lies in the monoclinic system. The number of materials in the triclinic system alone is greater than that for the cubic, hexagonal, and tetragonal systems combined. The orientation of low symmetry anisotropic compounds, therefore, will become increasingly important as the unique physical properties of these materials are investigated.

Research on the physical properties of anisotropic compounds is predicated upon a systematic procedure for determination of the orientation of single crystals and the description of their orientation in the formal notation of crystallography. The science and symbolism of crystallography are well known, and many systematic compilations of crystal data are available in the literature (6, 7, 9, 11, 17). However, the actual laboratory procedure which is required to reliably identify the orientation of a test specimen is not so well systemized, and considerable time and effort usually must be expended in trial and error before one can arrive at a plausible identification.

A pragmatic approach to the problem is described on the following pages. It organizes the crystallographic information of a specific crystalline substance in a table of triads of indices and interplanar angles for expedient comparison with diffraction data and systematic indexing of that data.

Theory

This work is not concerned with the diffraction geometry of various experimental methods, but several fundamental relationships must be realized before the concept of a table of triads of interplanar angles and indices can be adequately comprehended. The relative positions of spots on an X-ray film depend on the spectrum of X-rays incident on a specific single crystal and the relative spatial location and orientation of the crystal, X-ray source and film.

A crystal is a complex arrangement of atoms which is periodic in three dimensions. Diffraction occurs when the atoms scatter waves which interfere constructively to form new wave fronts. The relative directions of these diffraction maxima depend only on the elements of the unit cell. The axial lengths (a_o, b_o, and c_o) and the axial angles (α, β, and γ) describe the size and shape of the unit cell. A unit stereographic triangle exists for each unit cell which completely describes the diffraction symmetry of the crystal. The original X-ray method used by von Laue[1] displays this symmetry quite vividly. Friedel (4) first noted that the diffraction symmetry corresponds to one of the 11 types of centrosymmetry given in Table 1. These 11 groups of crystal classes are called Laue classes and the diffraction symmetry is called Laue symmetry. Lattice planes are defined by Miller indices (hkl), which are the reciprocals of the intercepts of the planes on the crystallographic axes. Hexagonal and trigonal crystals are referenced to the Bravais axes with the associated Bravais indices (hkil). Since the sum of the first three indices is always zero, the third index is frequently replaced by a dot (hk.l). Lattice directions are identified

---

[1]The first Laue photograph ever taken was made in 1912 at Munich by W. Friedrich and P. Knipping on the basis of M. von Laue's theoretical considerations.

Table 1. The 11 Laue symmetries

Crystal system	Laue symmetry	Characteristic symmetry[a]
Triclinic	$\bar{1}$	One fold symmetry only
Monoclinic	2/m	A single 2 or $\bar{2}$ symmetry axis
Orthorhombic	mmm	Three mutually perpendicular nonequivalent twofold axes
Tetragonal	4/m 4/mmm	A single 4 or $\bar{4}$ symmetry axis
Trigonal	$\bar{3}$ $\bar{3}$m	A single 3 or $\bar{3}$ symmetry axis without a mirror plane normal to it
Hexagonal	6/m 6/mmm	A single 6 or $\bar{6}$ symmetry axis
Cubic	m3 m3m	Four equivalent 3 or $\bar{3}$ symmetry axes

[a]Descriptions are given according to Nuffield (10).

by brackets around the three integer indices which are the coordinates of the second point on a line through the origin. A set of planes whose intersecting edges are parallel make up a zone (uvw). The direction defined by the edges is called the zone axis [uvw]. Any two planes describe a zone according to APPENDIX, page 91, and inscribe an acute interplanar angle, ϕ, which is calculated from the other expressions in that section.

A particular Laue symmetry is displayed on the film by the presence and arrangement of hyperbolas which are comprised of individual spots. Each spot represents one or more parallel diffracting planes, and each hyperbola represents a zone whose prominence is noted by the number of spots which describe it. Prominent poles (a pole is the normal to a given set of diffracting crystallographic planes) which describe the symmetry in the unit stereographic triangle occur at the intersections of several prominent zones. Prominence in most cases implies high symmetry and low indices. The interplanar angle between two prominent poles can be deduced from the position of the spots on the film and the distance from the film to the crystal. Therefore it is possible to identify the poles on the film by measuring the interplanar angles between two or more prominent poles and comparing the observed values with the theoretically calculated values in a table of indices and interplanar angles. Unfortunately, measurement of the interplanar angles by means of a Greninger Chart (5) and their transfer to a stereographic projection by means of a stereographic net usually results in an experimental uncertainty of approximately 2 degrees. Even if all angles could be known accurately to the nearest tenth of a degree, two angles must be measured to positively identify any given pole. It is apparent that the inaccuracy of angular measurement confounds the problem by presenting multiple choices of pairs of indices which could be related to the imprecisely known identity angles.

Many contrivances have been invented to make an identification more systematic, but most of these devices require considerable investment of time and equipment and usually place some limitation on the size of the single crystal. An over-all, systematic method, one which requires minimum mechanization, sample preparation and manipulation, and which does not depend upon the user's familiarity with diffraction patterns, is still needed to increase the efficiency of single crystal research. Ideally the method should be applicable to most diffraction data originating from optical, X-ray, electron, or even neutron wave forms. The crystallographic information (in essence, a tensor capable of predicting the diffraction characteristics of the material in any given diffraction geometry) should be organized in tables or graphs which can be scanned in a simple systematic way to derive the correct identification of the spots on the diffraction pattern. Several such recent organizations of orientation information are of interest and are cited below.

"A systematic method of indexing spots of single crystals in Laue X-ray photographs" was suggested by Konnan (8). The method is graphical and uses a specially compiled table of pairs of crystallographic angles. It is not dependent on the complexity of the structure of the crystal or its symmetry, but with more complex crystal structures the method becomes very laborious and the help of a computer is suggested.

A "New technique for orientation of crystal from Laue back-reflection photographs" has been proposed by Riddhagni and Asimow (12). A Laue photograph is examined and an important hyperbola (one which contains some spots through which several hyperbolas pass) is selected. The interplanar angles for the planes causing all of the spots on this hyperbola can be read directly from the Greninger chart in the usual manner. These related angular measurements are plotted on a strip of paper and this strip is compared with a similar theoretical plot of the angular location of poles on various zones. Upon finding a favorable fit, the indices of each spot are given at the appropriate positions on the chart and thus the orientation problem is solved.

A "Computer program for calculating interplanar angles and indexing back-reflection Laue data in an arbitrary crystal system" was developed by Camp and Clum (2). A digital computer program calculates and prints out a table of interplanar angles for an arbitrary crystal structure and then systematically searches through this table, attempting to obtain a consistent fit to back-reflection Laue data supplied by the user. All combinations of poles whose interplanar angles are compatible with the associated indices (and are within the specified angular tolerance) are printed out; and, if no adequate fit is found, a negative indication is printed.

ORIENTATION OF SINGLE CRYSTALS

None of the cited procedures provides a complete, readily available answer to the full range of practical orientation problems. Therefore this new approach is presented to provide the solutions. A tabulation of indices and interplanar angles based upon triads of related angles for any given Laue symmetry is developed by computer methods, compiled in booklet form, and used for rapid identification of crystallographic planes.

This method allows even a novice crystallographer (*e.g.*, a geologist, metallurgist, ceramist, physicist, or chemist) to unambiguously identify crystallographic directions in an unfamiliar X-ray pattern by (1) measuring three angles between prominent zone intersections, (2) ranking the angles in decreasing order, and (3) matching the angles with similarly ranked angles in the computer-produced table which contains triads of angles for the particular crystalline substance. Many of the advantages of the techniques described in the previous paragraphs are combined in this procedure, which relies on the speed and accuracy of a digital computer to calculate and tabulate unique triads for specific Laue symmetries. The paragraphs which follow illustrate the problem, the classical solution, and demonstrate this new systematic procedure of pole identification.

The Problem

Consider a typical single crystal orientation problem presented by the Laue back-reflection X-ray pattern of sapphire (monocrystalline α -

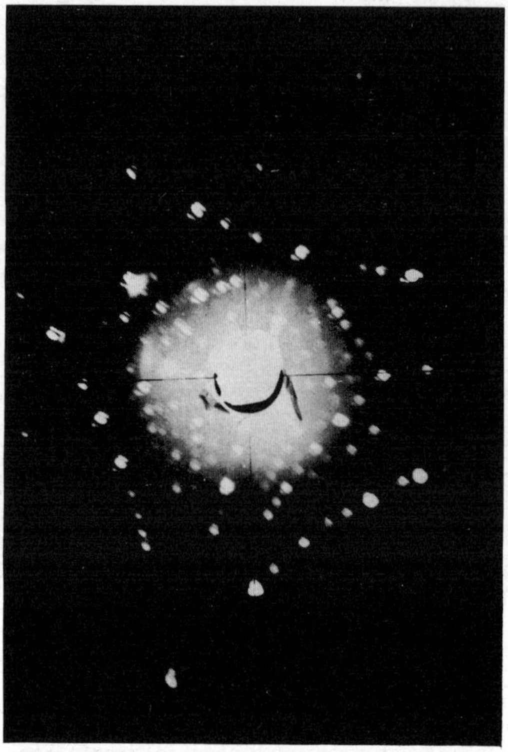

Fig. 1. Laue pattern of a randomly set sapphire crystal.

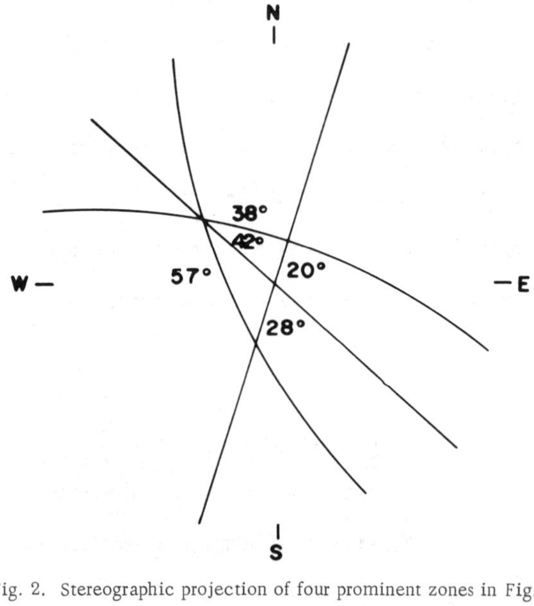

Fig. 2. Stereographic projection of four prominent zones in Fig. 1.

Al_2O_3) in Figure 1 and abstracted in the stereographic projection of four prominent zones in Figure 2.

The Classical Solution

To resolve the ambiguity of identification, three adjacent angles must be measured to define a possible arrangement of indices. Usually a closed loop of three angles is established by selecting the unmeasured angle between two of the poles, or alternatively, in the special case where all three poles are located on one zone, one which is the sum of two previously measured angles. Once three related angles have been found in a table of interplanar angles, a procedure of trial and error is used to determine which, if any, possible combination of pairs of indices can serve to identify the closed loop of angles. In any event, other angles must be measured and the associated poles identified by trial and error until all poles of the projection have been identified without contradiction.

A New Solution

A computer program called PIMAX (an acronymn for *P*ole *I*dentification by *M*aximization of *A*ngles in "*X*tals") has been developed to calculate the interplanar angles for all possible combinations of indices in triads which describe low index zones. The triads are ranked according to the magnitude of their interplanar angles, and tabulated in an order of decreasing magnitude. In effect, the unique triads of angles are organized for easy comparison with triads observed on the film. All similar triads appear in close proximity in the table which permits immediate scrutiny between possible solutions and allows immediate verification with adjacent triads. It is only necessary to assume that the zones selected from the film can be identified by small integer indices and that the intersection of these zones are prominent poles which were considered in the tabulation.

INDEX 1			ANGL 1,2	INDEX 2			ANGL 2,3	INDEX 3			ANGL 3,1	Z	
-1	1	4	59.482	-4	4	1	48.760	-2	2	5	10.723	1	
1	0	4	59.482	4	0	1	48.760	2	0	5	10.723	1	
-4	4	1	59.482	-1	1	4	36.099	-1	1	1	23.383	1	
4	0	1	59.482	1	0	4	36.099	1	0	1	23.383	1	
4	0	1	58.796	2	-2	1	55.637	1	0	1	23.383	3	
2	-2	1	58.796	4	0	1	34.406	4	-2	3	29.648	3	
4	0	1	58.796	2	-2	1	34.363	2	-1	0	31.204	3	
-2	2	5	57.774	-1	1	0	54.351	-1	2	3	20.527	3	
-1	1	0	57.774	-2	2	5	48.760	-4	4	1	9.014	1	
1	0	0	57.774	2	0	5	48.760	4	0	1	9.014	1	
-2	2	5	57.774	-1	1	0	32.397	-1	1	1	25.377	1	
2	0	5	57.774	1	0	0	32.397	1	0	1	25.377	1	
0	0	1	57.603	-1	1	1	46.990	0	1	2	38.237	3	A
0	0	1	57.603	1	0	1	46.990	1	-1	2	38.237	3	A
0	0	1	57.603	1	0	1	46.990	0	1	2	38.237	3	A
1	0	1	57.603	0	0	1	42.298	2	-1	3	27.326	3	B
1	0	1	57.603	0	0	1	42.298	1	1	3	27.326	3	B
-1	1	1	57.603	0	0	1	42.298	-1	2	3	27.326	3	B
0	0	1	57.603	1	0	1	37.769	2	-1	6	24.462	3	
0	0	1	57.603	1	0	1	36.099	1	0	4	21.503	1	

Fig. 3. Triads of indices and interplanar angles showing groups A and B.

After obtaining three angles from a stereographic projection or directly from a Laue film, one consults a table of triads of indices and interplanar angles similar to that illustrated in Figure 3. The largest angle is tabulated in the column titled "ANGL 1,2"; the intermediate angle is found adjacent to the largest angle in column "ANGL 2,3"; and the smallest angle is found adjacent to the previous angles under "ANGL 3,1". The first index identifies the plane represented by the intersection of the zones which include the largest and smallest angles. Similarly, the second and third indices identify the planes represented by the intersection of the zones which include the largest and intermediate angles and the smallest and intermediate angles, respectively. For additional information about the relationship of the three poles, a "1" appears under "Z" if all three poles occur on one zone and a "3" is present if the three poles occur at the intersections of three zones. If a selection of angles from the film cannot be matched with a combination of angles in the table, then another combination of poles must be selected until a *prominent* triad of poles can be identified in this manner. An identification may be verified by the selection of additional poles and comparing the resultant triads with those in the table.

Illustrative Example

The problem defined in Figure 1 will illustrate the use of the PIMAX tables in positively identifying the poles of the projection. Consider the three angles, 57°, 38°, and 48° (20° + 28°), which first are ranked in decreasing order: 57°, 48°, and 38°. They are then compared with the table and are found to favorably correspond with the group of angles marked "A" in Figure 3.

A group of triads is listed, since three similar triads occur within the portion of the stereographic projection selected for processing in the PIMAX program. Next, consider the triad 57°, 42°, and 28° in Figure 1, which is comparable to the group of indices marked "B" in Figure 3.

It is apparent that groups "A" and "B" are compatible since similar pairs of indices, {10$\bar{1}$1} and {0001}, exist in each group. Further verification can be obtained by considering the triad formed by the three poles on one group. Their angles, ranked as 48°, 28°, and 20°, correspond to group "C" in Figure 4, confirming the two previous

INDEX 1			ANGL 1,2	INDEX 2			ANGL 2,3	INDEX 3			ANGL 3,1	Z	
2	-1	3	47.702	2	-1	0	43.010	1	0	1	27.326	3	
-1	2	3	47.702	-1	2	0	28.790	-2	4	3	18.912	1	
1	1	3	47.702	1	1	0	28.790	2	2	3	18.912	1	
1	0	1	46.990	0	1	2	44.360	2	-1	6	37.769	3	
1	-1	2	46.990	1	0	1	37.769	2	-1	6	20.462	3	
0	1	2	46.990	-1	1	1	36.099	-1	1	4	32.411	3	
0	1	2	46.990	1	0	1	36.099	1	0	4	32.411	3	
1	-1	2	46.990	1	0	1	36.099	1	0	4	32.411	3	
-1	1	1	46.990	0	1	2	33.954	-2	2	5	25.377	3	
1	0	1	46.990	0	1	2	32.003	2	2	3	25.989	3	
1	-1	2	46.990	1	0	1	27.326	2	-1	3	19.663	1	C
0	1	2	46.990	1	0	1	27.326	1	1	3	19.663	1	C
0	1	2	46.990	-1	1	1	27.326	-1	2	3	19.663	1	C
1	1	6	44.360	1	-1	2	38.237	0	0	1	24.462	3	
2	-1	6	44.360	0	1	2	38.237	0	0	1	24.462	3	

Fig. 4. Triads of indices and interplanar angles showing group C.

comparisons. A brief mental or graphical summary of the information supplied by the three groups of angles yields three identical solutions involving planes of the form {0001}, {10$\bar{1}$1}, {11$\bar{2}$3}, and {01$\bar{1}$2}. Recent revisions of the input data for this compilation have eliminated some of the redundant information expressed in these groups of triads and compressed the table without excluding unique triads. Another selection of angles (42°, 38°, and 20°) is available and is present in the table, but it was not necessary to utilize this triad in identifying the orientation; however, it could be useful for additional confirmation.

The classical method for verifying an orientation is to make the necessary rotations on the specimen goniometer, bringing the point in question (in this case, the pole of the basal planes) to the center of the Laue film. Figure 5 shows such a verification photograph for this example. It illustrates the trigonal symmetry which in sapphire is characteristic only of the basal plane.

By these straightforward steps, the orientation of this sapphire crystal has been positively identified without relying on any special familiarity with the material and without reference to a standard stereographic projection of its prominent poles.

DESCRIPTION OF PIMAX AND ITS FUNCTIONS

PIMAX is the current result of a series of revisions which have embraced the idea that this program could be used by any person interested in indexing a well characterized single crystal who has access to adequate computer facilities. The computer card deck was prepared for Model 75 of International Business Machine's Operating System 360. The deck (APPENDIX, page 91, Appendix Figures 1 through 11) is composed of three steps involving a Fortran IV source program, a

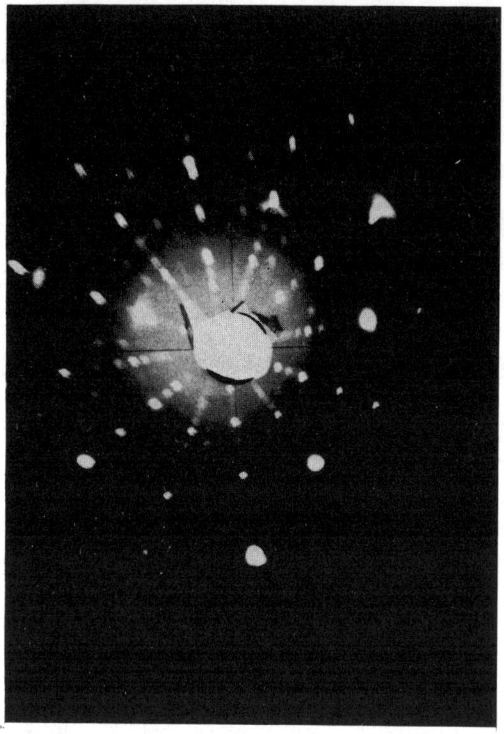

Fig. 5. Laue pattern of basal plane with its trigonal symmetry.

sorting routine, and a printing routine. The Fortran program performs all of the operations on the input data and prints Tables I and II. The other routines are called from the machine library to sort and print Table III. Other lesser machines with a Fortran IV translator could be used if adequate sorting and printing routines were available.

Program Logic

The logic of the program (APPENDIX, page 91, Appendix Figures 12, 13, and 14) is mainly concerned with three operations: (1) the interplanar angles for all possible pairs of indices are computed; (2) allowed pairs of indices are selected according to the low index (ultimate) zone criteria; and, (3) all possible triads of these allowed pairs are made with the criteria of keeping one pole within the unit triangle and the other poles are kept within multiples of the unit triangle on a characteristic portion of the stereographic projection. A computerized sorting routine is called from the machine library and the triads are sorted according to the magnitude of the interplanar angles. The resulting printout is a set of tables (spaced and paged for direct reproduction in booklet form) which can be easily scanned to solve problems relating to diffraction from a particular Laue symmetry.

The Laue symmetry of the crystal in question is expressed to PIMAX in terms of the axial angles and axial ratios, and in the selection of indices of prominent poles within unit triangles on the stereographic projection. The ultimate zone is important to limit the size of the table and to select only those zones which might be prominent.

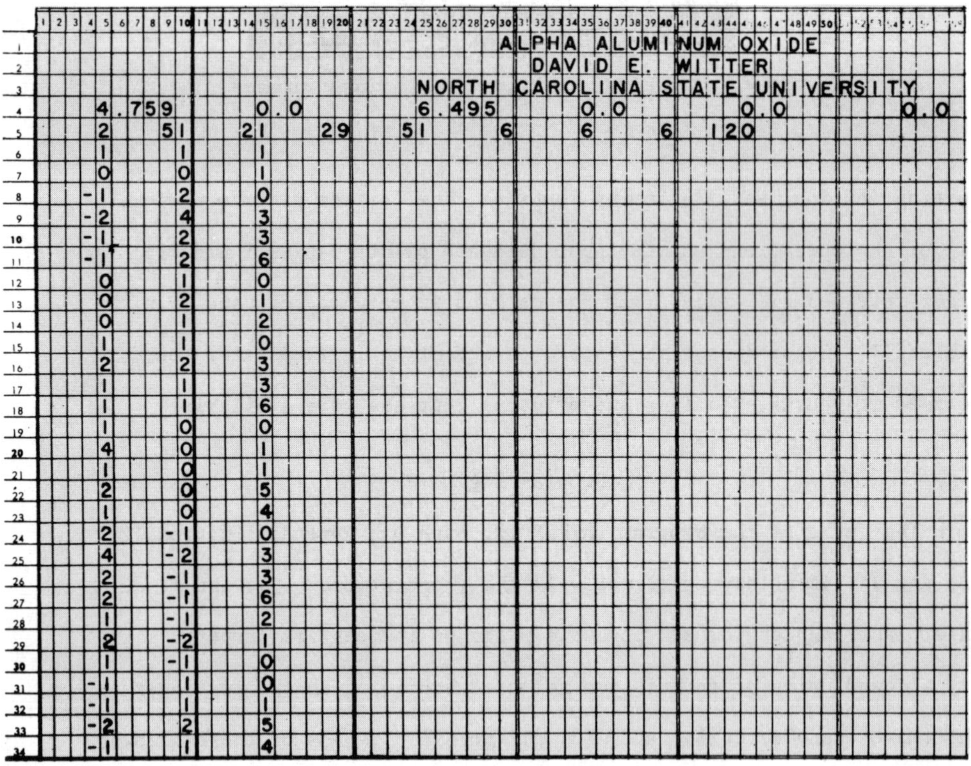

Fig. 6. Typical input data card.

A TABULATION OF INDICES AND INTERPLANAR ANGLES
FOR RAPID IDENTIFICATION OF CRYSTALLOGRAPHIC PLANES

FOR

ALPHA ALUMINUM OXIDE

COMPILED BY

DAVID E. WITTER
NORTH CAROLINA STATE UNIVERSITY

THIS PROGRAM (PIMAX) WAS PREPARED
IN
MARCH, 1967
BY
DAVID E. WITTER
AT
NORTH CAROLINA STATE UNIVERSITY
OF THE
CONSOLIDATED UNIVERSITY OF NORTH CAROLINA
WITH THE
TRIANGLE UNIVERSITIES COMPUTATION CENTER
USING
FORTRAN IV AND JOB CONTROL LANGUAGE FOR IBM 360 MODEL 75

Fig. 7. Sample title page

Data Definition and Preparation

Figure 6 shows some typical input data which were used to compile the table in the previous example. The input data are placed in the deck after the Fortran program and immediately before the sorting step. The format for all data is very simple. The following straightforward instructions will permit any interested person to prepare his own input cards, insert them in the proper positions in the deck, submit it for computation, and index his diffraction data with the resultant output.

Alphameric Cards

Three alphameric cards must be used. These cards contain information for the title page: more specifically, crystalline substance, investigator, and sponsor. A sample title page appears as Figure 7 which shows how this information is displayed. All of the information must be centered about column 40 on each card. The third card may be used for any additional information or may be a blank card.

Cell Card

The fourth card (Figure 6) contains the elements of the unit cell a_o, b_o, c_o, α, β, and γ, respectively. Each element appears in this order within a field which is ten columns wide. A decimal point must appear at the appropriate position in the number, but it need not be in any particular column. Since this compilation for sapphire was based on poles designated by abbreviated Bravais indices, the a_o and c_o values are the only values required. If sapphire were treated as a rhombohedral crystal in Miller indices, all of the values would need to be expressed.

Parameter Card

The fifth card supplies the parameters which determine the logic of the program. Table 2 gives the position and meaning of each parameter.

All possible combinations of the first NI planes with the first NJ planes and with the complete list of N planes form a list of triads of indices. This list is reduced in size by limiting the zone indices of each pair of planes in a triad to the values (ULTU, ULTV, ULTW). Further reduction of the list may be accomplished by limiting all of the interplanar angles to values less than MXPHI. After discarding all of the triads which do not fit these criteria, the list is sorted and printed as Table III.

Table 2. The position, name and meaning of parameters

Columns	Name	Description
1-5	MB	Type of indices used (1 = Miller, 2 = Bravais)
6-10	N	Number of planes to be submitted
11-15	NI	The first NI planes describe a unit triangle
16-20	NJ	The first NJ planes describe another characteristic portion of the stereographic projection
21-25	INT	The first INT planes are included in Table II
26-30	ULTU	The ultimate first zone index
31-35	ULTV	The ultimate second zone index
36-40	ULTW	The ultimate third zone index
41-45	MXPHI	The maximum value of ϕ in Table III

Control Card

The sixth card allows the user to select which of the three tables will be printed. A "1" in all three fields of the sixth card indicates that all three tables are desired; a "2" in the first field and "1's" in the other two fields would mean that only Tables II and III are desired. A "2" in all three fields would be nonsensical, since no output would be desired.

Data Cards

The remaining cards (numbering 51 for this example) give a three integer index (Miller or abbreviated Bravais) for the prominent poles to be included in the tables. Note that the order of these cards is important in compiling the tables. A unit triangle for sapphire is expressed by the first 21 indices. The first 29 indices comprise half of a stereographic projection, and the complete list of 51 indices are

```
INPUT DATA

A ZERO =     4.7590        ALPHA =     0.0

B ZERO =     0.0           BETA =      0.0

C ZERO =     6.4950        GAMMA =     0.0


THE NUMBER OF BRAVAIS, HK.L, INDICES OF PROMINENT POLES
IS EQUAL TO   51.

THE COMPLETE LIST OF INDICES IS GIVEN IN THE FIRST LOOP OF
TABLE I WHICH TABULATES THE ZONE AND INTERPLANAR ANGLE FOR
THE  1275. POSSIBLE COMBINATIONS OF   51 PLANES TAKEN
TWO AT A TIME.  THE INTERPLANAR ANGLE IS SET EQUAL TO ZERO WHEN
THE ZONE IS NOT CONSIDERED TO BE PROMINENT.


INTERPLANAR ANGLES ARE TABULATED IN TABLE II FOR THE FIRST   51
INDICES.


ALL COMBINATIONS OF THREE PLANES WITH THE ABSOLUTE VALUES OF
ZONE INDICES LESS THAN 4. 4. 6. WILL BE INCLUDED IN TABLE III,
WHICH TABULATES INDICES AND INTERPLANAR ANGLES FOR TRIADS
OF INDICES.  ONE PLANE IS SELECTED FROM THE FIRST   22
INDICES, A SECOND PLANE IS SELECTED FROM THE FIRST   29
INDICES, AND THE THIRD PLANE IS SELECTED FROM THE
COMPLETE LIST OF   51 INDICES.
```

Fig. 8. Typical input data page.

prominent poles on a complete ($11\bar{2}0$) standard projection. The entire projection was compiled merely to demonstrate this capability for possible application in transmission photographs. Practical considerations make those angles less than 90° most useful, since all angles greater than 90° can be expressed by a supplementary acute angle.

DESCRIPTION OF OUTPUT

A compilation of tables begins with the previously mentioned title page (Figure 7). All of the input data except the list of planes appears on the input data page illustrated as Figure 8. Table I (Figure 9) lists all possible combinations of pairs of indices, the indices of the zone which includes the two poles, and the interplanar angle. The interplanar angle is expressed as zero and a pair of indices is excluded

TABLE I PAGE 25

INDEX 1	ANGL 1,2	INDEX 2	ZONE 1,2
1 0 -2	46.990	1 -1 -1	-2 -1 -1
1 0 -2	33.954	2 -2 -5	-4 1 -2
1 0 -2	32.411	1 -1 -4	-2 2 -1
1 0 -2	64.822	-1 1 -2	2 4 1
1 0 -2	0.0	-2 2 -1	4 5 2
1 0 -2	38.237	0 0 -1	0 1 0
4 -2 -3	18.912	2 -1 -3	1 2 0
4 -2 -3	36.747	2 -1 -6	1 2 0
4 -2 -3	0.0	4 -4 -1	-5 -4 -4
4 -2 -3	25.989	1 -1 -1	-1 1 -2
4 -2 -3	0.0	2 -2 -5	2 7 -2
4 -2 -3	0.0	1 -1 -4	5 13 -2
4 -2 -3	0.0	-1 1 -2	7 11 2
4 -2 -3	0.0	-2 2 -1	4 5 2
4 -2 -3	61.209	0 0 -1	1 2 0
2 -1 -3	17.836	2 -1 -6	1 2 0
2 -1 -3	0.0	4 -4 -1	-11-10 -4
2 -1 -3	27.326	1 -1 -1	-2 -1 -1
2 -1 -3	20.527	2 -2 -5	-1 4 -2
2 -1 -3	0.0	1 -1 -4	1 5 -1
2 -1 -3	0.0	-1 1 -2	5 7 1
2 -1 -3	0.0	-2 2 -1	7 8 2
2 -1 -3	42.298	0 0 -1	1 2 0
2 -1 -6	0.0	4 -4 -1	-23-22 -4
2 -1 -6	0.0	1 -1 -1	-5 -4 -1
2 -1 -6	0.0	2 -2 -5	-7 -2 -2
2 -1 -6	11.949	1 -1 -4	-2 2 -1
2 -1 -6	0.0	-1 1 -2	8 10 1
2 -1 -6	0.0	-2 2 -1	13 14 2
2 -1 -6	24.462	0 0 -1	1 2 0
4 -4 -1	23.383	1 -1 -1	1 1 0
4 -4 -1	48.760	2 -2 -5	1 1 0
4 -4 -1	59.482	1 -1 -4	1 1 0
4 -4 -1	119.223	-1 1 -2	1 1 0
4 -4 -1	153.383	-2 2 -1	1 1 0
4 -4 -1	80.986	0 0 -1	1 1 0
1 -1 -1	25.377	2 -2 -5	1 1 0
1 -1 -1	36.099	1 -1 -4	1 1 0
1 -1 -1	95.839	-1 1 -2	1 1 0
1 -1 -1	130.000	-2 2 -1	1 1 0
1 -1 -1	57.603	0 0 -1	1 1 0
2 -2 -5	10.723	1 -1 -4	1 1 0
2 -2 -5	70.463	-1 1 -2	1 1 0
2 -2 -5	104.623	-2 2 -1	1 1 0
2 -2 -5	32.226	0 0 -1	1 1 0
1 -1 -4	59.740	-1 1 -2	1 1 0
1 -1 -4	93.900	-2 2 -1	1 1 0
1 -1 -4	21.503	0 0 -1	1 1 0
-1 1 -2	34.160	-2 2 -1	1 1 0
-1 1 -2	38.237	0 0 -1	-1 -1 0
-2 2 -1	72.397	0 0 -1	-1 -1 0

Fig. 9. A sample page from Table I.

from the list of triads if the imposed zone or angle criteria are not realized. The complete list of N planes appears in the first N-1 lines of Table I. Table II (Figure 10) provides a table of interplanar angles, similar to those which have been the sole reference for most crystal orienters to date. Given two indices, the interplanar angle can be easily located in this table. Table III (Figure 11) is the tabulation of sorted triads of indices and interplanar angles and is the principal output of this program.

APPLICATIONS

Use

For this procedure to be useful in the laboratory, a library of tables is necessary. A single table would identify the orientation of

TABLE II PAGE 3

			1 0 -2	2 0 -1	1 1 -6	1 1 -3	2 2 -3
0	0	1	141.763	107.603	155.538	137.702	118.791
-1	1	0	108.026	118.463	90.000	90.000	90.000
-1	1	1	133.010	124.363	119.189	113.347	104.954
-2	2	5	146.046	120.662	140.355	128.735	114.043
-1	1	4	147.589	117.134	147.874	133.486	116.621
-4	4	1	115.385	121.204	98.199	96.655	94.328
-1	2	0	90.000	90.000	78.051	70.337	64.011
-2	4	3	112.227	98.375	104.888	93.516	81.252
-1	2	3	125.519	102.926	122.271	108.702	93.516
-1	2	6	135.640	105.978	137.970	122.271	104.888
0	1	0	71.974	61.537	68.985	54.351	40.626
0	2	1	86.708	68.727	86.184	70.619	54.705
0	1	2	115.178	86.708	119.538	102.724	84.753
1	1	0	57.589	34.363	65.538	47.702	28.790
2	2	3	84.753	54.705	94.328	76.493	57.581
1	1	3	102.724	70.619	113.240	95.405	76.493
1	1	6	119.538	86.184	131.076	113.240	94.328
1	0	0	51.763	17.603	68.985	54.351	40.626
4	0	1	60.777	26.617	77.786	62.630	47.612
1	0	1	84.161	50.000	100.655	84.502	67.493
2	0	5	109.537	75.377	125.365	108.356	90.154
1	0	4	120.260	86.100	135.678	118.329	99.781
2	-1	0	57.589	34.363	78.051	70.337	64.011
4	-2	3	84.753	54.705	104.888	93.516	81.252
2	-1	3	102.724	70.619	122.271	108.702	93.516
2	-1	6	119.538	86.184	137.970	122.271	104.888
1	-1	2	115.178	86.708	135.640	125.519	112.228
2	-2	1	86.708	68.727	105.978	102.926	98.375
1	-1	0	71.974	61.537	90.000	90.000	90.000
-2	4	-3	67.773	81.625	51.696	49.374	51.977
-1	2	-3	54.481	77.074	35.649	39.326	49.374
-1	2	-6	44.360	74.022	23.898	35.649	51.696
0	4	-1	64.615	58.796	60.211	46.250	34.406
0	1	-1	46.990	55.637	37.769	27.326	25.989
0	2	-5	33.954	59.338	16.003	20.527	35.694
0	1	-4	32.411	62.866	11.949	25.603	43.424
2	2	-3	32.003	29.648	36.747	18.912	
1	1	-3	19.663	38.811	17.836		
1	1	-6	20.462	51.896			
2	0	-1	34.160				

Fig. 10. A sample page from Table II.

INDEX 1			ANGL 1,2	INDEX 2			ANGL 2,3	INDEX 3			ANGL 3,1	Z
2	2	3	40.626	1	0	0	32.397	1	0	1	25.989	3
-2	4	3	40.626	-1	1	0	32.397	-1	1	1	25.989	3
4	-2	3	40.626	1	0	0	32.397	1	0	1	25.989	3
2	2	3	40.626	1	0	0	30.000	1	1	0	28.790	3
4	-2	-3	40.626	1	0	0	30.000	2	-1	0	28.790	3
-2	4	3	40.626	-1	1	0	30.000	-1	2	0	28.790	3
-2	4	-3	40.626	-1	1	0	30.000	-1	2	0	28.790	3
-2	4	-3	40.626	0	1	0	30.000	-1	2	0	28.790	3
-2	4	3	40.626	0	1	0	30.000	-1	2	0	28.790	3
2	2	-3	40.626	1	0	0	30.000	1	1	0	28.790	3
2	2	-3	40.626	0	1	0	30.000	1	1	0	28.790	3
2	2	3	40.626	0	1	0	30.000	1	1	0	28.790	3
4	-2	3	40.626	1	0	0	30.000	2	-1	0	28.790	3
0	1	0	40.626	2	2	3	29.648	0	2	1	17.603	3
0	1	0	40.626	-2	4	3	29.648	0	2	1	17.603	3
1	1	3	39.326	-1	2	3	35.649	1	1	6	17.836	3
-1	2	3	39.326	1	1	3	35.649	-1	2	6	17.836	3
1	1	3	39.326	2	-1	3	35.649	1	1	6	17.836	3
2	-1	3	39.326	1	1	3	35.649	2	-1	6	17.836	3
1	1	3	39.326	2	-1	3	27.326	1	0	1	27.326	3
1	1	3	39.326	-1	2	3	19.663	0	1	2	19.663	1
-1	2	3	38.811	0	2	1	34.160	0	1	2	19.663	3
-1	2	3	38.811	0	2	1	29.648	-2	4	3	18.912	3
0	0	1	38.237	0	1	2	33.954	-2	2	5	32.226	3
0	0	1	38.237	0	1	2	32.411	-1	1	4	21.503	3
0	0	1	38.237	1	-1	2	32.411	1	0	4	21.503	3
0	0	1	38.237	0	1	2	32.411	1	0	4	21.503	3
0	1	2	38.237	0	0	1	24.462	1	1	6	20.462	3
0	1	2	38.237	0	0	1	24.462	-1	2	6	20.462	3
1	-1	2	38.237	0	0	1	24.462	2	-1	6	20.462	3
1	0	1	37.769	2	-1	6	36.747	4	-2	3	25.989	3
2	-1	6	37.769	1	0	1	36.099	1	0	4	11.949	3
1	0	1	37.769	2	-1	6	35.649	1	1	3	27.326	3
2	-1	6	37.769	1	0	1	27.326	2	-1	3	17.836	3
1	1	6	36.747	2	2	3	32.003	0	1	2	20.462	3
1	1	6	36.747	2	2	3	18.912	1	1	3	17.836	1
-1	2	6	36.747	-2	4	3	18.912	-1	2	3	17.836	1
1	0	4	36.099	1	0	1	27.326	1	1	3	25.603	3
1	0	4	36.099	1	0	1	25.377	2	0	5	10.723	1
-1	1	4	36.099	-1	1	1	25.377	-2	2	5	10.723	1
2	-1	6	35.649	1	1	3	25.603	1	0	4	11.949	3
1	1	3	35.649	2	-1	6	23.898	1	1	6	17.836	3
2	-1	3	35.649	1	1	6	23.898	2	-1	6	17.836	3
-1	2	3	35.649	1	1	6	23.898	-1	2	6	17.836	3
1	1	3	35.649	-1	2	6	23.898	1	1	6	17.836	3
-1	2	3	35.649	1	1	6	20.462	0	1	2	19.663	3
1	1	3	35.649	-1	2	6	20.462	0	1	2	19.663	3
4	-2	3	34.406	4	0	1	31.204	2	-1	0	28.790	3
4	0	1	34.406	4	-2	3	25.989	1	0	1	23.383	3
2	0	-1	34.363	2	-1	0	31.204	4	0	1	26.617	3

Fig. 11. A sample page from Table III.

all materials with cubic symmetry, however, most other materials would require an individual tabulation. The compilation of all these tables may seem to be an awesome task; but in this era of high speed digital computers, it represents just another small contribution to the information explosion. Tables for cubic symmetry and the more common engineering materials could be printed and distributed for the cost of reproduction. The computer program itself is meant to be used by the novice crystallographer-computer programmer in compiling tables pertinent to less common materials which require orientation. Once a tabulation is compiled, it is not necessary to make graphical artifices or to continually beseige the local computer complex with crystal orientation problems. Given a tried and true table, indexing becomes a simple task, easily executed by an inexperienced technician.

Several tabulations have been prepared for specific application. The author has experienced gratifying success with the tables for alpha alumina in the orientation of sapphire boules which are sectioned for bicrystal fabrication. Also, the cubic tables have been used successfully for the orientation of antimony doped silicon crystals for the Nuclear Engineering Department of North Carolina State University. Cadmium sulfide crystals with the wurtzite structure were oriented with only moderate success due to a misconception in the selection of the appropriate planes. Tables for an orthorhombic crystal, cobalt II silicide, have been compiled and sent to Northwestern University for the orientation of this lower symmetry anisotropic crystal. The author has recently prepared tables for cupric sulfite penta hydrate of the triclinic crystal system. The results with this system are not complete because of the lack of adequate diffraction data.

Several publications (14, 15, 16) will be available for wide distribution which will make future applications possible in other materials research laboratories.

Performance

A typical compilation of 4,500 triads and the tabulation of all three tables will consume only 5 or 6 minutes on Model 75 of this machine. (Five dollars a minute implies a thirty dollar investment.). The most efficient use of the computer was attempted to minimize expenses. During the course of compilation, the program computes and prints zone indices and interplanar angles according to equations in APPENDIX, page 35, for about 1,500 pairs of indices in Table I. All or some fraction of the interplanar angles are printed in the format of Table II, which was designed to consider any number of planes. The program then selects the 4,500 triads from a possible 25,000. The program does all of these operation in about one minute. Another two minutes are used to sort the 4,500 triads according to the 18 significant digits (six digits per interplanar angle) in each triad. The remaining two or three minutes are consumed in printing Table III.

Modifications

It has been suggested[2] that applications of this new indexing method to electron diffraction data could be accomplished better if only those combinations of three or more prominent poles on one zone were printed together with the d-spacings of the appropriate planes. Simple modifications in the program could produce an orientation library specifically tailored to the needs of electron diffraction.

RECOMMENDATIONS

The tremendous scope of current research on single crystals warrants further consideration of techniques to make the indexing of diffraction data as efficient as possible. It is hoped that PIMAX will help researchers characterize the anisotropic behavior of crystalline materials

---

[2]Personal communication, J. B. Wachtman, Jr., and N. Tighe, National Bureau of Standards, Washington, D. C., May 1967.

in order that new applications of these materials will advance technology.

More theoretical considerations of a scheme for selection and combination of indices should be conducted in order to provide an appropriate list for each of the 11 Laue symmetries.

CONCLUSIONS

A systematic method of indexing diffraction data has been established which reduces most of the crystallographic data describing a particular crystal structure to a simple, common, tabular form based upon unique triads of interplanar angles. This method has been successfully used to identify the orientation of single crystals in the cubic and hexagonal systems. Other symmetries will be compiled according to the interests of current research on single crystals.

LIST OF REFERENCES

1. Barrett, C. S. 1943. Structure of Metals. Pp. 511-515. McGraw-Hill Book Company, Inc., New York.

2. Camp, D. T., and J. A. Clum. 1966. Computer program for calculating interplanar angles and indexing back-reflection Laue data in an arbitrary crystal system. Transactions A.I.M.E., Vol. 236, p. 1752.

3. Cullity, B. D. 1959. Elements of X-ray Diffraction. P. 459. Addison-Wesley Publishing Company, Reading, Massachusetts.

4. Friedel, G. 1913. *Sur les symetries cristallines que peut reveler la diffraction des rayons Rontgen.* *Comptes Rendus*, Vol. 157, p. 1533. Paris, France.

5. Greninger, A. B. 1935. A back-reflection Laue method for determining crystal orientation. *Zeitschrift fur Kristallographie*, Vol. 91, p. 424. Frankfurt, Germany.

6. Groth, P. V. (first editor). 1877 to 1967. *Zeitschrift fur Kristallographie*, Vols. 1 through 123. Frankfurt, Germany.

7. Joint Committee on Chemical Analysis by X-ray Diffraction Methods. 1950. X-ray Diffraction Data Cards, Vols. 1 through 13. The American Society for Testing and Materials, Philadelphia.

8. Konnan, Y. A. 1963. A systematic method for indexing spots of single crystals in Laue X-ray photographs. *In* W. M. Mueller, G. Mallett, and F. Marie (editors), Advances in X-ray Analysis, Vol. 7, pp. 107-116. Plenum Press, New York.

9. Lonsdale, K. 1959. International Tables for X-ray Crystallography, Vols. 1, 2, and 3. The Kynoch Press, Birmingham, England.

10. Nuffield, E. W. 1966. X-ray Diffraction Methods, pp. 20-21, 351-384. John Wiley and Sons, Inc., New York.

11. Pearson, W. B. 1913 to 1959. Structure Reports, Vols. 1 through 23. *N.V.A. Oosthoek's Vitgevers Mij*, Utrecht, The Netherlands.

12. Riddhagni, B., and R. Asimow. 1966. New technique for orientation of crystal from Laue back-reflection photographs. Transactions A.I.M.E., Vol. 236, p. 1761.

13. Terpstra, P., and L. Codd. 1961. Crystallometry. Pp. 39-50, 209. Academic Press, New York.

14. Witter, D. E., and H. Palmour, III. 1967. Tabulation of indices and interplanar angles for rapid identification of crystallographic planes. Proceedings of an International Symposium on Anisotropy in Single Crystal Refractory Compounds, sponsored by Air Force Materials Laboratory, Wright Patterson Air Force Base, Ohio.

15. Witter, D. E., and H. Palmour, III. 1967. PIMAX tables for alpha alumina. Engineering Research Bulletin No. 84, North Carolina State University at Raleigh, Raleigh, North Carolina.

16. Witter, D. E., and H. Palmour, III. 1967. PIMAX tables for all cubic crystals. Engineering Research Bulletin No. 85, North Carolina State University at Raleigh, Raleigh, North Carolina.

17. Wyckoff, R. W. G. 1962 to 1967. Crystal Structures. Interscience Publishers, New York, Vols. 1 through 5.

APPENDIX

This section provides a card deck listing, a flow diagram, and equations to aid in the future use and further adaptation of PIMAX.

Listing of Card Deck

Appendix Figures 1 through 11 list a complete example card deck for PIMAX, including data and job control language.

```
//SAPPH  JOB NCS.MI.G949770,WITTER-DE,MSGLEVEL=0 TIME75=6
//  EXEC  PGM=MSG,PARM='TWO TAPES BOTH REL,SL,SCR,RING IN'
//  EXEC  PGM=MSG,PARM='ONE FOR FORTOUT AND SORTIN'
//  EXEC  PGM=MSG,PARM='ONE FOR SORTOUT AND UTILITY PRINT'
//  EXEC  FGLNKGO TIME=4
//FORT,SYSIN  DD *
C      A GENERAL PROGRAM FOR ALL CRYSTAL STRUCTURES
C       GIVEN ALPHA,BETA,GAMA,AND THE THREE DIMENSIONS OF THE UNIT CELL
C        THE ANGLES BETWEEN THREE PROMINENT POLES ON LOW INDEX ZONES ARE
C         TABULATED ACCORDING TO THE LARGEST INTERPLANAR ANGLE
C          WITH ASSOCIATED INDICES
      INTEGER H(100),ULTU,ULTV,ULTW
      DIMENSION K(100),L(100),PHI(100,100)
      READ (1,1)
1     FORMAT ('NAME OF MATERIAL FOR TITLE PAGE CENTERED ABOUT COLUMN
     1NUMBER FORTY ON CARD     ')
      READ (1,2)
2     FORMAT ('NAME OF INVESTIGATOR SAME AS ABOVE ON NEXT CARD
     1                          ')
      READ (1,3)
3     FORMAT ('ADDITIONAL INFORMATION - DATE, SPONCER, INSTITUTION, ETC
     1OR BLANK CARD INCLUDED ')
      WRITE (3,5)
5     FORMAT ('1',//////////17X,'A TABULATION OF INDICES AND INTERPLANA
     1R ANGLES',//14X,'FOR RAPID IDENTIFICATION OF CRYSTALLOGRAPHIC PLAN
     2ES',///37X,'FOR'/)
      WRITE (3,1)
      WRITE (3,7)
7     FORMAT (//34X,'COMPILED BY'/)
      WRITE (3,2)
      WRITE (3,3)
      WRITE (3,8)
8     FORMAT (///////////////24X,'THIS PROGRAM (PIMAX) WAS PREPARED',
```

Appendix Fig. 1. PIMAX card deck listing – part one.

```
     1/39X,'IN',
     2/35X,'MARCH, 1967',
     3/39X,'BY',
     4/32X,'DAVID E. WITTER',
     5/39X,'AT',
     6/25X,'NORTH CAROLINA STATE UNIVERSITY',
     7/37X,'OF THE',
     8/20X,'CONSOLIDATED UNIVERSITY OF NORTH CAROLINA',
     9/36X,'WITH THE',
     1/20X,'TRIANGLE UNIVERSITIES COMPUTATION CENTER',
     3/37X,'USING',
     3/15X,'FORTRAN IV AND JOB CONTROL LANGUAGE FOR IBM 360 MODEL 75')
      READ (1,100) A,B,C,AA,BB,CC
100   FORMAT (6F10.4)
      READ (1,104) MB,N,NI,NJ,INT,ULTU,ULTV,ULTW,MXPHI
104   FORMAT (9I5)
      READ (1,106) IUNO,IDUO,ITRE
106   FORMAT (3I5)
      COMB=(N*(N-1))/2.0
      WRITE (3,500) A,AA,B,BB,C,CC
500   FORMAT ('1',//27X,'INPUT DATA',//11X,8HA ZERO =,F10.4,6X,7HALPHA =
     1,F10.4,//11X,8HB ZERO =,F10.4,6X,7HBETA =  ,F10.4,//11X,8HC ZERO =,
     2F10.4,6X,7HGAMMA =,F10.4///)
      GO TO (58,59),MB
58    WRITE (3,501) N
      GO TO 57
59    WRITE (3,502) N
501   FORMAT (11X,'THE NUMBER OF MILLER, HKL, INDICES OF PROMINENT POLES
     1',/11X,'IS EQUAL TO',I5,'.'/)
502   FORMAT (11X,'THE NUMBER OF BRAVAIS, HK.L, INDICES OF PROMINENT POL
     1ES',/11X,'IS EQUAL TO',I5,'.'/)
57    WRITE (3,503) COMB,N,INT,ULTU,ULTV,ULTW,NI,NJ,N
503   FORMAT (11X,'THE COMPLETE LIST OF INDICES IS GIVEN IN THE FIRST LO
```

Appendix Fig. 2. PIMAX card deck listing – part two.

```
     1OP OF',/11X,'TABLE I WHICH TABULATES THE ZONE AND INTERPLANAR ANGL
     2E FOR',/11X,'THE',F7.0,' POSSIBLE COMBINATIONS OF',I5,' PLANES TAK
     3EN',/11X,'TWO AT A TIME.  THE INTERPLANAR ANGLE IS SET EQUAL TO ZE
     4RO WHEN',/11X,'THE ZONE IS NOT CONSIDERED TO BE PROMINENT.',///11X
     5,'INTERPLANAR ANGLES ARE TABULATED IN TABLE II FOR THE FIRST',I5,
     6/11X,'INDICES.',///11X,'ALL COMBINATIONS OF THREE PLANES WITH THE
     7ABSULUTE VALUES OF',/11X,'ZONE INDICES LESS THAN',3I3  ,' WILL BE
     8INCLUDED IN TABLE III,',/11X,'WHICH TABULATES INDICES AND INTERPLA
     9NAR ANGLES FOR TRIADS',/11X,'OF INDICES.  ONE PLANE IS SELECTED FR
     AOM THE FIRST',I5,/11X,'INDICES, A SECOND PLANE IS SELECTED FROM TH
     BE FIRST',I5,/11X,'INDICES, AND THE THIRD PLANE IS SELECTED FROM TH
     CE',/11X,'COMPLETE LIST OF',I5,' INDICES.')
      READ (1,101) (H(I),K(I),L(I),I=1,N)
101   FORMAT (3I5)
      AA=AA/57.29578
      BB=BB/57.29578
      CC=CC/57.29578
      S11=B*B*C*C*SIN(AA)**2
      S22=A*A*C*C*SIN(BB)**2
      S33=A*A*B*B*SIN(CC)**2
      S12=A*B*C*C*(COS(AA)*COS(BB)-COS(CC))
      S23=A*A*B*C*(COS(BB)*COS(CC)-COS(AA))
      S13=A*B*B*C*(COS(CC)*COS(AA)-COS(BB))
      ACS=0.75*((A*A)/(C*C))
      N1=N-1
      M=50
      IP=1
      DO 12 I=1,N1
      I1=I+1
      DO 12 J=I1,N
      M=M+1
      U=FLOAT(K(I)*L(J)-L(I)*K(J))
      V=FLOAT(L(I)*H(J)-H(I)*L(J))
```

Appendix Fig. 3. PIMAX card deck listing – part three.

```
      W=FLOAT(H(I)*K(J)-K(I)*H(J))
      E=ABS(U)
      F=ABS(V)
      G=ABS(W)
      IF (U) 40,41,40
40    X=E
      GO TO 46
41    X=1000.0
46    IF (V) 42,43,42
42    Y=F
      GO TO 47
43    Y=1000.0
47    IF (W) 44,45,44
44    Z=G
      GO TO 48
45    Z=1000.0
48    PRIME=AMIN1(X,Y,Z)
60    X=E/PRIME
      Y=F/PRIME
      Z=G/PRIME
      TEST=X+Y+Z-IFIX(X)-IFIX(Y)-IFIX(Z)
      IF (TEST) 49,54,49
49    PRIME=PRIME-1.0
      GO TO 60
54    JE=IFIX(U/PRIME)
      JF=IFIX(V/PRIME)
      JG=IFIX(W/PRIME)
      GO TO (55,56),MB
55    DI=SQRT(ABS(1.0/(S11*H(I)*H(I)+S22*K(I)*K(I)+S33*L(I)*L(I)+
     12.0*S12*H(I)*K(I)+2.0*S23*K(I)*L(I)+2.0*S13*H(I)*L(I))))
      DJ=SQRT(ABS(1.0/(S11*H(J)*H(J)+S22*K(J)*K(J)+S33*L(J)*L(J)+
     12.0*S12*H(J)*K(J)+2.0*S23*K(J)*L(J)+2.0*S13*H(J)*L(J))))
      COPHI=(DI*DJ)*(S11*H(I)*H(J)+S22*K(I)*K(J)+S33*L(I)*L(J)+S23*
```

Appendix Fig. 4. PIMAX card deck listing – part four.

```
      1(K(I)*L(J)+K(J)*L(I))+S13*(L(I)*H(J)+L(J)*H(I))+S12*(H(I)*K(J)+
      2K(I)*H(J)))
       GO TO 6
56     COPHI=(H(I)*H(J)+K(I)*K(J)+0.5*(H(I)*K(J)+H(J)*K(I))+ACS*L(I)*
      1L(J))/(SQRT(ABS((H(I)*H(I)+K(I)*K(I)+H(I)*K(I)+ACS*L(I)*L(I))*
      2(H(J)*H(J)+K(J)*K(J)+H(J)*K(J)+ACS*L(J)*L(J)))))
6      IF (COPHI) 25,26,27
26     PHI(I,J)=90.0
       GO TO 50
25     IF (COPHI+1.0) 29,28,29
28     PHI(I,J)=180.0
       GO TO 50
27     IF (COPHI-1.0) 29,31,29
31     PHI(I,J)=0.0
       GO TO 50
29     PHI(I,J)=ATAN(SQRT(1.0-COPHI*COPHI)/COPHI)*57.29578
       IF (PHI(I,J)) 24,50,50
24     PHI(I,J)=180.0+PHI(I,J)
50     IF (X-ULTU) 51,51,52
51     IF (Y-ULTV) 53,53,52
53     IF (Z-ULTW) 11,11,52
11     IF (PHI(I,J)-MXPHI) 64,64,52
52     PHI(J,I)=0.0
       GO TO 9
64     PHI(J,I)=PHI(I,J)
9      GO TO (80,12),IUNO
80     IF (M-51) 83,13,13
13     WRITE (3,105) IP
105    FORMAT ('1',//28X,'TABLE I',12X,'PAGE',I4,
      1//12X,'INDEX 1',4X,'ANGL 1,2',4X,'INDEX 2',5X,'ZONE 1,2'/)
       IP=IP+1
       M=0
83     WRITE (3,103) H(I),K(I),L(I),PHI(J,I),H(J),K(J),L(J),JE,JF,JG
```

Appendix Fig. 5. PIMAX card deck listing – part five.

```
103    FORMAT (10X,3I3,4X,F8.3,2X,3I3,4X,3I3)
12     CONTINUE
       GO TO (81,82),IDUO
81     J=INT
       IP=0
       GO TO 402
400    J=J-5
402    I=0
311    M=0
       IP=IP+1
       JILT=J+1
401    GO TO (302,302,303,304,305,306),JILT
301    WRITE (3,200) IP,H(J),K(J),L(J),H(J-1),K(J-1),L(J-1),H(J-2),K(J-2)
      1,L(J-2),H(J-3),K(J-3),L(J-3),H(J-4),K(J-4),L(J-4)
200    FORMAT ('1',33X,'TABLE II',20X,'PAGE',I4,
      1//21X,3I3,1X,3I3,1X,3I3,1X,3I3,1X,3I3/)
310    I=I+1
       M=M+1
300    WRITE (3,201) H(I),K(I),L(I),PHI(I,J),PHI(I,J-1),PHI(I,J-2),
      1PHI(I,J-3),PHI(I,J-4)
201    FORMAT (11X,3I3,5F10.3)
       IF (I+5-J) 360,600,600
360    IF (M-50) 310,311,311
306    WRITE (3,200) IP,H(J),K(J),L(J),H(J-1),K(J-1),L(J-1),
      1H(J-2),K(J-2),L(J-2),H(J-3),K(J-3),L(J-3)
600    I=I+1
403    WRITE (3,201) H(I),K(I),L(I),PHI(I,J),PHI(I,J-1),PHI(I,J-2),
      1PHI(I,J-3)
       GO TO 601
305    WRITE (3,200) IP,H(J),K(J),L(J),H(J-1),K(J-1),L(J-1),
      1H(J-2),K(J-2),L(J-2)
601    I=I+1
```

Appendix Fig. 6. PIMAX card deck listing – part six.

```
404    WRITE (3,201) H(I),K(I),L(I),PHI(I,J),PHI(I,J-1),PHI(I,J-2)
       GO TO 602
304    WRITE (3,200) IP,H(J),K(J),L(J),H(J-1),K(J-1),L(J-1)
602    I=I+1
405    WRITE (3,201) H(I),K(I),L(I),PHI(I,J),PHI(I,J-1)
       GO TO 603
303    WRITE (3,200) IP,H(J),K(J),L(J)
603    I=I+1
       WRITE (3,201) H(I),K(I),L(I),PHI(I,J)
       IF (J-5) 302,400,400
82     GO TO (302,4),ITRE
302    DO 4 I=1,N1
       I1=I+1
       DO 4 J=I1,NJ
       IF (PHI(J,I)) 30,4,30
30     J1=J+1
       DO 4 M=J1,N
       IF (PHI(M,J)) 32,4,32
32     IF (PHI(M,I)) 34,4,34
34     TEST=H(I)*K(J)*L(M)+K(I)*L(J)*H(M)+L(I)*H(J)*K(M)-
      1L(I)*K(J)*H(M)-K(I)*H(J)*L(M)-H(I)*L(J)*K(M)
       IF (TEST) 90,91,90
91     INDEX=1
       GO TO 92
90     INDEX=3
92     IF (PHI(I,J)-PHI(J,M)) 14,14,15
15     IF (PHI(J,M)-PHI(I,M)) 16,16,17
16     IF (PHI(I,M)-PHI(I,J)) 18,18,19
14     IF (PHI(J,M)-PHI(I,M)) 20,20,21
21     IF (PHI(I,M)-PHI(I,J)) 22,22,23
17     WRITE (4,102) H(I),K(I),L(I),PHI(I,J),H(J),K(J),L(J),PHI(J,M),
      1H(M),K(M),L(M),PHI(I,M),INDEX
       GO TO 4
18     WRITE (4,102) H(J),K(J),L(J),PHI(I,J),H(I),K(I),L(I),PHI(I,M),
```

Appendix Fig. 7. PIMAX card deck listing – part seven.

```
      1H(M),K(M),L(M),PHI(J,M),INDEX
       GO TO 4
22     WRITE (4,102) H(M),K(M),L(M),PHI(J,M),H(J),K(J),L(J),PHI(I,J),
      1H(I),K(I),L(I),PHI(I,M),INDEX
       GO TO 4
23     WRITE (4,102) H(J),K(J),L(J),PHI(J,M),H(M),K(M),L(M),PHI(I,M),
      1H(I),K(I),L(I),PHI(I,J),INDEX
       GO TO 4
20     WRITE (4,102) H(I),K(I),L(I),PHI(I,M),H(M),K(M),L(M),PHI(J,M),
      1H(J),K(J),L(J),PHI(I,J),INDEX
       GO TO 4
19     WRITE (4,102) H(M),K(M),L(M),PHI(I,M),H(I),K(I),L(I),PHI(I,J),
      1H(J),K(J),L(J),PHI(J,M),INDEX
102    FORMAT (3I3,F8.3,3I3,F8.3,3I3,F8.3,I2)
4      CONTINUE
       STOP
       END
/*
//GO.FT04F001  DD  UNIT=TAPE,DISP=(NEW,PASS),                        *
//               VOLUME=(,RETAIN),                                    *
//               DCB=(RECFM=FB,LRECL=53,BLKSIZE=2120)
//GO.SYSIN  DD *
                               ALPHA ALUMINUM OXIDE
                                 DAVID E. WITTER
                           NORTH CAROLINA STATE UNIVERSITY
   4.759     0.0        6.495      0.0        0.0        0.0
   2   51   21   29   51    6    6    6  120
   1    1    1
   0    0    1
  -1    2    0
  -2    4    3
  -1    2    3
  -1    2    6
```

Appendix Fig. 8. PIMAX card deck listing – part eight.

```
   0    1    0
   0    2    1
   0    1    2
   1    1    0
   2    2    3
   1    1    3
   1    1    6
   1    0    0
   4    0    1
   1    0    1
   2    0    5
   1    0    4
   2   -1    0
   4   -2    3
   2   -1    3
   2   -1    6
   1   -1    2
   2   -2    1
   1   -1    0
  -1    1    0
  -1    1    1
  -2    2    5
  -1    1    4
  -4    4    1
  -2    4   -3
  -1    2   -3
  -1    2   -6
   0    4   -1
   0    1   -1
   0    2   -5
   0    1   -4
   2    2   -3
   1    1   -3
```

Appendix Fig. 9. PIMAX card deck listing – part nine.

```
     1    1   -6
     2    0   -1
     1    0   -2
     4   -2   -3
     2   -1   -3
     2   -1   -6
     4   -4   -1
     1   -1   -1
     2   -2   -5
     1   -1   -4
    -1    1   -2
    -2    2   -1
     0    0   -1
/*
//        EXEC   PROC=SORT TIME=2
//SORT.SORTIN   DD   UNIT=TAPE,DISP=(OLD,DELETE),DSNAME=*.GO.FT04F001,     *
//               DCB=(RECFM=FB,LRECL=53,BLKSIZE=2120)
//SORT.SORTOUT   DD   UNIT=TAPE,DISP=(NEW,PASS),                           *
//               VOLUME=(,RETAIN),                                          *
//               DCB=(RECFM=FB,LRECL=53,BLKSIZE=2120)
//SORT.SORTWK01   DD   UNIT=DISK,SPACE=(TRK,(50),,CONTIG)
//SORT.SORTWK02   DD   UNIT=DISK,SPACE=(TRK,(50),,CONTIG)
//SORT.SORTWK03   DD   UNIT=DISK,SPACE=(TRK,(50),,CONTIG)
//SORT.SORTWK04   DD   UNIT=DISK,SPACE=(TRK,(50),,CONTIG)
//SORT.SORTWK05   DD   UNIT=DISK,SPACE=(TRK,(50),,CONTIG)
//SORT.SORTWK06   DD   UNIT=DISK,SPACE=(TRK,(50),,CONTIG)
//SORT.SYSIN   DD  *
 SORT    FIELDS=(11,3,CH,D,15,3,CH,D,28,3,CH,D,32,3,CH,D,                 *
               45,3,CH,D,49,3,CH,D),SIZE=E10000
 RECORD   TYPE=F,LENGTH=53
 END
/*
```

Appendix Fig. 10. PIMAX card deck listing – part ten.

```
//   EXEC   PGM=IEBPTPCH TIME=2
//SYSPRINT DD DUMMY
//SYSUT1     DD   UNIT=TAPE,DISP=(OLD,DELETE),DSNAME=*.SORT.SORTOUT,          *
//                DCB=(RECFM=FB,LRECL=53,BLKSIZE=2120)
//SYSUT2    DD    SYSOUT=A,                                                   *
//                DCB=(RECFM=F,BLKSIZE=132)
//SYSIN   DD   *
    PRINT MAXFLDS=1,MAXLINE=53
    TITLE   ITEM=('INDEX 1              INDEX 2',12),                         *
                ITEM=('INDEX 3          Z',46)
    TITLE   ITEM=('ANGL 1,2            ANGL 2,3',20),                         *
                ITEM=('ANGL 3,1',54)
    RECORD FIELD=(53,1,,10)
 *
```

Appendix Fig. 11. PIMAX card deck listing – part eleven.

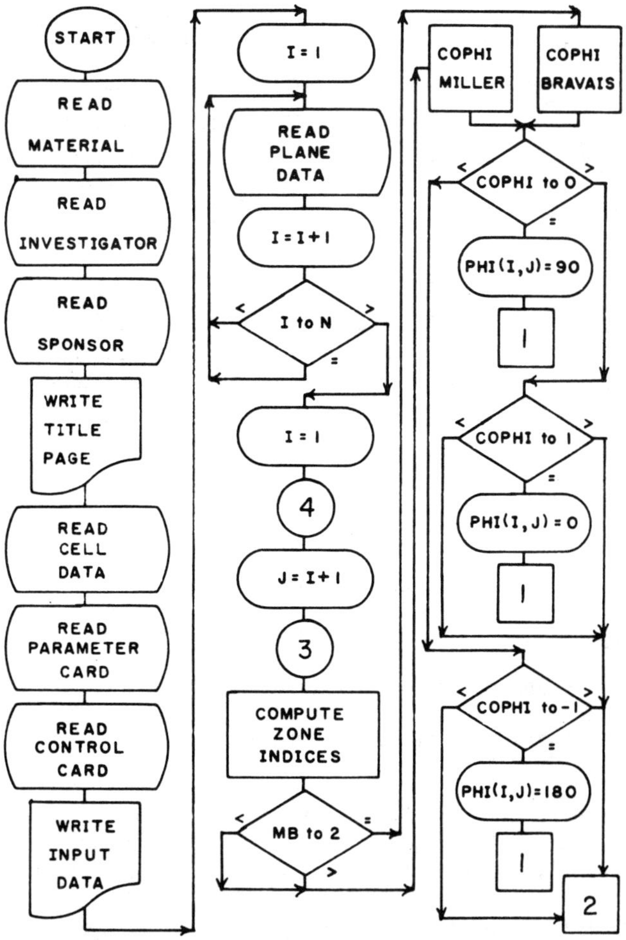

Appendix Fig. 12. Flow diagram for PIMAX – part one.

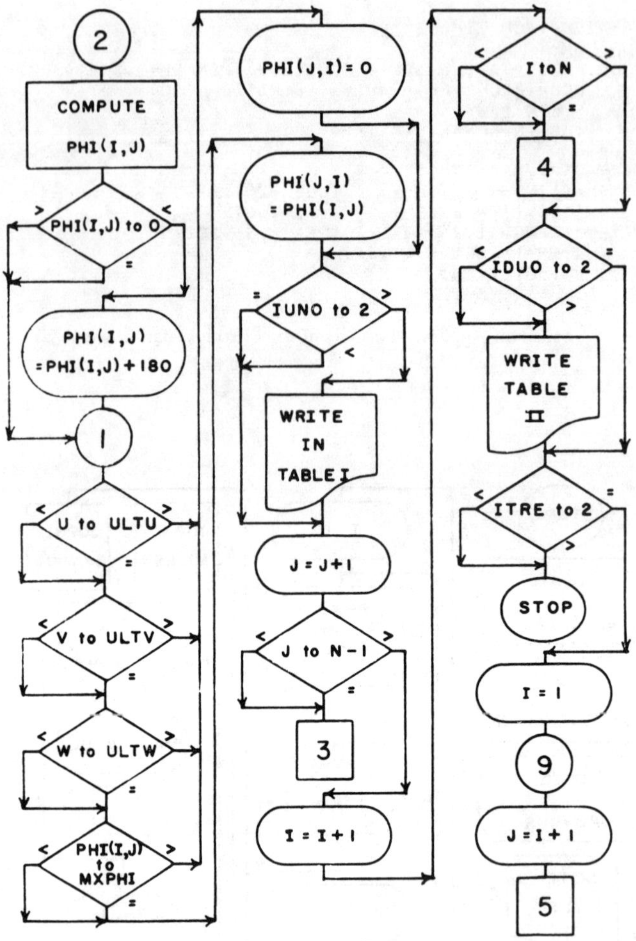

Appendix Fig. 13. Flow diagram for PIMAX — part two.

Flow Diagram

Appendix Figures 12, 13, and 14 graphically illustrate the main logic of PIMAX. A sequence which ends in a numbered square begins again in a circle containing the same numeral.

Equations

The equations (1, 3) considered by PIMAX are expressed in standard mathematical form.

Zone Indices

All possible combinations of indices of prominent poles taken two at a time are used to compute the zone indices according to the following relationships:

$$u = (k_1 l_2) - (l_1 k_2)$$

$$v = (l_1 h_2) - (h_1 l_2)$$

$$w = (h_1 k_2) - (k_1 h_2)$$

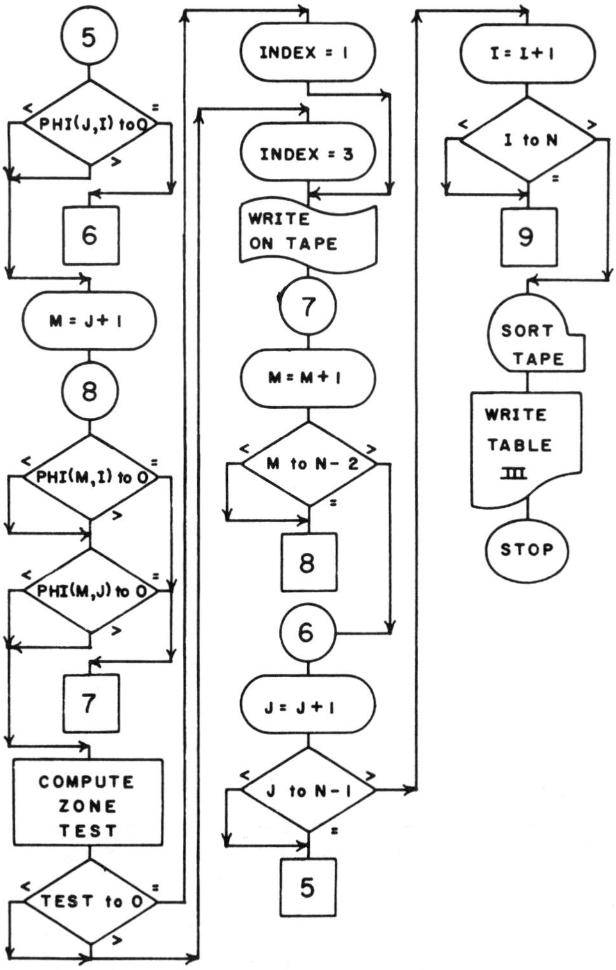

Appendix Fig. 14. Flow diagram for PIMAX – part three.

where,

$$uvw = \text{zone indices}$$

$$h_1k_1l_1 = \text{indices of one pole}$$

$$h_2k_2l_2 = \text{indices of other pole}$$

Interplanar Angles

If Miller indices are given, the following equations are used:

$$\cos\phi = \frac{d_1d_2}{V^2}\,[S_{11}h_1h_2 + S_{22}k_1k_2 + S_{33}l_1l_2 + S_{23}(k_1l_2 + k_2l_1)$$
$$+ S_{13}(l_1h_2 + l_2h_1) + S_{12}(h_1k_2 + h_2k_1)]$$

$$\phi = \arctan\frac{\sqrt{1 - \cos^2\phi}}{\cos\phi}$$

where,

a,b,c = axial ratios

α,β,γ = axial angles

$S_{11} = b^2c^2\sin^2\alpha$

$S_{22} = a^2c^2\sin^2\beta$

$S_{33} = a^2b^2\sin^2\gamma$

$S_{12} = abc^2(\cos\alpha\,\cos\beta - \cos\gamma)$

$S_{23} = a^2bc(\cos\beta\,\cos\gamma - \cos\alpha)$

$S_{13} = ab^2c(\cos\gamma\,\cos\alpha - \cos\beta)$

V = volume of unit cell

$V = abc\,\sqrt{1 - \cos^2\alpha - \cos^2\beta - \cos^2\gamma + 2\cos\alpha\,\cos\beta\,\cos\gamma}$

d = d-spacing

$$\frac{1}{d_1} = \frac{1}{V^2}\,(S_{11}h_1^2 + S_{22}k_1^2 + S_{33}l_1^2 + 2S_{12}h_1k_1 + 2S_{23}k_1l_1 + 2S_{13}h_1l_1)$$

$\frac{1}{d_2}$ = same with subscript 2

ϕ = interplanar angle

If abbreviated Bravais indices are given, the following equation is substituted for cos ϕ:

$$\cos \phi = \frac{h_1h_2 + k_1k_2 + \frac{1}{2}(h_1k_2 + h_2k_1) + \frac{3a^2}{4c^2} l_1l_2}{\sqrt{(h_1^2 + k_1^2 + h_1k_1 + \frac{3a^2}{4c^2} l_1^2)(h_2^2 + k_2^2 + h_2k_2 + \frac{3a^2}{4c^2} l_2^2)}}$$

Tautozonal Indices

Three indices are determined to be on one zone if the following condition exists:

$$h_1k_2l_3 + k_1l_2h_3 + l_1h_2k_3 - l_1k_2h_3 - k_1h_2l_3 - h_1l_2k_3 = 0$$

SESSION II

Electronic Structure and Bonding

Chairman: Bertil Aronsson
Swedish Institute for Metals Research
Stockholm, Sweden

Co-Chairman: L. E. Toth
University of Minnesota
Minneapolis, Minnesota

THE ELECTRONIC STRUCTURE OF REFRACTORY METALS, COMPOUNDS, AND ALLOYS

D. F. Gibbons
The Center for the Study of Materials
Case Institute of Technology
Cleveland, Ohio

Introduction

I would like to make it clear that I am approaching this subject as an experimentalist, not a theoretician. Because of their success in the case of closed d shell elements, we are developing confidence in using the results of APW calculations for the interpretation of experimental data on the transition metals and their compounds. I feel that, at the present time, the theoreticians are too far ahead and that we are in dire need of clear experimental data on those materials before the theoreticians can revise their calculations in a meaningful manner.

The relative closeness in energy of the s and d states in the early groups of the transition metal series makes these refractory metals ideal systems in which to study the interactions between these states. However, the hybridizing of the s and d states manifests itself in such a complex array of transitions and defect structures that we have some difficulty in sorting out the wheat from the chaff.

I will present a brief review of the various methods which are used to calculate the band structure of transition metals and compounds. I will then present and discuss some data on the metals and alloys of Groups IV and V of the first two transition metal series -- in particular, zirconium and its alloys with niobium. Finally, I would like to discuss some recent progress which is being made in determining the electronic configuration of molecules and speculate on how these data may help us gain a better insight into the antiferromagnetic behavior of some of the oxides.

Band Structure Calculations

The problems of the precise characterization of the electronic structure of solids which depart from the perfect ionic model -- for example, NaCl -- or the ideal free electron metal -- for example, Na -- provide a theoretically intractable problem. In order to make the problem solveable by present-day mathematical techniques, certain assumptions have to be made. These assumptions include, the assignment of the electronic configuration of the individual atoms, usually using the free atom configurations as a basis; the degree of ionization of the ions on the lattice; the amount of charge transfer between lattice sites and the omission of nuclear motion (thermal vibrations). These are some examples of the decisions which are made before setting down the problem in mathematical form.

The first successful attempt to modify the free electron model, taking into account bonding and hybridization of the wave functions for metal-like systems, was achieved by Cohen, Falicov, and Golin[1] by their use of the pseudo-potential method. They applied this technique to Bi, a Group V element. This method has been very successful, and has enabled them to make band calculations which agree remarkedly well with the experimental observations of the band gaps, etc. for many semimetals and semiconductors.[2,3] This technique, however, is less satisfying when attempts are made to apply it to ions with incomplete shells, such as the d shells in the transition metals although, recently, attempts are being made to use this technique in such cases by V. Heine and others.

The difficult problem of treating d bands by the OPW and pseudo-potential techniques, together with the emergence of relatively easy access to large-scale computers, has led to the increasing use of the augmented plane wave method[4] (APW), based on the muffin-tin model of the crystal potential. This method allows all core electrons outside the last inert gas shell to be treated together from Hartree-Fock-Slater functions within an arbitrarily-chosen sphere about the nucleus, the remaining volume of the unit cell being

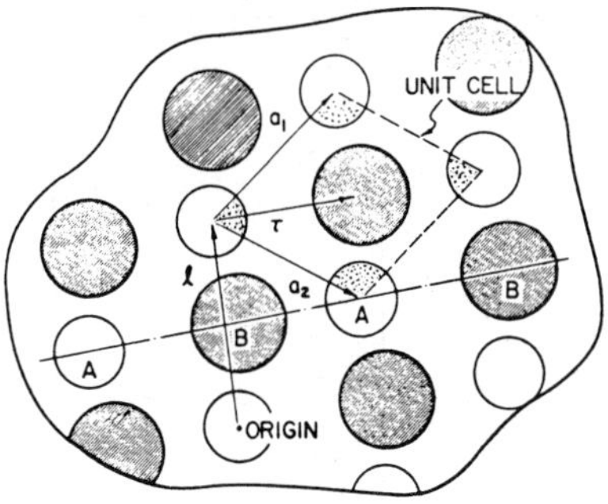

Fig. 1. Unit cell for a two-dimensional crystal showing sphere centered on the lattice sites in which the potential is spherically symetric. Outside spheres potential is taken as a constant.

occupied by plane "free electron-like" waves (Figure 1). These plane waves are matched at the surface of the sphere with the core functions in a self-consistent manner. This is a relatively simple procedure, but exceedingly tedious, and ideally suited to high-speed, large-memory computers.

Although the APW method does not have to make a choice between assigning certain of the electrons as core electrons and others as itinerant, it does have to make arbitrary decisions regarding the radius of the sphere, the muffin-tin radius, and the electron configuration of the core states which, in the case of the compounds, includes the degree of ionization. Also, whenever directional bonding becomes important, the spherically symmetrical muffin-tin model is not so good a choice, and leads to inaccuracies. In spite of these difficulties, the APW method has been successfully used in making meaningful band calculations of the transition metals[5] and their compounds.[6]

Figure 2 shows an example of such a calculation made across the elements of the first transition metal series[7] and demonstrates the progressive suppression of the d band as the series is traversed from Ti to Zn. Figure 3 demonstrates quite clearly that the decision regarding the precise core configuration is critical in defining the d band. Figure 4 shows the results of an APW calculation of the band structure for the compound titanium oxide, by Ern and Switendick.[6] As the authors point out in their paper, the decision of the degree of ionization is exceedingly important as the electronegativity of the anion increases. They assume a core configuration of Ti^+ and O^- for titanium oxide, but assume initially no charge transfer for the nitride. However, summing up the total number of electrons in all states, in both the

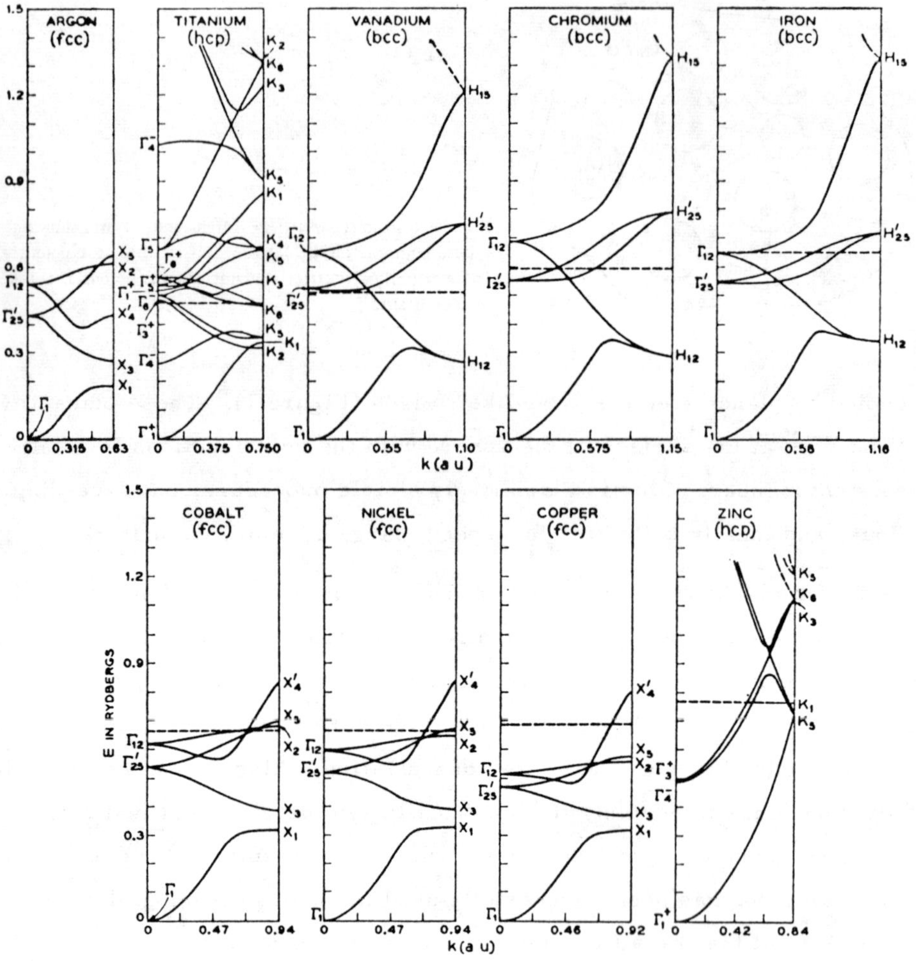

Fig. 2. Energy bands for the series argon through zinc, as function of wave vector (L. F. Mattheiss).

spheres and plane wave band, shows that, in the nitride, the APW model had enforced some charge transfer. This demonstrates that it is rather critical, if accurate band structure calculations are to be made, that all possible data which may give a clue as to the initial electron configuration of the atoms should be obtained.

Transition Metals and Alloys

I am going to focus my attention on one particular aspect of the metals and alloys, first because I believe that it demonstrates clearly some aspects of the bonding problem and, second, because we have some data which can

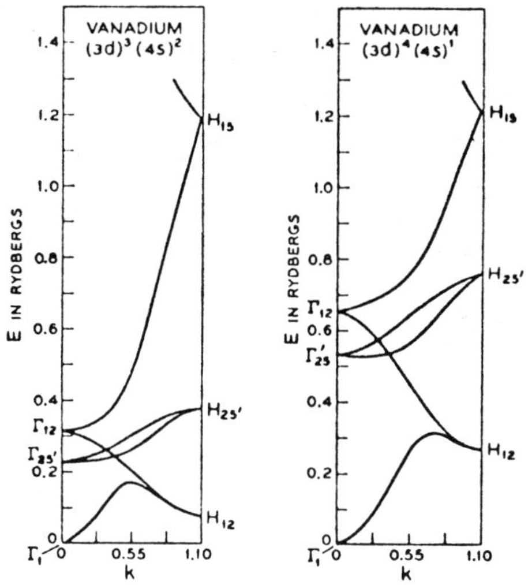

Fig. 3. Effect of choice of atomic configuration, for vanadium, on width and energy of bands (L. F. Mattheiss).

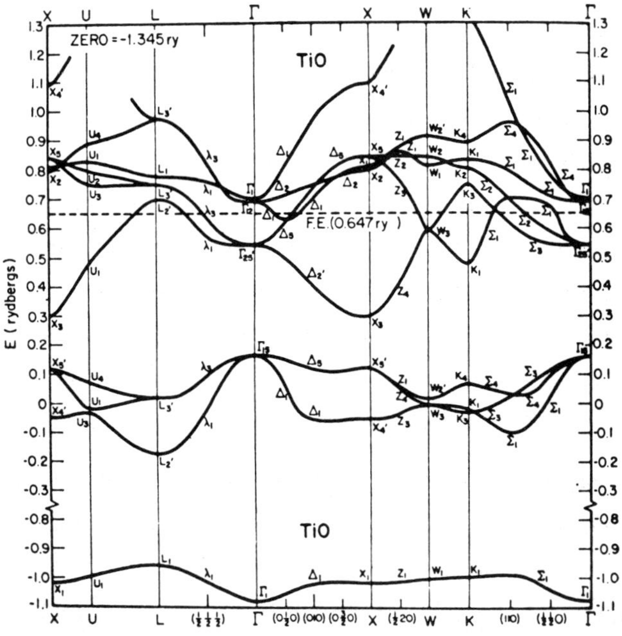

Fig. 4. Energy bands for cubic TiO as a function of wave vector (Ern and Switendick).

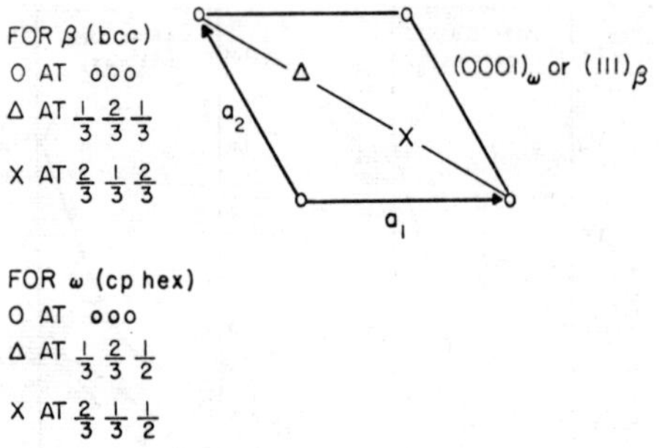

Fig. 5. Unit cell for the omega phase.

help to establish a model. The particular phenomenon is the so-called omega "ω" transformation which occurs in pure zirconium and quenched transition metal alloys of many Group V A elements as solutes in Group IV A solvents.

The omega phase is a hexagonal phase (D^1_{6h}) containing three atoms per unit cell (Figure 5). In pure zirconium, the omega phase is a high-pressure phase, produced at approximately 60 k.bar.[8] The transformation is unusual, however, in that it is retained as a metastable phase after reducing the pressure to atmospheric. There are very few high-pressure phases which are retained at atmospheric pressure; the cubic sodium chloride structure, InTe,[9] is another example. The stability of the omega phase at atmospheric pressure is strong evidence that it is the result of a change in the character of the bonding orbitals, rather than a Fermi surface-Brillouin zone interaction. Evidence that this is the correct model to use to explain the transformation is given by the fact that the superconducting transition temperature is, to a first approximation, unaffected by the transformation.[10] Since the expression for the energy gap of a superconductor involves the density of states at the Fermi surface, any interaction with the zone energy gap would be expected to significantly change the transition temperature.

The omega phase can also be produced by quenching the b.c.c. high temperature phase of titanium and zirconium alloys containing more than 7% of Group V A solutes, such as vanadium or niobium.[11] The restrictions on the alloying element are, in fact, much less stringent, and the omega phase can be produced in alloys where the solute comes from Group V A or higher; that is, the electronic structure of the free atom contains four

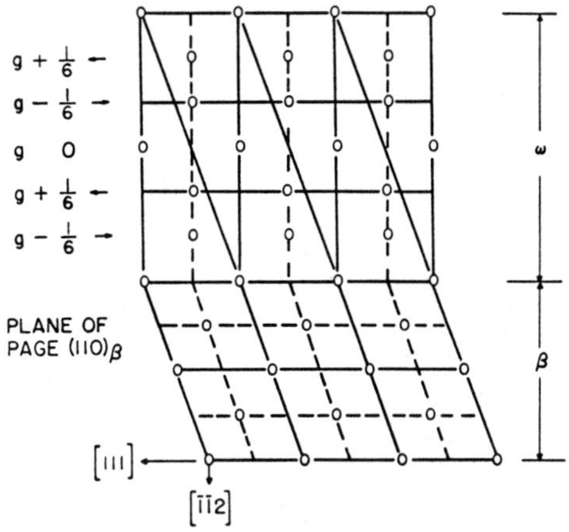

Fig. 6. Combination of shears which formally transform the b.c.c. phase to the omega phase.

or more d electrons. The omega phase is produced athermally from the b.c.c. phase. The transformation can be formally described by the two atoms at $\pm(\frac{2}{3}, \frac{1}{3}, \frac{1}{3})$ in the b.c.c. lattice, when indexed on a hexagonal basis $\{(111)_c$ // $(0001)_{hex}$ and $[10\bar{1}]_c$ // $[11\bar{2}0]_{hex}\}$, moving in $[111]_c$ directions to $\pm(\frac{2}{3}, \frac{1}{3}, \frac{1}{2})$ (Figure 6). Crystallographically, this can be described as $R\bar{3}m(D^5_{3d}) \rightarrow P\,6/mmm(D^1_{6h})$. The omega phase is brittle, which again indicates that the bonding has changed to one in which the bonds are more localized, having developed some of the character of covalent bonds.

TABLE I

Elastic moduli of b.c.c. Zr-25% Nb and Nb at 300°K.

	S_{11}	S_{12}	S_{44}	Anisotropy Constant $A = \frac{2(S_{11}-S_{12})}{S_{44}}$
Zr-25% Nb	2.22	-1.03	3.13	2.06
Nb	0.60	-0.233	3.481	0.51

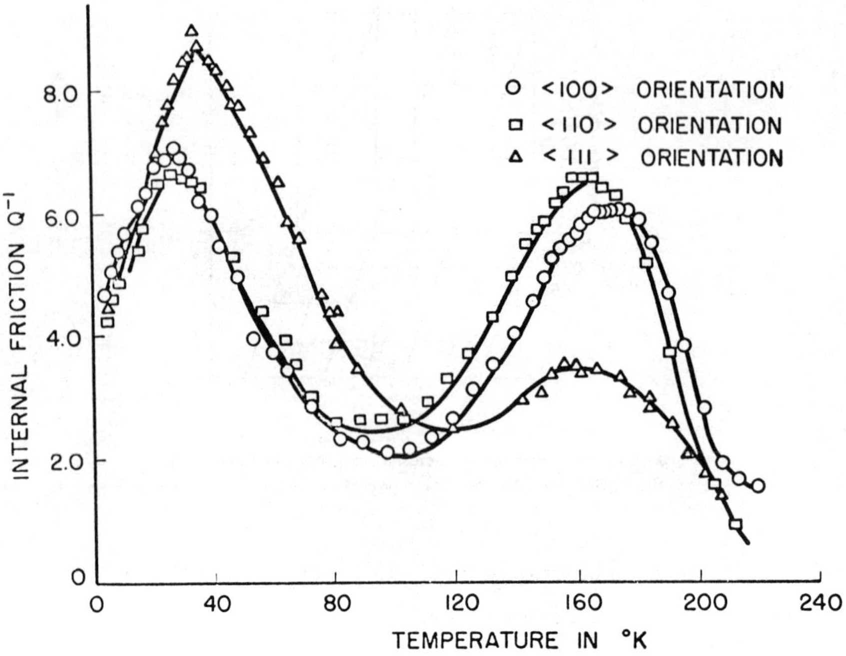

Fig. 7. Temperature variation of the internal friction for longitudinal waves propagated along <100>, <110>, and <111> directions of Zr–25% Nb single crystals.

We have measured the elastic moduli of single crystals of Zr-25% Nb alloys. Table 1 lists these data. The most significant aspect of these data is that the anisotropy constant is approximately 2.0, in contrast to the b.c.c. niobium for which A = 0.50. This indicates that the bonding in the b.c.c. phase is undergoing a radical change as we move toward the zirconium-rich alloys.

Acoustic measurements[12] on zirconium-base alloys containing from 5-25% Nb have demonstrated a low temperature acoustic relaxation peak (40°K) associated with the omega phase. Figure 7 shows this relaxation for different orientations of single crystals of Zr-25% Nb.* The activation energy for this relaxation process is 0.05 eV/atom (1.1 k.cal/mole). This activation energy is so small that it almost certainly is associated with some electronic relaxation phenomenon.

---

* The relaxation occurring at 160°K is the result of another dissimilar process and will not be discussed further in this paper.

The orientation dependence of the magnitude of the relaxation peak for the Zr-Nb alloys is identical to that observed in the ferrites[13] where the relaxation process is known to have as its origin an electronic re-arrangement. The activation energy is also of the same order of magnitude for the two systems. It is, therefore, very tempting to assign the origin of the relaxation process in Zr-Nb alloys to an electronic process, where the application of an external strain changes the electronic population between two orbitals.

At this point, I would like to draw an analogy between the local site symmetry in the omega phase and that of the $(ReH_9)^{--}$ ion.[14] In each case, they have a lattice site with three nearest neighbors in a planar triangular array with six other neighbors, whose separation distance is only about 2% greater, at the corners of a triangular prism. This is illustrated in Figure 8 for the $(ReH_9)^{2-}$ molecule. There is no doubt, in the case of $(ReH_9)^{2-}$, that the bonding is covalent and that the symmetry is dictated by the sd molecular orbitals.

We may speculate that the origin of the stability of the omega phase in the alloys and pure zirconium may reside in the formation of sd hybrid bonding orbitals. If the orbitals responsible for the bonding in c. p. hexagonal and

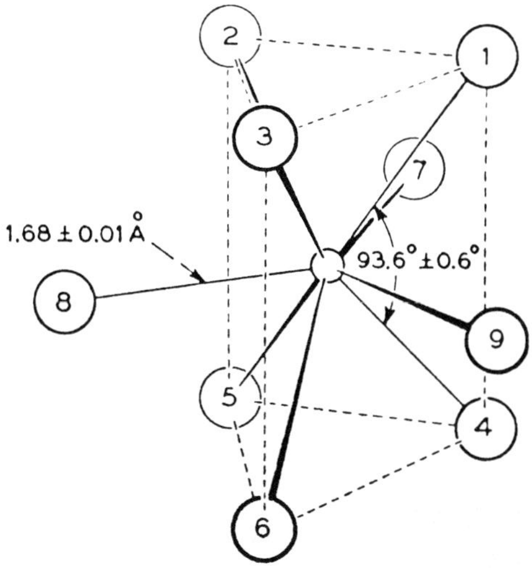

Fig. 8. Configuration of nearest neighbor (7-9), and next nearest neighbor sites (1-6) for $(ReH_9)^{2-}$ and the omega phase. Distances and angles are those for the $(ReH_9)^{2-}$.

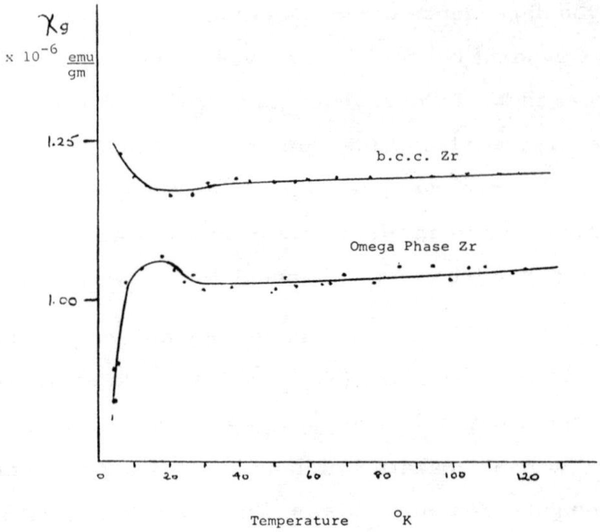

Fig. 9. Magnetic susceptibility of zirconium in the c. p. hexagonal and omega phases.

b. c. c. phases are primarily sp or pd hybrids, then it is possible to obtain a strain induced relaxation between these levels and the s-d omega levels. This is possible if the energy difference between the gerade sd orbitals and ungerade sp or pd orbitals is large enough to prevent mixing.[15]

Some of the consequences of this model for the bonding involved in the omega transformation are presently being tested. The low temperature acoustic relaxation should occur in the pure zirconium omega phase. A change in the bonding orbital structure should manifest itself as a change in the magnetic susceptibility and N. M. R. spectrum. Preliminary data on the magnetic susceptibility is shown in Figure 9. It can be seen that, below about 40°K, the susceptibility of the Zr omega phase shows a sharp decrease, the phase becoming more diamagnetic. Formation of localized orbitals would produce such a change; however, more experimental work has to be done before we can be certain that this is the correct explanation.

Transition Metal Compounds

Since it is becoming evident that the bonding orbitals play an important role in the structure and properties of the transition metals and their alloys, it is not surprising that they are even more important in the case of their compounds, especially with carbon, nitrogen, and oxygen. My previous comments in the section on band structure pointed out the weakness of the muffin-tin, spherically-symmetrical potential model when localized bonding orbitals occur. In order to realistically take such bonding into account, we should be able to make use, as a starting point, of the data from molecular orbital calculations.

During the last few years, again as a result of the introduction of high-speed computers, it has become possible to carry out calculations on diatomic molecules in which all the electrons are treated simultaneously, using the many-electron Hamiltonian in the Hartree-Fock approximation.[16,17] These calculations require no initial assumptions because of the simplicity of the molecules and the relatively small number -- of the order of 30 or 40 -- of electrons. They provide reliable, quantitative data on the lowest energy ground state configuration of the molecular orbits. The internuclear spacing of such molecules is only 20 to 30 percent different from that in the solids and, therefore, these molecular calculations should provide an excellent basis for purturbation calculations of an APW model.

Calculations of the electronic configuration of the series scandium, titanium, and vanadium oxide molecules have been made, and it is possible to correctly calculate a ground state of these molecules which agrees with spectrographic data wherever such data is available. It is also possible to observe the trends in these molecular configurations across the series and down the columns through increasing nuclear charge. Table II quite clearly demonstrates these data. Figure 10 shows the distribution of the four outermost electronic states for titanium oxide, TiO.

From the contours of $|\Psi_i|^2$, it can be seen that the 7σ consists mainly of the 2s (O), with a small mixture of $3p\sigma$ and $3d\sigma$ from the titanium, and the 9σ is mainly the 4s (Ti). The two main bonding orbitals are the 8σ, which is a mixture of $3d\sigma$ (Ti) and 2 $p\sigma$ (O), and the 3π, which consists of a mixture of $3d\pi$ (Ti) $2p\pi$ (O). Thus, the ordering of the levels is very much in line with the bands calculated for TiO (Figure 4).

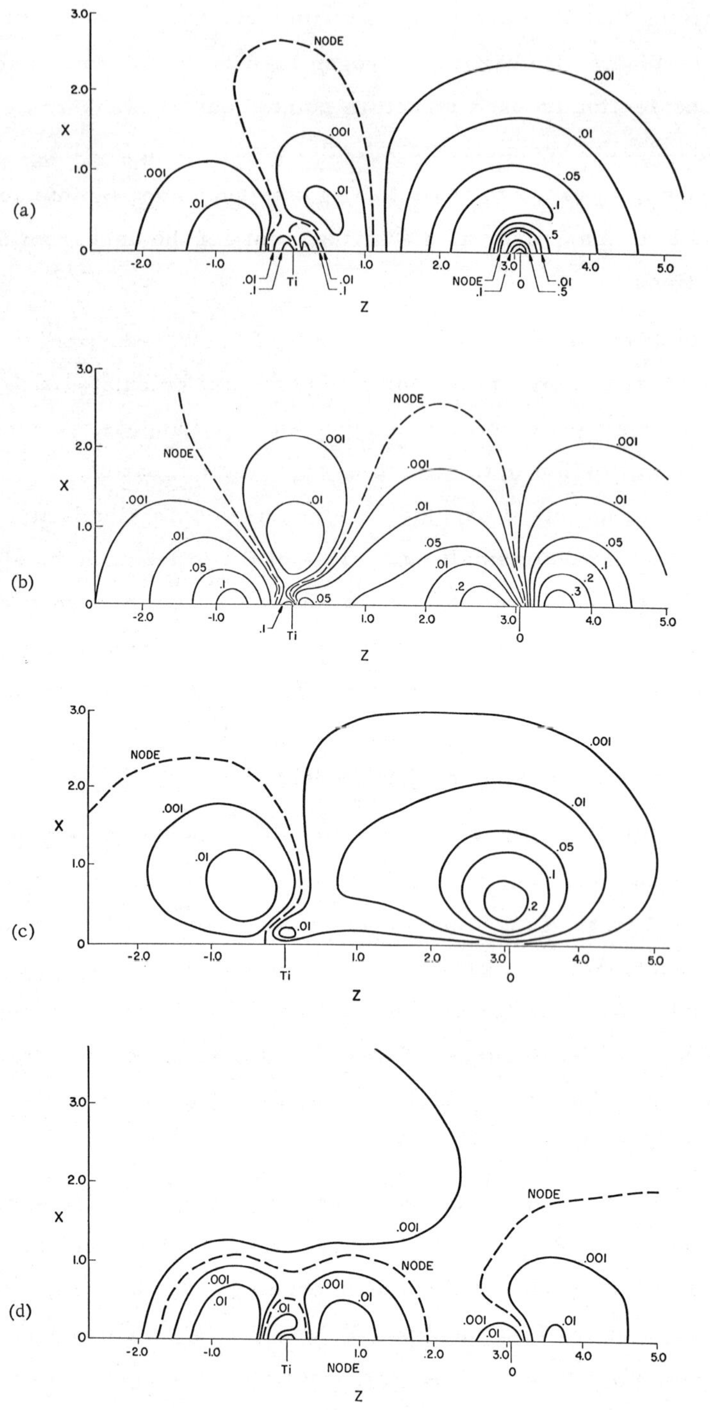

Fig. 10. Contours of constant $|\psi_i|^2$ for (a) 7σ, (b) 8σ, (c) 3π, (d) 9σ orbitals of TiO ($^1\Sigma^+$) (after I. Cohen).

TABLE II
Ground state orbital configurations of the monoxides of Sc, Ti, and V family elements.

ScO	$\sigma(^2\Sigma^+)$	TiO	$\delta\sigma(^3\Delta_r)$	VO	$\delta^2\sigma(^4\Sigma^-)$
YO	$\sigma(^2\Sigma^+)$	ZrO	$\sigma^2(^1\Sigma^+)$	NbO	$\delta\sigma^2(^2\Delta_r)$
LaO	$\sigma(^2\Sigma^+)$	HfO	$\sigma^2(^1\Sigma^+)$	TaO	$\delta\sigma^2(^2\Delta_r)$

Calculations on vanadium oxide, however, demonstrate that the ordering of the 8σ and 3π is reversed. Thus, energies of these two levels are very close. The relative energies of these two levels in the solid is very important with respect to the magnitude of the super-exchange term in those

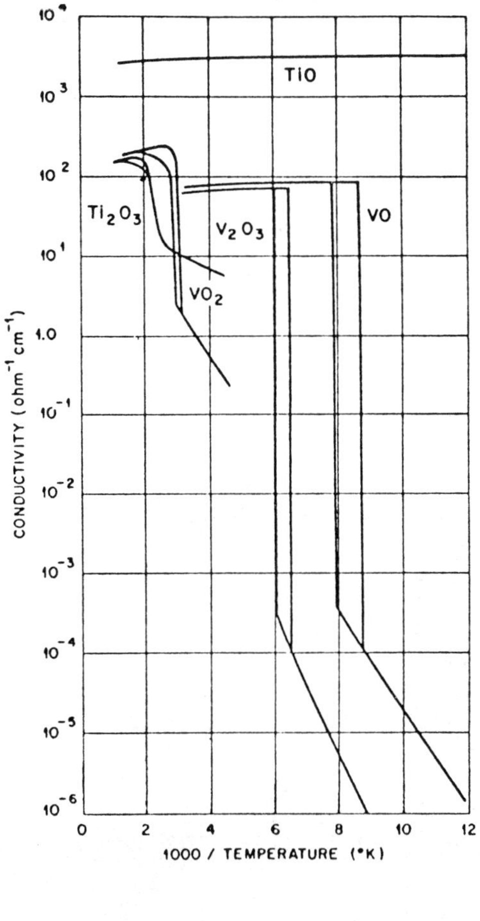

Fig. 11. Conductivity as a function of reciprocal temperature for oxides of titanium and vanadium (after F. J. Morin).

compounds, the kinetic exchange term of P. W. Anderson.[18] Figure 11 represents the data of Morin on the conductivity versus temperature for a number of the vanadium and titanium oxides. The transition in conductivity with temperature coincides with the Neél temperature. TiO does not have an antiferromagnetic transition; however, the other oxides of titanium and those of vanadium do.

To a first approximation, we may expect the same ordering of the levels to hold over into the solid state, with the higher energy states being the first to be broadened into band states. We may expect the 3π orbital to play an important role in super-exchange. In TiO, this orbital may become a band state whereas, in the other oxides of titanium and the vanadium oxides at low temperatures, it may remain a localized bond orbital responsible for the large kinetic exchange term. At higher temperatures, however, it may well become nonlocalized and mix with the band states and remove the super-exchange. In order to test this speculation further, it will be necessary to determine precisely how these molecular orbitals transform when inserted into the octahedral crystal symmetry.

I would like to thank Professor D. Carlson for illuminating discussions regarding the *ab initio* many electron molecular orbital calculations and Professor R. Hehemann for discussions on the omega phase studies in Zr-Nb alloys.

References

1. M. H. Cohen, L. Falicov, and S. Golin, *IBM J. Res. and Dev.*, *8*, 215 (1964).

2. G. Weisz, *Phys. Rev.*, *149,* 504 (1966).

3. W. A. Harrison, *Pseudo-potentials in the Theory of Metals*. W. A. Benjamin, Inc., New York, (1966).

4. See, for example, T. Loudes, *Augmented Plane Wave Method*. W. A. Benjamin, Inc., New York (1967).

5. L. F. Mattheiss, *Phys. Rev.*, *139,* 236 (1965).

6. V. Ern and A. C. Switendick, *Phys. Rev.*, *137*, 202 (1965).

7. L. F. Mattheiss, Phys. Rev., 134, 192 (1964).

8. J. C. Jamieson, Science, 140, 72 (1963).

9. A. J. Darnell, W. F. Libby, A. J. Yencha, Science, 141, 713 (1963).

10. B. Tittmann, D. Hamilton, and A. Jayaraman, J. Appl. Phys., 35, 732 (1964).

11. P. D. Frost, W. M. Parris, L. L. Hirsch, J. R. Doig, and C. M. Schwartz, Trans. Am. Soc. Metals, 46, 231 (1954).

12. C. W. Nelson, D. F. Gibbons, and R. F. Hehemann, J. Appl. Phys., 37, 4677 (1966).

13. D. F. Gibbons, J. Appl. Phys., 28, 810 (1957).

14. S. C. Abrahams, A. P. Ginsberg, and K. Knox, Inorg. Chem., 3, 558, (1964).

15. S. L. Altman, C. A. Coulson, and W. Hume-Rothery, Proc. Roy. Soc. (London), A240, 145 (1957).

16. K. D. Carlson and C. R. Claydon, Advances in High Temperature Chemistry, Vol. 1. Academic Press, New York. To be published.

17. K. D. Carlson and R. K. Nesbet, J. Chem. Phys., 41, 1051 (1964).

18. P. W. Anderson, Proceedings of the Buhl International Conference on Transition Metal Compounds. Gordon & Breach, NY (1964).

BONDING, IMPERFECT STRUCTURE, AND PROPERTIES OF THE REFRACTORY NITRIDES OF TITANIUM, ZIRCONIUM, AND HAFNIUM

M. E. Straumanis
Graduate Center for Materials Research*
University of Missouri
Rolla, Missouri

Abstract

Titanium, zirconium, and hafnium nitrides were investigated by chemical means and by X-ray diffraction concerning their structure. From the dissolution of the nitrides in HF (under formation of NH_4^+, N_2, and H_2), from N_2 production at elevated temperatures in vacuum and furthermore, from the heats of formation, lattice energies, hardness, and high melting points, the conclusion was drawn that the bonding in the three nitrides is of a mixed type. The bonding appears to be predominantly ionic and to some extent metallic; the cationic sublattice consists throughout of the trivalent ions (Ti^{3+}, Zr^{3+}, and Hf^{3+}) and there is a nitrogen equilibrium ($N \rightleftharpoons N^{3-} + 3e$) in the anionic sublattice. The partially metallic character of the bonding accounts for the high electrical conductivity, magnetic properties, metallic luster of the nitrides. Since the N atoms do not accept all the available valence electrons from the metal atoms, the average radius of the N anions must be smaller, from 1.4 to 1.425 Å,

*Contribution No. 24 from the Center

(depending upon the degree of ionization of N) than that calculated for N^{3-}. Using these values for the N ionic radius and the radii of the three trivalent cations, a good agreement between calculated and experimental lattice constants can be reached. There are vacant sides (cationic and anionic) in the three nitrides: 5.14 in the TiN, 4.59 in the ZrN (both containing an excess of N) and 12.62% in the stoichiometric HfN.

Introduction

The nitrides of the transition metals Ti, Zr, and Hf of the IV B group have unusual properties. At first, there is the high melting point of these nitrides (Table I), which, however, can be utilized only under conditions of absence of oxygen or in a neutral atmosphere, because the three nitrides are easily converted (at or even below 1000°C) into the dioxides, e.g.,

$$HfN + O_2 \longrightarrow HfO_2 + 0.5\ N_2 \tag{1}$$

In vacuum a slight dissociation of the nitrides occurs (liberation of N_2) far below their melting points. All three nitrides are very hard and brittle.

A further, very significant property of the nitrides is their high electrical conductivity (Table II) and their metallic

Table I

Melting point of nitrides of the IV B group

Subst.	Melt. pt. in °C	Literat.
TiN	2930-50±50°	1,2
ZrN	2930-2980±50°	1,2
HfN	3310	3

Table II

Specific electrical conductivity (in $\Omega^{-1}cm^{-1}$) of the IV B group metals and nitrides [4)]

	Ti	Zr	Hf
Metals	1.11×10^4	2.24×10^4	3.07×10^4
Nitrides	4.6×10^4	7.4×10^4	*

*The conductivity of HfN is very high according to the qualitative tests by the author.

luster. The conductivity is even larger than that of the metals. To the presence of free electrons points also the metallic luster of the nitrides. Although TiN in powder form is brown, ZrN grayish brown, and HfN olive green, the particles of all three display under high magnification and incident light a metallic, more or less golden color. [5,6,7)] The magnetic properties of these nitrides also partially confirm their metallic nature.[8)]

However, in disagreement with the metallic structure of the nitrides are the high heats of formation which amount to 26.8 for TiN, 27.4 for ZrN, and 29.4 kcal/g.equiv. for HfN being the highest of all the nitrides.[5)] Furthermore, Baughan found from his calculations employing the Born-Haber cycle that the lattice energies are very close to those of the alkali nitrides, which, surely have an ionic structure.[9)] In agreement with this structure is also the very low vapor pressure of the three nitrides.[10)] There is also some direct evidence for the ionic structure: Holliday concluded from his measurements of the shift in wavelengths of X-ray emission bands that there is an ionic structure in Ti- and ZrN.[11)] Mössbauer spectroscopy as performed by Gielen and Kaplow with Fe-N alloys also confirms the existence of a nitrogen negative ion behavior in both martensite and austenite.[12)]

Although there is strong evidence for both, the metallic and the ionic structure, the first one is more favored.[13,14,15)] A mixture of a homopolar, metallic, and to a very slight extent of anionic structure is assumed by Biltz.[16)] Samsonov et. al are of the opinion that the bonding is in part ionic.[17)] Clearly, new information is necessary to make a decision concerning the nature of bonding in the three nitrides.

This information based on chemical evidence was collected by the author and his coworker Mr. C. A. Faunce (in his thesis for the M.S. degree at the Graduate School of the University of Missouri at Rolla) in collaboration with Dr. W. J. James (now Director of Graduate Center for Materials Research).

The chemical behavior of the 3 nitrides

TiN, ZrN, and HfN are very resistive against the action of acids, except HF, in which ZrN and HfN dissolve easily, but TiN only under special conditions. As reaction products appear the fluorides of the respective metals, ammonium fluoride, small amounts of N_2 and H_2.

Quantitative determinations showed [5,6,7)] that the reactions of dissolution are as follows:

$$\text{ZrN or HfN} + 3\text{HF} \longrightarrow \text{HfF}_3 + \text{NH}_3 \qquad (2)$$

then rapidly
$$\text{HfF}_3 + \text{HF} \longrightarrow \text{HfF}_4 + 0.5\ \text{H}_2 \qquad (3)$$

and
$$\text{NH}_3 + \text{HF} \longrightarrow \text{NH}_4\text{F} \qquad (4)$$

Since both Zr^{3+} and Hf^{3+} are unstable (strong reducers) they react rapidly with the acid according to Eq. (3). Hydrogen is developed, the amount of which was measured and compared with that calculated from Eq. (2) and (3). The agreement within the error limits proved that Zr and Hf in the nitride are in the trivalent form, as otherwise 0.5 H_2 would not have been developed, and there would be no reaction (3).

The Ti^{3+} ion is much more stable than the respective ions of Zr and Hf. Unfortunately, TiN does not dissolve to any appreciable extent in HF, evidently because no TiF_4 is formed as it is in the reactions 2 and 3. The reaction (3) working with Ti was studied separately and it was found to be very slow.[18] However, it should be expected that upon addition of a strong oxidizer, which would rapidly oxidize Ti^{3+} to Ti^{4+}, the rate of the dissolution reaction would strongly be enhanced. Indeed, oxidizers and as HNO_3, H_2CrO_4, H_2O_2, $KMnO_4$[19] if added to HF increase the rate of the reaction (5)

$$TiN + 3HF \longrightarrow TiF_3 + NH_3 \quad (5)$$

$$5Ti^{3+} + MnO_4^- + 8H^+ \longrightarrow 5Ti^{4+} + Mn^{2+} + 4H_2O \quad (6)$$

The nitride dissolves, although still under evolution of gases (see below). The reason of inactivity of TiN in pure HF may be formation of protective layers, consisting probably of low solubility oxifluorides of 3-valent Ti. As soon as the Ti of the protective film is oxidized to Ti^{4+}, the film dissolves. Thus TiN, if attacked by HF, also produces in the initial stage Ti^{3+} like Ti metal [20] and the nitrides of Zr and Hf, while Zr and Hf metals go into solution in the final step (3) in form of 4-valent ions.[21]

The quantitative determination of NH_3 and of N in the nitrides was interesting in so far as there were differences de-

Table III

N(in weight %) of the nitrides from combustion and Kjeldahl analyses. x of the formula Me_xN.

Nitride	Combustion	Kjeldahl	Δ	x(from comb. an.)
TiN	24.38	20.22	4.16	0.907
ZrN*	13.86	12.38	1.48	0.963
HfN**	7.52_5	7.35_5	0.17	1.0

*Containing 2 wt % Hf

**Containing 4.16 wt % Zr

pending on the method of determination: The combustion analyses according to eq. (1) and the Dumas method gave larger results than the Kjeldahl method (determination of NH_3 in the acid after the dissolution, eq. 3, 4, and 5), as shown in Table III.
The difference Δ between the N-analyses as obtained by these two methods increased from HfN to TiN. In the case of ZrN the Dumas method gave the same results as the combustion analyses. This behavior of the nitrides was noticed already previously.[22,23) Therefore, the statement as follows can be made: part of the nitrogen present in the solid escapes during its dissolution in HF.

There are two explanations for this behavior: 1. The bonding in the nitrides is atomic (homopolar) or metallic. In such a case the N and Me atoms would exist beside each other in the NaCl type cubic lattice. Upon the attack of HF the following reactions would occur, e.g. for HfN:

$$\text{Hf (of the lattice)} + 3H^+ \text{ (of HF)} \longrightarrow Hf^{3+} + 3H \quad (7)$$

adsorption of the atomic H by the nitride surface (8)

then, 3H + N (of the lattice) $\longrightarrow$ NH_3 (and solution in HF) (9)

Reaction (9) may be incomplete, therefore, part of N escapes from the solution and the Kjeldahl analysis shows a smaller N content than the solid.

A further possibility is that the nitrides have a predominantly ionic structure in which the N is in equilibrium with the ammonia ions N^{3-} and the free electrons:

$$N + 3e \rightleftharpoons (N^{2-} + e) \rightleftharpoons N^{3-} \qquad (10)$$

where the electrons are supplied by the metal atoms, e.g., by Zr

$$Zr \rightleftharpoons Zr^{3+} + 3e \qquad (11)$$

Upon dissolution in HF, the H^+ of the acid are simply attracted by the N^{3-} already present in the structure, and the preformed Zr^{3+} ions go into solution in a simple exchange reaction:

$$N^{3-} \text{ of the solid} + 3H^+ \text{ (from HF)} \longrightarrow NH_3 \qquad (12)$$

and Zr^{3+} of the solid $\longrightarrow$ Zr^{3+} (into the electrolyte) (13)

Of course the N, present in form of atoms in the solid structure escapes as the gas N_2, which can be collected (see Table IV).

Assuming that the second concept is the more probable one (see discussion), two kinds of nitrogen have to be present in the three nitrides: one which is converted into NH_3 and the other, more loosely bound by the lattice, and which is liberated as gas upon dissolving the nitrides in HF. The ionic and atomic forms of N should be in equilibrium (10) with each other in the solid nitrides. A shift to the left with N_2 release must, therefore, occur upon heating the nitrides even far below their melting

Table IV

N_2 in wt % resulting from dissolution of TiN, ZrN, and HfN in HF and from extraction in vacuum at 1250°C

Nitride	N_2 from diss. in HF	N_2 from extraction	Calcl.
TiN	4.3	1.9	2.3
ZrN	0.4	1.1	0.67
HfN	0.17*	1.1	-

*taken from Table III, no direct determination

point. Hence, degassing experiments were performed at 1250°C in vacuum (to prevent release of N_2 according to eq. 1): all three nitrides liberated considerable amounts of N_2 as shown in Table IV. Special tests were performed to assure that the N_2 collected was not a consequence of absorption, but that the N was constitutional. Table IV shows clearly the comparatively large amounts of nitrogen present in the compounds in form of atoms. The agreement between the two methods of determination is, however, poor, and is explained by analytical difficulties.[5,6,7]

Imperfect structure of the nitrides from X-ray and density data

It can be seen from Table III that the three nitride preparations used for the experimental work were, with the exception of HfN, nonstoichiometric:

$$Ti_{0.904}N,\ (Zr,2wt\%Hf)_{0.95}N \text{ and } (Hf,4.16wt\%Zr)\ N$$

The last two compounds were solid solutions.[24] The first two contained excess N, meaning that the nitrides must have an imperfect structure involving vacant sites. Whether the vacancies are in the cationic or anionic lattice or in both can be determined only from lattice parameter and density measurements. The latter were made with all the precautions not to obtain smaller densities (due to the presence of cavities in the grains or due to surface adsorbed air or both) than are the actual ones.

The lattice parameter determinations were made at constant temperatures of the sample. Asymmetric powder patterns were obtained in cylindrical cameras (64mm in dia.) which were placed into thermostatted jackets (constancy of temperature ±0.05°C). Such patterns do not require the knowledge of the diameter of the camera or the amount of film shrinkage for the calculation of the Bragg angles. As the powder mounts were very thin, about 0.2 mm in diameter, no absorption correction was necessary. The refraction correction was added to the calculated constants The whole procedure has been described previously.[25)]

The comparison of the density d_x, calculated from X-ray data, with that of the experimental density d gives the possibility to decide whether or not compounds contain imperfections. The densities of all three nitrides are listed in Table V, which indicates that even the stoichiometric HfN has vacant sites,

Table V

Experimental d_{25} and X-ray densities d_x of the Ti-, Zr-, and Hf-nitrides at 25°C

	d_{25} gcm^{-3}	d_x gcm^{-3}
$Ti_{.904}N$	4.975±0.004	5.394
$Zr_{.95}N$	6.884±0.003	7.284
HfN	11.696±0.002	13.386

assuming the real experimental density of the nitride was obtained correctly.

It is interesting to note that from the lattice parameter and experimental density of a non-stoichiometric compound its formula (and molecular wt.) can be estimated. Assuming, because of N-excess, a filling of all the four nitrogen sites in the unit cells of the first two compounds (Table V), the number of occupied metallic sites is then 4x, where x of the Me_xN is < 1. One thus obtains from the equation [26)]

$$n' = vdN_0/M , \tag{14}$$

the equation $$4.000 = vdN_0/Me_xN \tag{15}$$

where v is the volume of the unit cell, n', the actual number of Me_xN molecules per cell, d, the experimental density of the two nitrides at the respective temperature, N_0, Avogadro's number ($= 6.024 \times 10^{23}$) and $M = xA_{Me} + A_N$, A referring to the atomic weights of the metals and of nitrogen.

Solving equation (15) for x, 0.900 ± 0.002 is calculated for the Ti- and 0.95 ± 0.03 for the Zr-nitride, which is in reasonable agreement with the x obtained from analytical data (0.907 and 0.963 resp., Table III) by a completely different method.

Having the formula of the compound (from the chemical analysis) the actual number n' of molecules per unit cell can be calculated from equation (4), and, using n', the number of constitutional vacancies in the metallic and nitrogen sublattices can be found (Table VI). The table shows that the three nitrides are defective; the sum of the vacancies in the metallic and the N-sublattices amounts to 5.14, 4.59, and 12.62% respectively.

Table VI

Imperfect structure of the Ti-, Zr-, and Hf-nitrides in molecules, atoms (or ions), and vacant sites per unit cell (u.c.)

	x Me/N	n' molec/u.c.	Atoms or ions per u.c.	vacancies n'-4 in sites/u.c.	vacancies %***
Ti nitride	0.904	3.986	3.603*	-0.397	4.96
Zr nitride	0.95	3.913	3.72 *	-0.28	3.50
Hf nitride	1.0	3.495	3.495*	-0.505	6.31
N atoms(ions)=n'in	TiN	-	3.986**	-0.014	0.18
	ZrN	-	3.913**	-0.087	1.09
	HfN	-	3.495**	-0.505	6.31

For the calculation of n' the molecular weights of the solid solutions were used.
* - in the metallic sublattice: xn'
**- in the N sublattice
*** of the total number os sites = 8.

Discussion

An excess of N atoms over metal atoms follows from Table VI: 0.383 for TiN and 0.193 at/u.c. for ZrN. From these figures the amount of atomic nitrogen which is present in the nitrides at room temperature can be calculated (see Table IV). There is some agreement with the experimental data. Originally it was thought that only the excess nitrogen is driven out of the compounds at high temperatures in vacuum. However, the stoichiometric HfN also releases N_2 (Table IV), which again indicates that there is a N-equilibrium in the nitrides, as represented by equation (10). Thus, by far the largest part of the N in the nitride structures is in form of N^{3-}, the electrons being supplied by the d and s levels of the metal atoms (see eq. 11), which all are ionized and are present in the structure as trivalent ions: Ti^{3+}, Zr^{3+}, and Hf^{3+} (see equations 2,3,5). However, not all the electrons released according to eq. (11) are accepted by the N atoms, as indicated by eq. (10). Hence, there are many free electrons in the nitrides. As the metallic ions all are trivalent, one valence electron in the d level still remains, which may also contribute to the increase of free electron concentration.[27]

The latter explains the high electrical conductivity of the nitrides and their magnetic and metallic properties (luster) although it is difficult to understand why the conductivity is higher than that of the pure metals (Table II).

The bonding in the nitrides is, thus, of a mixed type[28]: it is mainly ionic and to a smaller extent metallic. Since N^{3-} (or even N^{2-}) is present in the nitride structure, there is no difficulty to explain the NH_3 or even the NH_2^- formation (eq. 12 and 10) and likewise the development of N_2 in vacuum at elevated temperatures.

Nitrogen is given up also by other nitrides, e.g. by NbN, only degassing starts at the much higher temperature of 1600°C,[29] meaning that in this nitride the equilibrium (10) is shifted still more to the right. Hence this nitride should be more ionic than the three under discussion. In comparison, oxygen in the solid solutions of Ti, Zr, and Hf, as well as in Nb is possibly nearly completely ionic[18] and, hence, is not given up upon heating in vacuum.[29] Conversely, hydrogen in the solid solutions of these metals can easily be extracted at elevated temperatures, meaning that the hydrogen equilibrium, which is similar to that of (10), is shifted to the left side. Hydrogen in these solid solutions, as it is in case of alkali hydrides[30] must be in form of negative ions H^- and of atoms (as in eq. 10).

With this concept of the structure of nitrides, where the atoms of the cationic sublattice are completely ionized, the nitrogen atoms of the anionic sublattice are in equilibrium with their ions, N^{3-}, and the free electrons are scattered throughout the lattice (forming the conductivity band), all the other properties and reactions of the nitrides are in accord. They are as follows:

1) The large amounts of NH_4^+ formed during digestion of the nitrides in HF, correspond to those found for the recognized ionic alkali and alkali earth nitrides. Conversely, NH_3 formation, assuming metallic bonding and N in form of atoms in the solid, by the reaction of active N and H atoms at the nitride surface [15] is improbable (see eq. 7,8, and 9), especially if one takes into consideration that TiN dissolves in HF only in the presence of large amounts of strong oxidizers (e.g. $KMnO_4$)[14,19];

2) The N development at elevated temperatures (but far below the melting point of the nitrides) and during dissolution of them in HF;

3) The simultaneous H_2 development in the last case (see eq. 3);

4) The heats of formation, which are the highest of all the nitrides and are difficult to explain on other than ionic models;

5) The lattice energies (-511 ± 57 Kcal for $N \longrightarrow N^{3-}$) calculated for the nitrides are in agreement with the ionic model;[9]

6) The hardness, brittleness, high melting point, and other properties, are, according to Pauling[10], characteristic of multivalent ionic crystals;

7) The shift of wavelengths of the X-ray emission bands[11];

8) The existence of nitrogen negative ions from Mössbauer spectroscopy[12];

9) The electrical and magnetic properties (explained above); and

10) The semiconductor properties, which are only apparent.[7)]

Concerning the formation of vacant sites in the nitride structures, the appearance of the entities in several valence states may be responsible for that. Nitrogen is present in form of N, N^{3-}, and maybe in some intermediate states. There is an analogy to TiO, which represents a solid solution of Ti_2O_3 in Ti[31)], and hence, Ti is in the oxidation states of 0 and 3. As a consequence there are 15.5% vacancies in TiO[32,33)]. In the HfN, where the atoms in various oxidation states are in the anionic sublattice, there are only 12.6% of all sites vacant.

One observation is seemingly in contradiction to the worked out concept of nitride structure: the lattice parameters of the three compounds, as calculated from their atomic diameters, agree much better with the measured values than those calculated from the ionic diameters.

Lattice parameters from ionic diameters and conclusion

The main factor which causes this disagreement is the ionic radius of N^{3-}, which is too large: 1.71 Å. The latter was calculated from Pauling's equation, using the ligancy n of 7:

$$R_Z = R_1 z^{-2/(n-1)} \tag{16}$$

where R_Z in the ionic (crystal) radius, R_1, the univalent radius (2.47 for N^{3-}), and z, the valency (=3). Since on the average, not all the three electrons are taken up by N, the ionic radius must be smaller. Instead of finding the average radius of the nitrogen ion from the lattice parameter, attempts were made to calculate it from Pauling's equation (16), using the ligancy number

Table VII

Lattice constants calculated from the diameter of the atoms and ions in Å. Radii ratios.

	Ti	Zr	Hf
	Atomic or homopolar bond		
Diam. of Me atoms	2.92	3.12	3.16
Diam. of N atoms	1.42	1.42	1.42
Σ	4.34	4.54	4.58
Constants measured	-4.24	-4.576	-4.512
Δ	0.1	-0.036	0.068
r_N/r_{Me}	0.486	0.455	0.449
	Mixed bonding		
Diam. of Me^{3+}	1.38	1.84*	1.76*
Diam. of N^{3-} = N + 3e	2.85	2.85	2.85
Σ	4.23	4.69	4.61
Constants measured**	-4.24	-4.576	-4.512
Δ	-0.01	0.114	0.098
r_{Me}/r_N	0.484	0.645	0.6175

*Estimated by interpolation
**The precise values of the constants <u>a</u> at 25°C are in Å:

4.23986±0.00007($Ti_{.904}N$); 4.57560±0.00005[$(Zr,2\%Hf)_{.95}N$]

and 4.5118±0.0001[(Hf,4.5%Zr)N].

of 5: a radius of about 1.425 Å is obtained. This radius which may correspond to a mixed bonding, ionic and metallic (of the anionic sublattice while the metals are all in form of three valent cations) gives lattice constants much closer to the measured ones, as shown in Table VII. The constants calculated for the three pure ionic nitrides are not mentioned in the table, because they are much larger than the experimental. The table shows that the constants calculated for a mixed bonding are nearly as close to the experimental constants, as those calculated for homopolar bonding. However, it is clear that the average diameter of N ions will depend upon the degree of ionization of N atoms and it will decrease with the shift of the equilibrium (10) to the left. If the N ion diameter were 2.80 Å a still better agreement between the computed and observed values could be attained. Furthermore, the concentration of vacancies may also influence the average diameter of the ions.

The ratios of the radii are in agreement with those found for the NaCl type structures. The relationship r_N/r_{Me} around 0.6 and 0.48 (for a NaCl type structure) is characteristic for the ionic NaCl type structures; however, if the bonding is regarded atomic in the three nitrides, the relation is reverse and around 0.47.

Thus, the three nitrides have a predominantly ionic character and to a smaller extent - a metallic. This statement also agrees with the preliminary results obtained in our laboratory in comparing the intensities of X-ray lines with those calculated. The study indicated electronic states intermediate between N and N^{3-}, thus excluding total N (atomic) and positive N ionic states, in agreement with some investigators of other nitrides.[34,35] However, it is difficult to estimate the degree of ionization of the solid nitride nitrogen (eq. 10), because the ionization may be much larger than given by the amount of free N developed during digestion of the nitrides in HF or by the difference in N between combustion and Kjeldahl analyses (see Table III). The reason may be reaction (9) which cannot be completely excluded, especially if oxidizers are absent.

References

1. C. Agte & K. Moers, Z. anorg. allgem. Chem. 198 233(1931).
2. E. Friederich & L. Sittig, ibid. 143, 293(1925).

3. A. E. van Arkel & J. H. de Boer, ibid. _148_, 345(1925).

4. H. Remy, "Inorganic Chemistry II," Akad. Verlagsges. Leipzig, 1959, p. 68.

5. M. E. Straumanis, C. A. Faunce, & W. J. James, Acta Met. _15_, 65(1967).

6. M. E. Straumanis, C. A. Faunce, & W. J. James, Inorgan. Chem. _5_, 2027(1966).

7. M. E. Straumanis & C. A. Faunce, Z. anorg. allgem. Chem. _353_, 329 (1967).

8. W. Klemm & W. Schüth, ibid. _201_, 24(1931).

9. E. D. Baughan, Trans. Farod. Soc. _55_, 736, 2025(1959).

10. L. Pauling, "The nature of the chemical bond," Cornell Univ. Press, 1960, pp. 513-538.

11. J. E. Holliday in "Advances in X-ray analysis," _9_, 365 371 (1966).

12. P. M. Grelen & R. Kaplow, Acta Met. _15_, 49(1967).

13. O. Schmitz-Dumont & K. Sternberg, Naturwiss. _41_, 117(1954).

14. W. H. Philipp, Acta Met. _10_, 583(1962); _12_, 740(1964).

15. W. B. Pearson, ibid. _10_, 1123(1962).

16. H. Biltz, Z. Physik _153_, 338, 344, 347(1958).

17. See e.g. G. V. Samsonov & T. S. Verkhoglijadova, Dopov. Akad. Nauk Ukr. RSR _1_, 48(1962).

18. M. E. Straumanis, C. H. Cheng, & A. W. Schlechten, J. Electrochem. Soc. _103_, 439 (1956); Analyt. Chem. _28_, 1883 (1956); J. W. Johnson, K. S. Chen and W. J. James, J. Less Comm. Metals (1968) in press.

19. W. J. James & M. E. Straumanis, Acta Met. _12_, 739(1964).

20. M. E. Straumanis & P. C. Chen, ibid. _98_, 234(1951).

21. M. E. Straumanis & J. I. Ballass, Z. anorg. allgem. Chem. 278, 33(1955).

22. G. Hägg, Z. physik. Chem.B 7 339(1930).

23. W. Kern & G. Brauer, Talanta 11, 1177 (1964).

24. See e.g. H. Nowotny, F. Benesowsky, & E. Rudy, Monatsh. Chem. 91, 349(1960).

25. M. E. Straumanis in G. L. Clark's "Encyclopedia of X-rays and Gamma Rays" p. 700, Reinhold, New York (1963).

26. M. E. Straumanis, ibid. p. 733.

27. A. E. van Arkel, Metallurg. 13, 511 (1934).

28. D. K. Sebera, "Electronic Structure and Chemical Bonding," Blaisdell, New York (1964) p.62.

29. E. Gebhardt, W. Dürrschnabel & H. Görz, J. Nuclear Mat. 18, 149(1966).

30. H. Remy, "Inorganic Chemistry I," Akad. Verlagsges., Leipzig, 1957, p. 207.

31. S. M. Arija & M. V. Golomolzina, Solid St. Phys. USSR, 4, 2921(1962).

32. P. Ehrlich, Z. Electrochem. 45, 362(1939); Z. anorg. allgem. Chem. 247, 53(1941).

33. M. E. Straumanis & H. W. Li, Z. allgem. anorg. Chem. 305, 143(1960).

34. W. J. Takei, R. R. Heikes, & G. Shirane, Phys. Rev. 125, 1893(1962).

35. M. Kuriyama, S. Hosoya, & T. Suzuki, ibid. 130, 898(1963).

THE NATURE OF THE ELECTRICAL CONDUCTION TRANSIENT BEHAVIOR OF RUTILE

R. N. Blumenthal, W. M. Hirthe, and B. A. Pinz
Metallurgy and Materials Science
College of Engineering
Marquette University
Milwaukee, Wisconsin

ABSTRACT

Single crystal and ceramic specimens of rutile were examined to determine the existence of conduction transients over the range 500° to 1000°C in air and argon by employing a constant current dc power supply and a six-probe high resistance input-isolated recording system. Voltage transients were observed only when the current was applied parallel to the "c" direction of single crystal rutile. Potential distribution curves at several different times revealed that the electrical conductivity increases with time at the negative end of the specimen and is larger in magnitude than at the positive end where the conductivity initially decreases with time. These results were interpreted in terms of a model that assumes blocking of titanium interstitials at the cathode and a redistribution of the electrons so as to preserve electroneutrality.

INTRODUCTION

Transient electrical conduction in rutile has been reported by a number of investigators.[1-4] Cardon[2] observed current changes in unetched rutile crystals at room temperature and attributed this effect to the injection of electrons at the negative electrode by field emission. For an etched crystal, Cardon considered the origin of the conductivity changes observed at room temperature to be the result of a small electrolytic conductivity and a discharge at the negative electrode. Greener and Whitmore[3] observed both potential and current transient effects at higher temperatures, 632° to 908°C, and they identified these effects with the existence of space-charge-limited-currents in the bulk of the crystal. They obtained a value of the order of 10^4 cm^2/volt-sec for the mobility. This value is high because all reported mobility values in rutile are of the order of 10^{-1} to 10 cm^2/volt-sec.[5-7] In a more recent investigation, VanRaalte[4] observed current transients in rutile at room temperature and explained his results in terms of an injection of positive space charge from the anode into the bulk of the crystal. It was the purpose of our investigation to examine the nature of the conductivity transients at high temperatures and low applied fields. On the basis of previous investigations, studies at high temperatures and low applied fields should have the following advantages (1) greater reproducibility (2) the impurities should play less of a role and (3) the possibility of quenched-in defects is minimized. In order to perform this type of experiment it was necessary to obtain the electrical conductivity as a function of temperature, oxygen partial pressure, time, and voltage distribution within the bulk of the specimen.

EXPERIMENTAL PROCEDURE

The "c" direction and "a" direction single crystal specimens used in this investigation were cut from boules supplied by the National Lead Company. These boules are about 99.9% minimum with respect to TiO_2. The dimensions of the single crystal specimens were approximately 0.2 cm x 0.4 cm x 2.0 cm. Johnson-Matthey high-purity powder was used to prepare

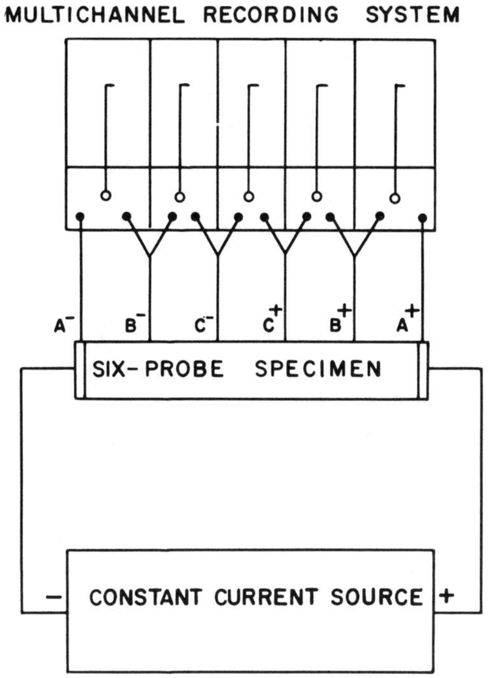

Fig. 1. High-input resistance-isolated measurement system.

the sintered specimens. The powder was reported to be 99.999% minimum with respect to TiO_2. The sintered specimens were prepared by cold-pressing the TiO_2 powder without a binder at 20,000 psi into a bar about 0.7 cm x 0.7 cm x 4.5 cm. A surface layer was scraped from the specimens to avoid contamination from the die walls. The specimens were then placed in a boat lined with a layer of high-purity TiO_2 powder and sintered at 1400° to 1500°C for at least two hours in a dynamic flow of oxygen. The sintered specimens were cut with a diamond saw and the final specimen size was about 0.6 cm x 0.6 cm x 2.0 cm in length.

In order to determine the voltage distribution within the bulk material, a six-probe specimen, shown in Fig. 1, was employed. The four potential probes, B−, C−, C+, and B+, were formed by pressing beaded 0.016 in diameter platinum wire into 0.035 in diameter holes drilled into

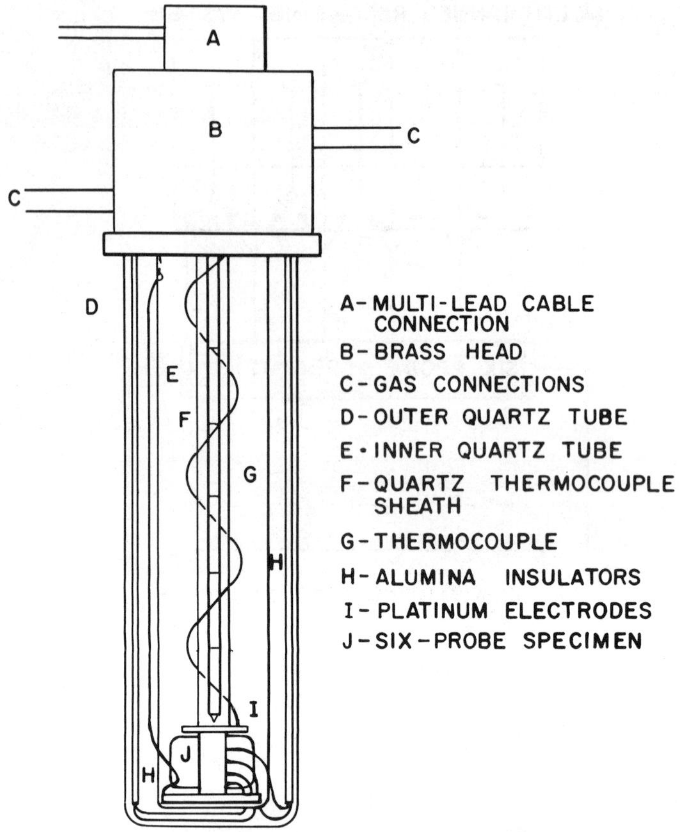

Fig. 2. High isolation six-probe specimen jig.

the material at intervals of 0.4 cm. Platinum paste was applied to the ends of the specimen and the current probes, A- and A+, were formed by pressing platinum foil against the ends of the specimen. The voltages across the probe pairs were measured by means of a multichannel recording system. The multichannel system used consisted of ultralinear Brush Oscillographs driven by Offner 462 preamplifiers using guarded differential input electrometer couplers. The maximum input voltage to the couplers is approximately 1 V with this system. The input resistance to the recording system was thus maintained at greater than 10^{11} ohms. Rather than allow both the current and voltage to vary as a result of the transient, a constant current dc source was used so that only the voltage varied with time. In order to provide a high isolation system, the six-probe specimen was mounted in a quartz jig as shown in Fig. 2. The jig was designed for high resistance (10^9) ohms and six-probe measurements. The problem of noise associated with this high impedance system was alleviated by (1) shielding and guarding the entire system including the power supply (2) coating the alumina tubes containing the platinum lead wires in the conductivity jig with platinum paste to protect the leads from stray fields inside the quartz jig and (3) enclosing the outer quartz tube with a nichrome screen to shield against fields produced by the furnace windings.

RESULTS AND DISCUSSION

The transient effects were investigated by means of the multichannel transient analysis system described. The voltages across the selected probe pairs were recorded as a function of time at constant temperature, partial pressure of oxygen, and current. The voltage versus time data

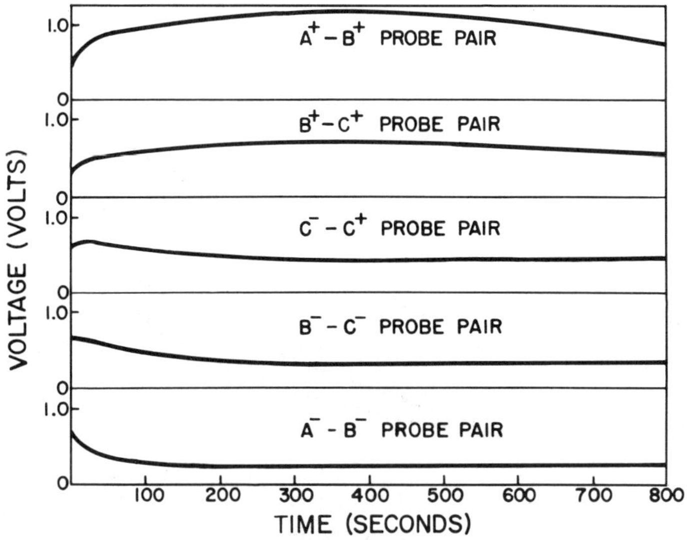

Fig. 3. Voltage versus time curves at constant current (500 μA) for the probe pairs at 900°C in air.

for the five distinct probe pairs were then used to obtain the voltage distribution within the specimen as a function of time. The V(x) curves at fixed times were obtained with the aid of a Gerber data reduction system by converting the initial results to a polynomial expansion ($V = a + bx + cx^2 + dx^3$) suitable for analysis with an IBM 7040 computer.

Typical results of the transient investigation are shown in Fig. 3 for a "c" direction specimen; i.e. the constant current is applied parallel to the "c" axis of the specimen. This is a plot of voltages across the individual probe pairs as a function of time at 900°C in air with a current of 500 microamperes. It is interesting to note that, initially, the voltage decreases with time at the negative end of the specimen, probe pair $A^- - B^-$, the voltage increases with time at the positive end of the specimen, probe pair $A^+ - B^+$, and the voltage first increases and then decreases at the center of the specimen, probe pair $C^- - C^+$. The corresponding plot of voltage as a function of distance from the negative end at various times is shown in Fig. 4. The calculated

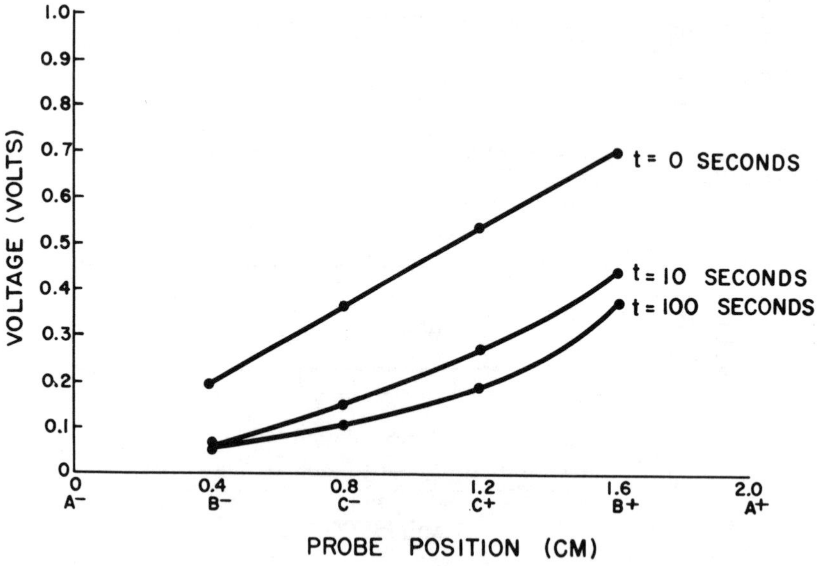

Fig. 4. Constant-current potential distribution curves at 900°C in air.

potential distributions are shown as solid lines for times equal to 0, 10, and 100 seconds. For comparative purposes the actual data points obtained from the previous curves are also shown. The agreement between the calculated curves and the data points is excellent. To avoid the effects of contact resistance at the A- and A+ probes on the potential distribution, the data were only analyzed between the B- and B+ probes. Since the voltage transient was measured with the current held constant, the slope of a potential-distance curve is inversely related to the electrical conductivity. With this in mind it can be seen that, since the initial voltage distribution is linear, the conductivity at the zero of time is independent of position. The absolute magnitude of the dc conductivity at $t = 0$ is equal to the value obtained from ac measurements. However, over short periods of time, the conductivity at the negative end is increasing with time and the conductivity at the positive end is decreasing with time. At a fixed time greater than $t = 0$, the electrical conductivity decreases with increasing distance from the cathode.

There are at least two possible explanations for these changes in electrical conductivity. First let us consider the possibility of the injection of electrons into the bulk of the material at the cathode. For this space charge model, the calculated value of mobility is about 10^4 cm^2/volt-sec. This value of mobility is at least four orders of magnitude greater than the value recently calculated by Blumenthal, Kirk, and Hirthe.[7] A charge carrier concentration of the order of 10^{17} to 10^{18} was also calculated from the electrical conductivity data and the thermogravimetric data obtained in our laboratory. This value seems high for the space charge model which assumes the bulk material to be an insulator. In addition, the dependence of current on voltage was obtained as part of a number of experiments. The relationship was of the form: $I \sim V^n$ but n was not equal to two as in the space charge model. The value of n obtained was less than two and it was dependent on time and the particular probe pair selected for the measurement of voltage. Therefore, electron injection does not appear to be a reasonable explanation for the observed voltage transients.

A second explanation of the changes in conductivity obtained involves the formation of an inhomogeneous distribution of point defects. When the single crystal is placed in an applied dc field, the point defects migrate under the influence of the field. The point defects with a negative effective charge migrate toward the anode and those with a positive effective charge migrate toward the cathode. If the rate of electrolysis is small, the charged point defects are essentially blocked at the electrodes. Because of the requirement of electroneutrality, the electronic defects are distributed so as to neutralize the charged point defects; i.e. there is a higher concentration of electrons at the cathode and a lower concentration of electrons at the anode. The electronic conductivity would then vary with distance. Thus, an inhomogeneous distribution of charged point defects produced by the application of a dc field would produce a distribution of electronic defects and hence a variation in the electrical conductivity with distance and time. The migrating ions could be impurity defects or nonstoichiometric point defects.

In order to elucidate the nature of the point defect involved, a high-purity (less than 10 ppm) sintered specimen was prepared and placed in the jig. No dependence of voltage on time was noted in the range 500° to 1000°C in air and argon. The conclusion drawn from the experiments on the high-purity sintered rutile was that the transients were due to impurities or the migration of point defects along a preferred crystallographic direction. In order to test the latter possibility, i.e. dependence on crystallographic direction, an "a" direction specimen was prepared in an identical fashion to the "c" direction specimen for which the voltage transients were

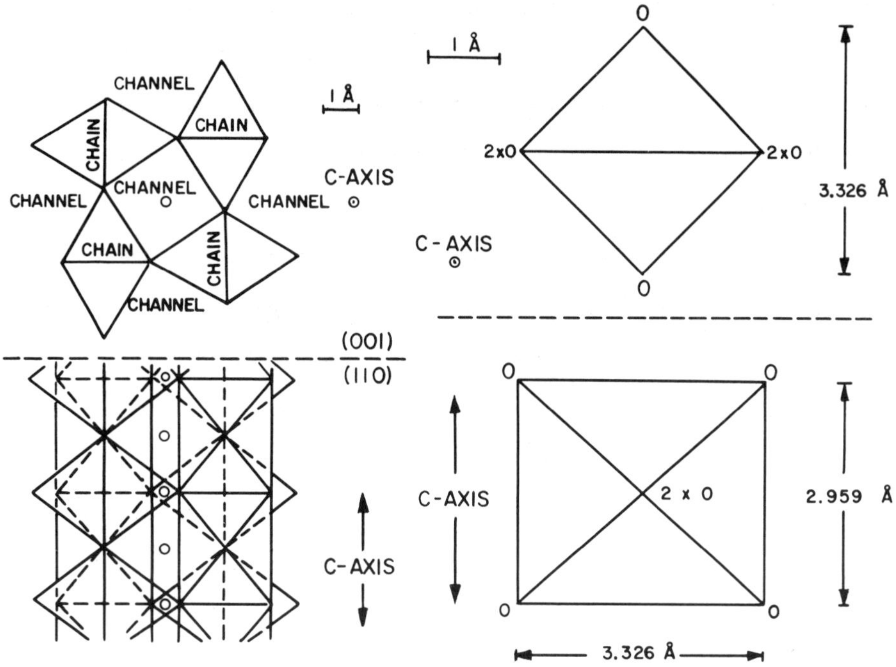

Fig. 5. (a) Octahedral interstitial positions in rutile. (b) The shape and size of the O_6-octahedra surrounding interstitial positions in rutile.

observed. The "a" direction specimen did not exhibit transient behavior over the range 500° to 1000°C in air and argon. It was concluded that the transient effects are dependent on crystallographic direction. The significance of this dependence can be seen if we examine the crystal structure of rutile as shown in Fig. 5.[8] It can be seen that large interstices in the rutile lattice are aligned in the "c" direction and have been referred to as the "c" channels for the migration of ions by a number of investigators. In addition, Bogomolov[9] clearly indicated the diffusion of boron into rutile along the "c" axis. Johnson[10] reported that the diffusion of interstitial Li in rutile was strongly anisotropic with the diffusion coefficient perpendicular to the "c" axis less than that parallel by at least eight orders of magnitude up to 550°C. Huntington and Sullivan[11] have pointed out that the strong anisotropy of diffusion in rutile is uniquely compatible with interstitial motion in this lattice. Therefore, it is reasonable to propose here the migration of interstitial point defects in the "c" direction channels as the mechanism responsible for the transients in voltage observed in our experiments. It now remains to identify these migrating ions.

Recent investigations[12,13] of equilibrium electrical conductivity over the range 1000° to 1500°C and 1 to 10^{-15} atm of oxygen have been interpreted in terms of a nonstoichiometric defect structure consisting of both triply and quadruply ionized titanium interstitials. In order to identify the migrating point defects, a 200 V dc potential was applied to a "c" direction specimen for several hours at 900°C. The specimen was quenched and then cut in half. The temperature dependence of the electrical conductivity was then measured for each half and found to be identical above 800°C. If the migrating point defects were impurities, the electrical conductivity for the two halves could not be the same because the conductivity in the low temperature region is dependent on the concentration of impurities. However, if the migrating point defects were titanium interstitials, the conductivity could be the same because of the equilibration of the non-stoichiometric defect structure with the atmosphere on heating during the course of the conductivity measurements. Thus it appears that the charged point defects responsible for the voltage transients observed in rutile are probably titanium interstitials.

SUMMARY

Single crystal and ceramic specimens of rutile were examined to determine the existence of conduction transients over the range 500° to 1000°C in air and argon by employing a constant current dc power supply and a six-probe high resistance input-isolated recording system. Voltage transients were observed only when the current was applied parallel to the "c" direction of single crystal rutile. Potential distribution curves at several different times revealed that the electrical conductivity increases with time at the negative end of the specimen and is larger in magnitude than at the positive end where the conductivity initially decreases with time. These results were interpreted in terms of a model that assumes blocking of titanium interstitials at the cathode and a redistribution of the electrons so as to preserve electroneutrality.

ACKNOWLEDGMENT

The authors gratefully acknowledge the support of Aerospace Research Laboratories, Office of Aerospace Research, United States Air Force, Contract No. AF 33(615)-1244.

REFERENCES

1. F. Cardon, J. Appl. Phys., 33, 3358 (1962).
2. F. Cardon, Phys. Stat. Sol., 3, 1415 (1963).
3. E. H. Greener and D. H. Whitmore, J. Appl. Phys., 32, 1320 (1961).
4. J. A. Van Raalte, J. Appl. Phys., 36, 3365 (1965).
5. R. G. Breckenridge and W. R. Hosler, Phys. Rev., 91, 793 (1953).
6. H. P. R. Frederikse, J. Appl. Phys., Suppl., 32, 2211 (1961).
7. R. N. Blumenthal, J. C. Kirk, Jr., and W. M. Hirthe, J. Phys. Chem. Solids, in press.
8. T. Hurlen, Acta Chem. Scand., 13, 365 (1959).
9. V. N. Bogomolov, Soviet Phys.-Solid State 5, 1468 (1964).
10. O. W. Johnson, Phys. Rev., 136, A 284 (1964).
11. H. B. Huntington and G. A. Sullivan, Phys. Rev. Letters, 14, 177 (1965).
12. R. N. Blumenthal, J. Coburn, J. Baukus, and W. M. Hirthe, J. Phys. Chem. Solids, 27, 643 (1966).
13. R. N. Blumenthal, J. Baukus, and W. M. Hirthe, J. Electrochem. Soc., 114, 172 (1967).

ELECTRONIC STRUCTURE OF TRANSITION CARBIDES AND NITRIDES

P. Costa
ONERA
Chatillon-sous-Bagneux, France

SUMMARY

Previous publications have suggested that the band structure of the transition carbides and nitrides might be very similar to the band structure of the transition metals with a narrow, predominantly d, band.

In this paper we want to show how this model can be modified to account for NMR experiments by Rossier. Ionic character, d-p hybridization are discussed.

Two other parts deal with:

1) the existence of ordered structures of interstitial atoms (V_8C_7).

2) the validity of the transition metal model for sub-stoichiometric compounds and with recent experiments on semi-carbides M_2C.

INTRODUCTION

I.

The theories which have been presented for the structure of transition elements compounds can be classified in three groups.

One of these groups includes the models (Bilz for instance (1)) which are related to the descriptions given for other types of compounds between metals and non-metals, and for which the band-states are derived from an hybridization between atomic states of the metal and the interstitial atom, with a similar contribution of both types of states. Such band models are evidently very close to the descriptions which could be given by chemists in terms of localized states. They can explain the metallic properties of the compounds by a slight overlap between a (s-p-d) bonding band and an almost unoccupied d band.

In a second group we find the "transition-metal-like" models which have been given by DEMPSEY (2), COSTA (3) and LYE (4). These authors have tried to present a model which might account for the striking similarities existing between the properties of these materials and the properties of transition metals near the middle of each period: maximum of the cohesive energy, minimum of the density of states. In this line, COSTA and LYE have tried to give, in a tight binding approximation, a description of the band states derived from the d states of the metal and they have shown that it is possible to account for the properties of these compounds under the assumption that the d population for the carbides and the nitrides would be larger than for the metal (an increase of roughly 1.5 electrons in the first case, 2 or 2.5 in the second case).

In a third group we find ERN's APW band calculations on TiC, TiN and TiO (5), which give a similar minimum of the density of states and might partially reconcile both types of descriptions, though it is probably closer to the first one.

All these models are in reasonable agreement with the experimental data, mainly specific heat and magnetic susceptibility (see Table 1). The purpose of this paper is not to try to compare them, but rather:

TABLE 1

	γ (mJ/OK^{2}mole)	ref.	$n(E_F)$*	χ (10^{-6}uem/mole)	ref.
ScC	6	(18)	2,7	-	
TiC	0,51	(3)	0,23	(+6	(22)
				(+8	(18)
ZrC	-		-	(-26	(22)
				(-13	(18)
$VC_{0,84}$	3,0	(21)	1,3	(+39	(22)
				(+28	(18)
NbC	2,6	(19)	1,12	(+18	(22)
				(+23	(18)
TaC	3,2	(3)	1,44	(+12	(22)
				(+20	(18)
LaN	3,5**	(20)	1,7	-	
TiN	2,5	(18)	1,05	(+40	(23)
				(+38	(18)
ZrN	-		-	+22	(23)
VN	4,5	(18)	2,1	+130**	(18)
NbN	2,6	(19)	1,12	(+31	(23)
				(+17	(18)
TaN	-		-	(+25	(23)
				(+10	(18)

*Values of number of states per electron-volt and per atom for both spin directions.

**Doubtful value - presence of a widespread superconductive transition.

***Doubtful value - given by a Honda extrapolation (1/H=0) on a specimen showing ferromagnetic impurities.

1) to show how the "transition-metal like" description can be modified to account for the hybrid character of the occupied states, as suggested both by Ern's model and by recent experiments carried out by Rossier at Orsay;

2) to make some remarks on ordered interstitial structures and the nature of interactions between interstitial vacancies;

3) to discuss the validity of the transition metal model, for MC_{1-x} compounds, to give some recent results on low temperature specific heat for M_2C type carbides and to try to interpret them.

II.

In the transition-metal like model, the various authors have made the assumption that the population of the d-states is larger than for the metal. On the other hand, it is of course not reasonable to assume any important charge effect produced by a positive ionization of the non-metal. In fact, these two statements are not quite contradictory,

1) because of the interstitial character of the compounds:

a) the distances between metal atoms are very close to the value they would have in a f.c.c. metal phase. As a consequence, the overlap between d orbitals in the non-metal sphere is important and it can give a good screening to the nucleus of the non-metal;

b) the 2p - and eventually 2s - orbitals of the interstitial atom have, by far, a smaller size than the d orbitals of the metal.* Therefore, it is possible to give locally to the binding a partially covalent character, and still to retain a dominant d character for the occupied band states, by a partial 2p hybridization in the vicinity of the non-metal.

2) because a very large part of the transfer effect which produces the increase of the d population can derive from the metal s-states.

These assumptions are much in the line of Ern's results: binding through hybrid states derived from 2p and metalstates, but for the latter, considerable decrease of the s population and increase of the d population.

Their validity should besides vary with the electronegativity of the two elements. Ern's calculations have shown that an ionic character, with <u>negative</u> ionization of the non-metal, might show up for nitrides; it is certain for monoxides. On the other side of the carbon, the <u>positive</u> ionization of hydrogen in hydrides seems well established (7), (8), (9); for the monoborides of the last elements of the first transition series, magnetic experiments indicate an increase of the number of occupied d states (10).

---

*This can be readily shown on TiC for instance. From Slater's empirical rule, the 3d wave functions of titanium have a radial variation given by: $r^2 e^{\frac{1,22r}{ao}}$, while for the 2s and 2p functions of carbon the law of variation is $re^{\frac{1,63r}{ao}}$.

As for the metal, things are more complex since there is a competition between two opposite effects:

a) within a transition period, the electronegativity is increased with increasing atomic number;

b) beyond the IVth or the Vth column, antibonding d states should appear.

The first effect should, though, be dominant.

We wish to check now how these assumptions fall in with Rossier's results.

In his work on the V^{51} resonance for various compositions of vanadium carbide, this author has shown that the Knight shift for VC is negative with short relaxation times ($T_1T \simeq 2$sec.°K). If one remembers that the K.S. for vanadium metal is strongly positive, this modification could be due to a violent decrease in the number of metal s-states. (The Knight-shift for d states K_d is negative, since it is caused by an exchange spin polarization of the inner s shells. This term might then be dominant for the carbide, as it is for platinum for instance.)

Besides, a more precise examination of the resonance line has proved that for each specimen one can detect several peaks with different intensities corresponding to different values of the Knight shift and different quadrupolar effects. For the $VC_{0,88}$ specimen, for instance, Rossier has found two peaks; the first one did not show any quadrupolar effect (no field dependence of the line shape); the second one corresponded to a higher Knight shift (-0,11% instead of -0,20%) and showed large second order quadrupolar effects.

Rossier explained these results in terms of different nearest neighbours distributions: the first peak would correspond to vanadium atoms surrounded by six carbon atoms; the second to vanadium atoms with one carbon vacancy occupying one of the six next neighbours positions. Figure (1) shows that

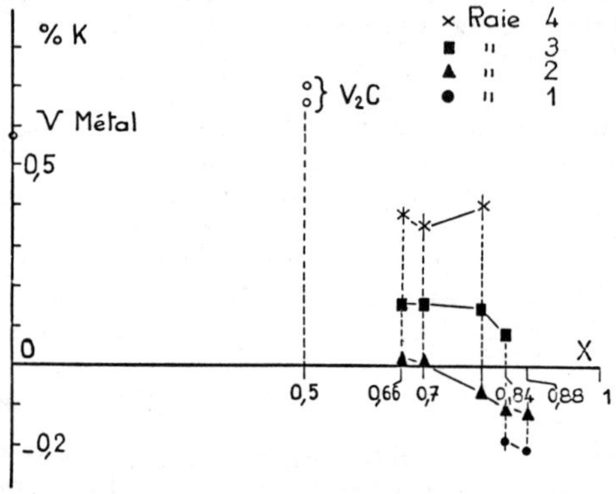

Fig. 1. Knight-shifts of V^{51} in vanadium carbides at 10 Kg and 4.2°K.

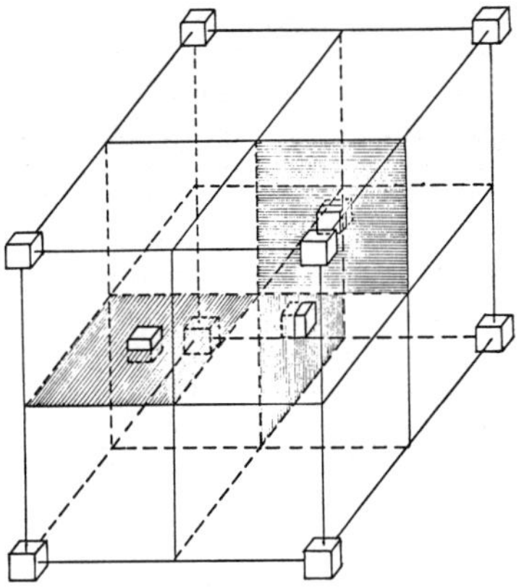

Fig. 2. Position of the carbon vacancies in the V_8C_7 carbide cell.

the Knight shift for the different local configurations varies with the overall carbon concentration.

Such a result would not have been expected in a good metallic phase: for metallic alloys, conduction electrons give rise to long range interactions with the nuclei moments. Therefore, the interactions with the nuclei would not have led to discrete resonances, but rather to a continuous wide-spread line; on the other hand, it is in agreement with:

1) the probable decrease of the free-electron like s population;

2) with the existence of important covalent effects leading to a binding between next neighbour metal and non-metal atoms.

Another point would confirm these conclusions. Rossier has checked that field gradients due to several carbon vacancies were additive, which cannot be explained unless the electrons are assumed to be distributed on fairly independent and, at least partially, covalent bonds. From the values of the field gradients, Rossier has tried to compute the rate of 2p hybridization of the d states, and suggests a figure of about 0.2.

Another important result which can be deduced from this work is that no important ionic character can be present in carbides: the observed quadrupolar frequency (40kc/sec) is much larger than the value computed for a charge $\pm e$ occupying one of the interstitial sites (890kc/sec).

III.

Most of the carbides and nitrides can be very highly nonstoichiometric. The question has arisen very early to know whether the interstitial vacancies were ordered or not.

For hexagonal M_2C carbides, this order has been reported by several authors, some compounds showing a Fe_2N ζ or ϵ type of ordering (V_2C (11), Nb_2C (12), Mo_2C (13)), some others giving rise to a phase with alternately empty and occupied (001) interstitial layers (Ta_2C (14)).

For cubic carbides there is some evidence now that the same type of effects can show up. Goretzki (15) has recently reported an ordering TiC_x with x = 0,5, which was discovered by neutron diffraction. We have on the same lines found (16) an ordering of carbon vacancies in VC_x for the upper limit of carbon concentration, which corresponds to the chemical composition V_8C_7.

This effect has been deduced from the study on an X-ray goniometer of extra-diffraction lines, which had already been reported by Volkova (17). It could be detected by X-rays since vanadium atoms are displaced in the vicinity of a carbon vacancy. It leads to a new cubic cell with a parameter, a = 8,334 A° double of the face centered cubic cell parameter. The structure corresponds to the symmetry groups ($P4_3$ 32 0^6) or ($P4_1$ 32 0^7) which are enianthomorphic. In fact, since these groups including helicoidal axes have never been reported yet for non-organic materials, the actual group might be more complicated, owing to a slight crystallographic distortion.

As it can be noticed on Fig. (2), the minimal distance between two carbon vacancies is equal to the distance between third neighbours. This result suggests that there is a mutual repulsion between carbon vacancies.

Such repulsions had previously been reported by Rossier for vanadium carbide, even in the region where the long range ordered phase does not appear. His experiments show indeed that the intensity of the NMR lines which are assumed to correspond to several local environments of vanadium atoms are not consistent with a random distribution of interstitial atoms. The observed deviations are interpretable under the assumption that vacancies would repel each other.

The nature of this repulsion is still to be clarified. It is doubtful that it could be due to ionic effects (see part II). Interactions through conduction electron gas charge oscillations are more likely to appear.

Calculations have been made, in order to try to fit the line intensities with the vanadium displacements parameters. They show that <u>vanadium atoms move away from the carbon vacancies</u>. The same effect had been reported by GORETZKI on $TiC_{0,5}$. In both cases, the interpretation of this phenomenon is still to be found.

IV.

If the transition metal-type model is correct, it can give information on what should be the electronic structure of non-stoichiometric compounds by assuming that the increase in the number of d states compared to the metal is roughly proportional to the interstitial atom concentration. The different behaviours of TiC_x and NbC_x susceptibilities have already been explained in this way (3).

Along the same lines, we thought that the minimum for the density of states which is observed for the NaCl type f.c.c. carbides (MC_x with x close to 1), near the fourth column (Ti, Zr, Hf), should be found near the fifth for the hexagonal semi-carbides.

Low temperature specific heat experiments have been carried out (Caudron, Costa, (24)) on V_2C, Nb_2C, Mo_2C and $(V,Mo)_2C$ specimens. ($(Nb,Mo)_2C$ samples could not be studied since they are not stable.) The results are shown on Fig. (3). The values for γ deduced from the law $Cp = \gamma T + \alpha T^3$, and proportional to the density of states at the Fermi level are collected in Table (2).

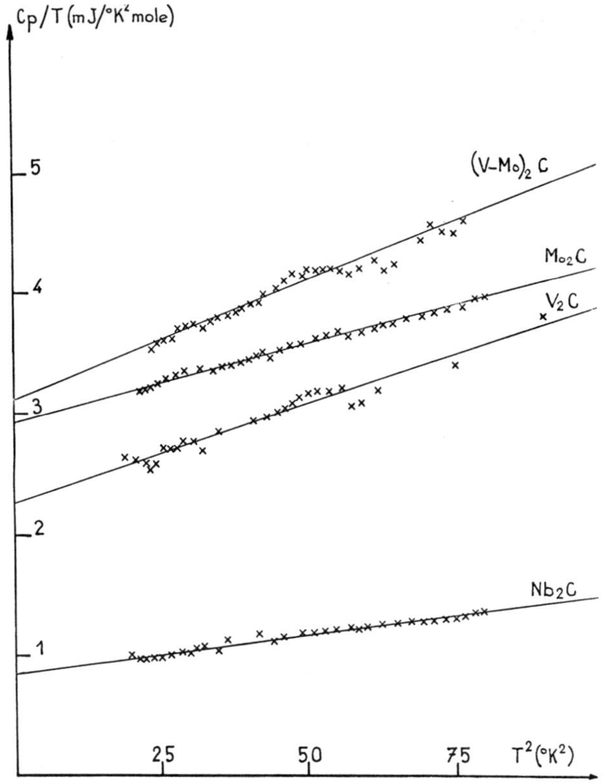

Fig. 3. Specific heat of semi-carbides.

Table 2

Low Temperature Electronic Specific Heats

	V_2C	Nb_2C	Mo_2C	MoVC
γ (mJ/$^oK^2$mole)	2,26±5%	0,84±5%	2,93±5%	3,16±5%

Table 3

Parameters of Semi-Carbides

	V_2C (orth*)	Nb_2C (orth*)	Mo_2C (Hex)	MoVC (Hex)
a(A°)	2,873±0,01	3,0955±0,001	3,007±0,001	2,9535±0,001A°
b(A°)	10,250±0,03	10,904±0,003	-	-
c(A°)	4,572±0,02	4,967±0,001	4,729±0,002	4,645±0,002A°

*The Nb_2C and V_2C phases are slight distortions of hexagonal phases (12).

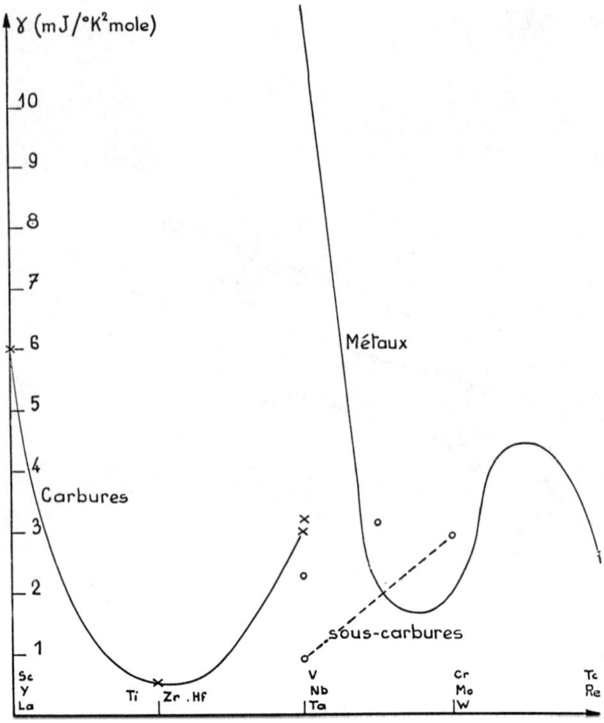

Fig. 4. Electronic specific heat of the transition metals and their carbides.

One can notice the drastic decrease from Mo_2C to Nb_2C. The difference is not so sharp between Mo_2C and V_2C, but this effect should be normal since the density of states associated with d-states is roughly 1.5 times higher for the first transition period than for the second one.

This result strongly suggests the existence of a minimum of the density of states for the fifth column or maybe for a slightly lower electron per atom ratio. It indicates a progressive displacement of this minimum when changing from the C.C. metal (VIth column) to the hexagonal semi-carbide (Vth column) and then to the f.c.c. monocarbide (IVth column). It would suggest a shape of the bands rather insensitive to the crystallographic structure. To confirm these results, we need to know the actual shape of the d band for the hexagonal close-packed structure, and in particular to check that it does show - like c.c. and NaCl f.c.c. bands - a minimum of the density of states corresponding to a half-filled d band.

References

1) H. Bilz. Zeits. für Phys., 153, 338, (1958).
2) E. Dempsey, Phil. Mag., 86, 8, 225, (1963).
3) P. Costa and R. R. Conte, Compounds of Interest in Nuclear Technology. AIME, Boulder 1964, 3.

4) R. G. Lye and E. M. Logothetis, Phys. Rev., 147, 622, (1966).
5) V. Ern and A. C. Switendick, Phys. Rev., 137A, 1927, (1965).
6) D. Rossier. Thesis, Orsay 1966.
7) B. Stalinski, C. K. Coogan and H. S. Gutowsky, J. Chem. Phys. 34, 1191, (1961).
8) D. W. Jones, J. of Less Com. Metals 6, 100, (1964).
9) T. Tsuchida, J. of Phys. Soc. of Japan 18, 1016, (1963).
10) M. C. Cadeville and A. J. P. Meyer, C. R. Ac.Sc., 255, 3391, (1962).
11) K. Yvon, W. Rieger and H. Nowotny, Mh. Chem. 97, 690, (1966).
12) K. Yvon, H. Nowotny and R. Kieffer, Mh. Chem. 98, 34, (1967).
13) E. Parthe and V. Sadagopan, Acta Cryst. 16, 202, (1963).
14) A. L. Bowman, T. C. Wallace, J. L. Yarnell, R. G. Wenzel and E. K. Storms, Acta Cryst. 19, 6, (1965).
15) H. Goretzki, Phys. Stat. Sol. 20, K141, (1967).
16) C. H. deNovion, R. Lorenzelli and P. Costa., C. R. Ac. Sc. 263, 775, (1966).
17) N. M. Volkova, S. I. Aljamovskji and P. V. Gel'd 5, 1934, (1963) Izv. Akad. Nauk SSSR, otd. Tekhn. Nauk, Met i Gorn Delo.
18) P. Costa. Thesis, Orsay.
19) T. H. Geballe and al. Physics 2, 293, (1966).
20) J. J. Veyssie, D. Brochier, A. Nemoz and J. Blanc, Phys. Letters 14, 261 (1965).
21) Bonnerot. Unpublished work. Orsay.
22) H. Bittner and H. Goretzki Mh. Chem. 93, 1000, (1962).
23) H. Bittner and H. Goretzki Mh. Chem. 94, 518, (1963).
24) R. Caudron, P, Costa and B. Saulgeot, 2nd Int. Conf. on Semi-Metallic Compounds of Trans. Elem. Enschede 1967.

SESSION III

Non-Stoichiometry and Phase Relations

Chairman: Donald F. Gibbons
Case Institute of Technology
Cleveland, Ohio

Co-Chairman: F. W. Vahldiek
Air Force Materials Laboratory
Wright-Patterson Air Force Base, Ohio

NON-STOICHIOMETRY AND BONDING IN REFRACTORY MONOCARBIDES

Michael Hoch
Materials Science Program
University of Cincinnati
Cincinnati, Ohio

ABSTRACT

The refractory monocarbides with the sodium-chloride type structure have a wide homogeneity range. Equations have been derived from statistical considerations to give the activity of each component as a function of composition and temperature based on the interaction energies between the various components.

From the activity measurements as a function of composition in various binary carbides and nitrides of the Groups IV and V transition elements and in the ternary zirconium-niobium-carbon system, the various interaction energies were calculated. The metal to metal interaction energy is the same in the carbides as it is in the pure metals and metallic solutions; therefore, the metal to metal bonding is the same in pure metals and in the carbides. The high melting point and other refractory properties of the carbide arise from the metal-carbon bond. The metal-carbon bond is stronger in Group IV carbides than in Group V carbides. The metal-carbon bond energy decreases as the atomic weight of the metal increases. The experimental results also indicate that the metal-carbon bond is such that the carbon atom is ionized--the ionization increasing with the bond strength.

INTRODUCTION

The borides, carbides, and nitrides of the transition metals of Groups IV, V, and VI of the Periodic Table are generally characterized by very high melting points, brittleness, and hardness, that is, properties which are generally ascribed to materials having ionic or strongly covalent bonding. However, the resistivity and temperature dependence of the resistivity indicate metallic behavior. For some of these compounds the resistance is lower than that of the parent metal. Also, in several of the compounds, large deviations from stoichiometry exist, which is also characteristic of metallic behavior.

There have been several theoretical investigations dealing with the bonding in these so-called hard metals, that is, the carbides, nitrides, and borides of the Groups IV, V, and VI, which are mostly based on the behavior and description of the electronic structure of the system. In the calculations of bonding by Bilz[1], interactions involving the d and s electrons of the metal atom and the 2p and 2s electrons of the metalloid atom were taken into account. Based on these calculations, Bilz[1] concludes that in the hard metals there is a metal-metalloid bond and also a metal-metal bond. Bilz[1] also concludes that the density of states curve undergoes the minimum somewhere between 8 to 9 electrons per "molecule" (TiC is counted as 8, 4 electrons contributed by Ti and 4 by C; whereas TiN is counted as 9, 4 contributed by Ti and 5 by N).

A somewhat different approach is taken by Dempsey[2] who concludes that the metalloid contributes electrons into the d band of the transition metal. This kind of treatment had already been proposed by Kiessling[3] who said that the nonmetal is an electron donor into the d band of the metal and that the metal-metalloid bonding decreases as the atomic number of the transition metal increases. According to Umanski[4], the metalloid is ionized to some extent, and the d band of the metal is an electron acceptor. To explain the relatively lower stability of the Group VI carbides, molybdenum carbide and tungsten carbide, Dempsey[2] indicates that bonding is increased until the transition metal d band contains around 6-1/2 electrons per atom, whereas if this number is surpassed, the bonding and stability decrease.

According to Samsonov[5], the electronic structure of the carbides is influenced by the unfilled d shell of the metal atom. The model of Lye[6] is very close to that described by Dempsey[2]. Ern's[7] calculations also show a minimum in the density of states curves in agreement with Bilz[1]. The experimental results of Costa and Conte[8] and of Piper[9], who measured resistivity, thermoelectric power, electronic specific heat, Hall coefficients, superconducting transition temperatures, and magnetic susceptibility, seem to be in agreement with the model of Dempsey[2]; but as the authors point out, the experimental results are not in any disagreement with the model of Bilz[1].

The present investigation attempts to obtain the various bond energies in the monocarbides from thermodynamic activity measurements. From statistical considerations, equations were derived to determine the activity of each component as a function of composition. In this model, the various interaction energies or bond energies between metal and nonmetal, metal and metal, and nonmetal-nonmetal are involved. By measuring the activity variation of each component with composition, the various interaction energies are obtained. From the interaction energies the type of bonding present in the various materials can be deduced.

THEORY

The statistical treatment given below corresponds to the "regular solution" treatment[10] of substitutional solid solutions. In the sodium-chloride type structure of the monocarbides, the metal atoms occupy one sublattice and the carbon atoms occupy the other. In the single-phase region of a ternary Me^{I}-Me^{II}-C system having the sodium-chloride type structure, the metal sites are randomly filled with Me^{I} and Me^{II} atoms, whereas some of the nonmetal sites are empty. Depending on the ratio of Me^{I} atoms to Me^{II} atoms and the fraction of the nonmetal sites which are empty, the composition of the material can vary.

The general formula of this material, therefore, can be written: $Me^{I}_{x}Me^{II}_{1-x}C_{y}$, where $0 < x < 1$, $0 < y < 1$.

From statistical considerations the grand partition function for this type of structure is

$$GPF = \sum \frac{N!}{N_1!(N-N_1)!} \cdot \frac{N!}{N_3!(N-N_3)!} [a_1K_1(T)]^{N_1} [a_2K_2(T)]^{N-N_1}$$

$$[a_3K_3(T)]^{N_3} \exp[-\frac{1}{kT}(N_1E_1 + N_2E_2 + N_3E_3 + \frac{N_1^2}{2N}E_{11} +$$

$$\frac{N_2^2}{2N}E_{22} + \frac{N_3^2}{2N}E_{33} + \frac{N_1N_2}{N}E_{12} + \frac{N_1N_3}{N}E_{13} + \frac{N_2N_3}{N}E_{23})] \quad (1)$$

where N is the number of available metal sites and N_i is the number of the various atoms present. Subscript 1 refers to Me^{I}, 2 to Me^{II}, and 3 to C. a_i is the activity of each component in the crystal phase; $K_i(T)$ is the contribution of an added atom to the partition function for the normal vibrational modes of the crystal; E_i is the energy required to bring an atom from its standard state into the sodium-chloride type lattice; and E_{ij} is the pairwise interaction energy. The small amount of Schottky defects and Frenkel disorders present in the metal and nonmetal sublattice is negligible.

By the standard treatment[10], the following equations are obtained:

$$RT\ln a_1 = RT\ln\frac{(N-N_2)(N-N_3)}{N^2} - RT\ln K_1(T) + [E_1 + \frac{N^2-N_2^2}{2N^2}E_{11} -$$

$$\frac{N_2^2}{2N^2} E_{22} - \frac{N_3^2}{2N^2} E_{33} + \frac{N_2^2}{N^2} E_{12} + \frac{N_2 N_3}{N^2} E_{13} - \frac{N_2 N_3}{N^2} E_{23}] \quad (2)$$

$$RT\ln a_2 = RT\ln \frac{(N-N_1)(N-N_3)}{N^2} - RT\ln K_2(T) + [E_2 + \frac{N^2-N_1^2}{2N^2} E_{22} -$$

$$\frac{N_1^2}{2N^2} E_{11} - \frac{N_3^2}{2N^2} E_{33} + \frac{N_1^2}{N^2} E_{12} + \frac{N_1 N_3}{N^2} E_{23} - \frac{N_1 N_3}{N^2} E_{13}] \quad (3)$$

$$RT\ln a_3 = RT\ln \frac{N_3}{N-N_3} - RT\ln K_3(T) + [E_3 + \frac{N_3}{N} E_{33} + \frac{N_1}{N} E_{13} + \frac{N_2}{N} E_{23}] \quad (4)$$

Introducing the equations

$$N_1 + N_2 = N$$

$$\frac{N_1}{N} = x$$

$$\frac{N_2}{N} = 1-x$$

$$\frac{N_3}{N} = y$$

gives

$$RT\ln a_1 \cdot \frac{1}{x(1-y)} = E_1 + (1-x)^2 [E_{12} - \frac{E_{11}}{2} - \frac{E_{22}}{2}] + \frac{E_{11}}{2} + (1-x)y\ E_{13} -$$

$$(1-x)y\ E_{23} - \frac{1}{2} y^2\ E_{33} - RT\ln K_1(T) \quad (5)$$

$$RT\ln a_2 \cdot \frac{1}{(1-x)(1-y)} = E_2 + x^2 [E_{12} - \frac{E_{11}}{2} - \frac{E_{22}}{2}] + \frac{E_{22}}{2} + xyE_{23} -$$

$$xyE_{13} - \frac{1}{2} y^2\ E_{33} - RT\ln K_2(T) \quad (6)$$

$$RT\ln a_3 \cdot \frac{(1-y)}{y} = E_3 + yE_{33} + xE_{13} + (1-x)E_{23} - RT\ln K_3(T) \quad (7)$$

By making y = 0, equations (5) and (6) reduce to respectively

$$RT\ln a_1\frac{1}{x} = E_1 + \frac{E_{11}}{2} + (1-x)^2[E_{12} - \frac{E_{11}}{2} - \frac{E_{22}}{2}] - RT\ln K_1(T) \tag{8}$$

$$RT\ln a_2\frac{1}{1-x} = E_2 + \frac{E_{22}}{2} + x^2[E_{12} - \frac{E_{11}+E_{22}}{2}] - RT\ln K_2(T) \tag{9}$$

Equations (8) and (9) are the equations for the substitutional regular solution model where the elements in the pure state have a different crystal structure from that of the solution. $E_1 + \frac{E_{11}}{2} - RT\ln K_1(T)$ is the free energy required to transform pure 1 from its standard state into the structure of the solution. Similarly, $E_2 + \frac{E_{22}}{2} - RT\ln K_2(T)$ is the corresponding term for component 2.

If x = 1, then the interstitial system $Me^{I}C_y$ is formed and equations (5) and (7) reduce to

$$RT\ln a_1\frac{1}{1-y} = E_1 + \frac{E_{11}}{2} - y^2\frac{E_{33}}{2} - RT\ln K_1(T) \tag{10}$$

$$RT\ln a_3\frac{1-y}{y} = E_3 + E_{13} + yE_{33} - RT\ln K_3(T) \tag{11}$$

Similarly, when x = 0, the interstitial system $Me^{II}C_y$ results, equations (6) and (7) reducing to

$$RT\ln a_2\frac{1}{1-y} = E_2 + \frac{E_{22}}{2} - y^2\frac{E_{33}}{2} - RT\ln K_2(T) \tag{12}$$

$$RT\ln a_3\frac{1-y}{y} = E_3 + E_{23} + yE_{33} - RT\ln K_3(T) \tag{13}$$

In equation (11), if y = 1, then one has the hypothetical stoichiometric crystal $Me^{I}C$ and therefore, $E_3 + E_{13} + E_{33} - RT\ln K_3(T)$ is the free energy difference between carbon in its standard state and the stoichiometric compound.

Because the pure metals at the temperatures in question are body centered cubic, the standard state for them in further calculations will be body centered cubic metal, whereas the standard state for carbon will be taken as graphite.

The pairwise interaction energies have been discussed in detail[11]. As pointed out, the interaction energy E_{ij} contains bonding not only from i and j atoms which are closest neighbors, but from those which are second, third, etc., neighbors. Furthermore, the interaction energy consists of a "chemical" energy and "strain" energy, which may be of opposite sign. Thus, it is very difficult to calculate the pairwise interaction energies from first principles.

EXPERIMENTAL RESULTS AND DISCUSSION

Binary Systems

Hoch[12] has evaluated the available data on binary carbides and nitrides of the Group IV and V transition metals and obtained the variation of the activity of each component as a function of composition and the various interaction energies. The results of interest here are as follows:

For ZrC_y

$$RT\ln a_{Zr} \cdot \frac{1}{1-y} = (1\pm1)-(y^2)(14\pm1.5) \text{ kcal/mole} \quad (14)$$

$$RT\ln a_C \cdot \frac{1-y}{y} = -(35.2\pm4)-(28\pm3)(1-y)+2.7\times10^{-3}T \text{ kcal/mole} \quad (15)$$

For NbC_y

$$RT\ln a_{Nb} \cdot \frac{1}{1-y} = -(0.05\pm1.0)-(y)^2(7.6\pm1.3) \text{ kcal/mole} \quad (16)$$

$$RT\ln a_C \cdot \frac{1-y}{y} = -(28.6+2.5)-(15.2\pm2.5)(1-y)+0.4\times10^{-3}T \text{ kcal/mole} \qquad (17)$$

Though the above data was deduced from experimental results obtained at very high temperatures, 2500-2700°K, one may compare the values with heat of formation data obtained by conventional calorimetry. Assuming that $\Delta H = \Delta E$, the heat of formation of ZrC_y can be expressed from equations (14) and (15) as

$$\Delta H_f = (1\pm1) - (63.2\pm5)y + (14\pm1.5)y^2 \text{ kcal/mole} \qquad (18)$$

This equation can be compared with the values obtained by Mah[13]. The above equation gives for $ZrC_{1.0}$, $\Delta H_f = -48\pm5$ kcal/mole, whereas Mah's value is -47.0 ± 0.6 kcal/mole. For $ZrC_{0.710}$, the above equation gives $\Delta H_f = -36.8\pm4$ kcal/mole, whereas Mah obtained -33.1 ± 0.8 kcal/mole. One can see that the agreement is good.

Similarly for NbC_y, the heat of formation calculated from equations (16) and (17) is

$$\Delta H_f = -(0.05\pm 1.0) - (43.8\pm3.5)y + (7.6\pm1.3)y^2 \text{ kcal/mole} \qquad (19)$$

This equation is compared with the combustion calorimetric measurements of Huber, et al.[14], Kusenko and Geld[15], Mah and Boyle[16], and Kornilov, et al.[17] In Figure 1 where $\frac{\Delta H_f}{y}$ is plotted versus y all the calorimetric data was corrected by using the heat of formation for Nb_2O_5 of -455.2 ± 0.6 kcal/mole. As can be seen from Figure 1 in the middle range, the agreement between the heat of formation obtained from the high-temperature vaporization data and from the combustion measurements is extremely good. In the low and high carbon regions there are discrepancies which may be due to the fact that when these samples are quenched for the calorimetric measurements considerable stresses may build up.

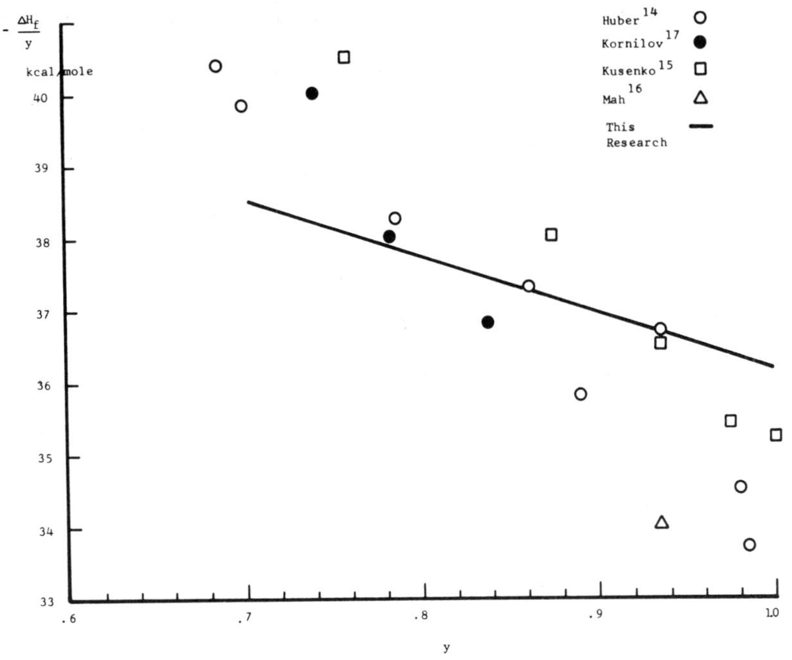

Fig. 1. Heat of formation of NbC_y.

TABLE 1

Energy Terms in Binary Nitrides and Carbides

Material		E_{33} kcal/mole	$E_3 + E_{13} + E_{33}$ kcal/mole
ZrC	-473	28. ± 3.	-35.2 ± 4.
HfC	-394	15. ± 8.	-47 ± 10.
TiN		78.2 ± 9.	-57.5 ± 9.0
ZrN		19.0	-72.1 ± 2.0
HfN		0	-72.4 ± 1.6
VC	-242	25 ± 4.6	-10.6 ± 4.6
NbC	-313	15.2 ± 2.5	-28.6 ± 2.5
TaC	-201	0	-27.0 ± 1.3

In the zirconium-niobium binary metallic solution which is bcc, Hoch[18] had derived from the phase diagram the metal-metal interaction energy $E_{12} - (\frac{E_{11}}{2} + \frac{E_{22}}{2}) = 4.9\pm0.6$ kcal/mole.

In the other binary carbides and nitrides, the energy terms derived by Hoch[12] are given in Table 1. This table shows certain regularities. The values of $E_3 + E_{13} + E_{33}$ (the energy difference between the nonmetal in its standard state and at the stoichiometric composition in the interstitial phase) are equal in the pairs of compounds that are in the same group and have the same lattice parameters, that is, ZrC-HfC, ZrN-HfN, and NbC-TaC. E_{33} is always positive and in the pairs of compounds decreases as the atomic weight of the metal increases.

Ternary Systems

Zr-Nb-C System

Jun[19], using a time-of-flight mass spectrometer, has measured the relative activities of the various components in the NaCl type structure single phase region of the zirconium-niobium-carbon system. From these activities, using equations (1), (2), and (3), he has deduced the various interaction energies in the system which are given in Table 2.

In the fcc single phase region of the ternary zirconium-niobium-carbon system, the metal atoms are randomly distributed on the metal sublattice and form a substitutional solution on this sublattice. $E_{12} - (\frac{E_{11}}{2} + \frac{E_{22}}{2})$ is the energy parameter on this fcc metal sublattice. The value of $E_{12} - (\frac{E_{11}}{2} + \frac{E_{22}}{2}) = 6.5\pm0.3$ kcal/mole is very close to that in the pure binary bcc Zr-Nb system, $E_{12} - (\frac{E_{11}}{2} + \frac{E_{22}}{2}) = 4.9\pm0.6$ kcal/mole[18]. The similarity of $E_{12} - (\frac{E_{11}}{2} + \frac{E_{22}}{2})$ in the two systems indicates that the bonding between metal atoms is the same in the metals and in the carbides.

The difference in bonding between zirconium to carbon and niobium to carbon is $E_{13} - E_{23} = -5.2\pm1.6$ kcal/mole. This shows, as expected, that the

TABLE 2

Energy Terms in $Zr_xNb_{1-x}C_y$

$[E_{12} - \frac{E_{11} + E_{22}}{2}]$	=	6.54 ± 0.27 kcal/mole
$[E_{13} - E_{23}]$	=	-5.16 ± 1.59 kcal/mole
E_{33}	=	15.62 + 9.37x - 29.5(1-x)x kcal/mole

Subscript 1 refers to Zr, 2 to Nb, 3 to C.

zirconium-carbon bond is stronger than the niobium-carbon bond. Table 1 gives $E_3 + E_{13} + E_{33}$, which is the energy difference between the nonmetal in its standard state and at the stoichiometric composition in the interstitial phase. The difference of this value between zirconium carbide and niobium carbide is -6.6±4.8 kcal/mole. This value is equal to $E_{13} - E_{23}$ measured in the ternary Zr-Nb-C system. Thus, the stronger bonding of carbon in stoichiometric zirconium carbide compared to niobium carbide (more negative $E_3 + E_{33} + E_{13}$) is due only to a stronger zirconium-carbon bond. This type of metal-carbon bond can be described by the model of Bilz[1] or that of Dempsey[2]. The metal to carbon bond probably ionizes the carbon atom; a stronger metal-carbon bond forms a more ionized carbon atom. As the ionization of the carbon increases, so does the carbon-carbon repulsion due to simple Coulomb-interaction. This explains the larger value of E_{33} in ZrC compared to that in NbC as shown in Table 1. An ionized carbon atom, coupled with Kiessling's[3] finding that the metal-nonmetal bond strength decreases as the atomic number of the metal increases, explains why E_{33} decreases in the same groups as shown in Table 1 (ZrC-HfC, ZrN-HfN, NbC-TaC). A weaker metal-nonmetal bond gives rise to a less ionized nonmetal, and therefore, a lower Coulombic repulsion between the nonmetal ions.

To explain the quadratic variation of E_{33} with metal composition in the ternary zirconium-niobium-carbon system, two types of carbon atoms must be postulated. One may describe the two types of carbon atoms by saying that one type, so to speak, belongs to niobium; the other type to zirconium. If these two types of carbon atoms interact with each other and are distributed randomly, three types of bonds will result--those which are formed between carbon atoms belonging to niobium, those which belong to zirconium, and those where one carbon atom belongs to zirconium and the other to niobium. The presence of three types of bonds gives rise to a quadratic term in the interaction energy as a function of composition. Since to have three types of bonds requires two kinds of carbon atoms, the metal to carbon bond must be localized in order to distinguish the carbon atoms.

REFERENCES

1. H. Bilz, Z. Physik 153, 338 (1958).

2. E. Dempsey, Phil. Mag. 8, 285 (1963).

3. R. Kiessling, Metall. Rev. 2, 77 (1957).

4. J. S. Umanski, Dokl. Akad. Nauk SSSR 26, 127 (1943).

5. G. V. Samsonov and I. I. Korenskii, Phys. Metals Metallog. (USSR), (English trans.) 16, 93 (1963).

6. R. G. Lye and E. M. Logothetis, Phys. Rev. 147, 622 (1966).

7. V. Ern and A. C. Switendick, Phys. Rev. 137A, 1927 (1965).

8. P. Costa and R. R. Conte, "Properties of the Carbides of the Transition Metals", In Compounds of Interest in Nuclear Reactor Technology, AIME, 1964.

9. J. Piper, "Electrical Properties of Some Transition Metal Carbides and Nitrides", In Compounds of Interest in Nuclear Reactor Technology, AIME, 1964.

10. R. Fowler and E. M. Guggenheim, Statistical Thermodynamics, University Press, Cambridge, 1949.

11. M. Hoch, Trans. Met. Soc. AIME 230, 138 (1964).

12. M. Hoch, "The Role of the Defect Interaction Energy on the Stability of Interstitial Phases", In Phase Stability in Metals and Alloys, Eds., P. S. Rudman, J. Stringer, and R. I. Jaffee, McGraw-Hill, New York, 1966.

13. A. D. Mah, USBM Rept. of Ind., 6518 (1964).

14. E. J. Huber, Jr., E. L. Head, C. E. Holley, Jr., E. K. Storms, and N. H. Krikorian, J. Phys. Chem. 65, 1846 (1961).

15. F. G. Kusenko and P. V. Geld, Izvest. Sibir. Otdel. Akad. Nauk SSSR 2, 46 (1960).

16. A. D. Mah and B. J. Boyle, J. Am. Chem. Soc. 77, 6512 (1955).

17. A. N. Kornilov, I. D. Zaikin, S. M. Skuratov, and G. P. Shveikin, Zh. Fiz. Khim. 40, 1070 (1966).

18. M. Hoch, Trans. Met. Soc. AIME 224, 379 (1962).

19. C. K. Jun, Ph.D. thesis, University of Cincinnati, 1967.

THE CONVERSION OF CUBIC AND HEXAGONAL SILICON CARBIDE AS A FUNCTION OF TEMPERATURE AND PRESSURE

C. E. Ryan, R. C. Marshall, J. J. Hawley, I. Berman, and D. P. Considine

Air Force Cambridge Research Laboratories
Office of Aerospace Research
L. G. Hanscom Field
Bedford, Massachusetts

Abstract

Polytypism in silicon carbide is reviewed. The geometric properties of close-packed hexagonal structures are illustrated by models of the stacking arrangements in the more common polytypes. The inadequacies of a number of theories on the origin of polytypes are briefly discussed, and it is concluded that no satisfactory theory exists.

From the results of recent AFCRL experiments in the temperature range of 2100° to 2800°C, it is concluded that the transformation of beta to alpha silicon carbide can be suppressed by 20 atm of inert gas, and consequently that the alpha silicon carbide is a defect structure probably involving deficiencies in the carbon sublattice. It is further concluded that both polytypism and the beta-alpha transformation are impurity-dominated, and that the most significant impurities are boron and nitrogen. Speculating on the origin of polytypes, it appears that any adequate theory must include the influence of chemical and physical imperfections, temperature, and ambient gas composition and pressure.

1. INTRODUCTION

Silicon carbide has unique chemical, physical, optical, thermal, and electrical properties that endow it with attractive possibilities as semiconductor material, particularly for difficult temperature and radiation environments and for electro-optical and high-power devices in normal environments. Despite substantial research efforts during the past two decades (O'Connor and Smiltens, Eds., 1960; J. J. O'Connor, 1963; Knippenberg, 1963, pp. 161-274; Neuberger, 1965; Verma and Krishna, 1966), silicon carbide has failed to live up to its promise, primarily because certain problems associated with the controlled and reproducible growth of single crystals of this material have not been clearly understood.

In the silicon carbon system, the only compound species that exists in the solid state is silicon carbide, SiC (Elliott, 1966). It does, however, exist in both beta and alpha modifications. The beta, or cubic, modification crystallizes in the zinc-blende or sphalerite (diamond cubic) form. A large number (approximately 45) of the alpha are hexagonal or rhombohedral forms known as polytypes (Parthé, 1964).

At one time it was thought that the beta (cubic) modification was thermodynamically stable at low temperatures and that it transformed irreversibly to the alpha modification by a solid state phase transformation at about 2100°C (Jagodzinski and Arnold, 1960). Above this temperature the alpha modification was considered stable. It was also thought that the alpha polytypes were associated with slight energy differences involving different third-nearest-neighbor interactions in hexagonal close-packed structures. We now know that there has been no adequate explanation either for the beta-to-alpha transformation or for the origin of polytypes in silicon carbide.

It is the purpose of this paper to briefly outline the nature and present status of these problems and indicate the inadequacy of current theories. A number of experiments recently conducted at AFCRL are described. From the results of these experiments and our interpretation of experiments published by other workers, we draw some conclusions and speculate on the trend of current research.

2. POLYMORPHISM AND POLYTYPISM

When the same chemical compound exists in two or more crystallographic forms, the forms are called polymorphs and the phenomenon is called polymorphism. Polymorphs were originally associated with thermodynamic phases in such a way that the crystallographic form having the least Gibbs free energy was considered to be the most stable polymorph in a given range of temperature and pressure. To explain kinetic aspects such as speeds of transition from one polymorphic form to another and polymorphs that are metastable, energy-barrier theories depending largely on details of bonding structure were developed.

In certain close-packed structures like SiC, ZnS, and CdI_2, there is a special one-dimensional sort of polymorphism called polytypism. Polytypes are alike in the two dimensions of the close-packed planes, differing only in the stacking sequence in the dimension perpendicular to the close-packed planes. Since all polytypes of SiC have close-packed structures, all have the same density (3.17 gm $\cdot$ cm^{-3}). Consequently, gross pressure effects cannot be considered to be the prime

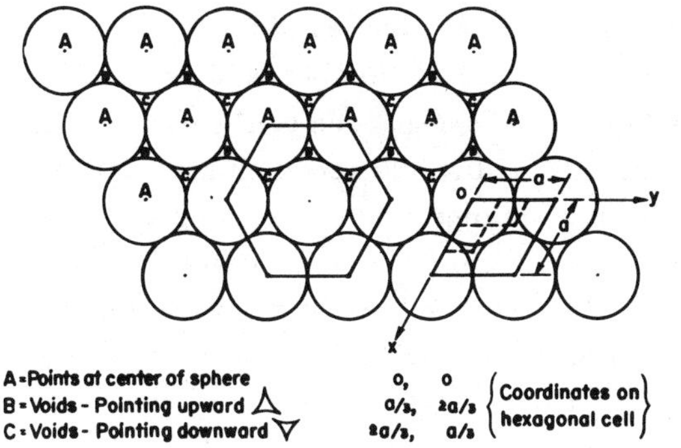

Fig. 1. Single layer of close-packed spheres of constant radius, with positions of centers of spheres and voids projected into plane of paper.

factor in polytype formation. Since polytypes differ geometrically only in third- and higher-nearest-neighbor spacing, all of them have very nearly the same energy. Nevertheless, temperature alone seems inadequate to explain the relative stability and occurrence of certain polytypes.

3. CLOSE-PACKED STRUCTURES AND NOMENCLATURE

In order to deal more precisely with polytypes in silicon carbide and be able to illustrate the more prevalent ones with models (Verma and Krishna, 1966, Chap. 4), we must first review close-packed structures and develop a consistent set of notations.

Figure 1 shows a single layer of close-packed spheres of constant radius. Each sphere is surrounded by six spheres whose centers A form the vertices of a regular hexagon. Half of the interstices (voids) between spheres, which look like triangles pointing upward (Δ), are labeled B. The other half, like triangles point-

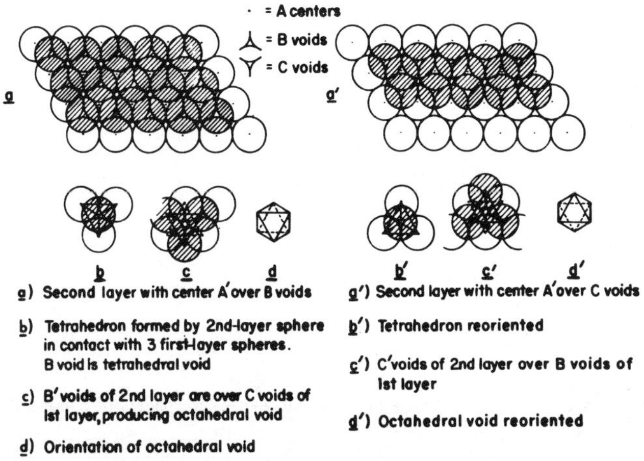

Fig. 2. Two layers of close-packed spheres of constant radius.

ing downward (▽), are labeled C. The hexagonal coordinates of the A, B, and C points projected into the plane of the paper are: A(0, 0); B(a/3, 2a/3); C(2a/3, a/3).

If we pack a second layer over the first in a close-packed configuration, all the centers A' of the second-layer spheres will be directly above either the B or the C voids. Figure 2 shows the two possible arrangements of packing a second layer above the first. In Figure 2(a) the second-layer centers A' are above the B voids of the first layer; in 2(a') they are above the C voids. Each sphere of the second layer touches three spheres of the first layer; similarly, each sphere of the first layer touches three spheres of the second layer. The centers of the four spheres in contact form the vertices of a regular tetrahedron. By comparing 2(a) and 2(a') it can be seen that the tetrahedrons surrounding the C voids are both translated and rotated with respect to those surrounding the B voids. Looking into the voids of the second layer in Figure 2(a), we see half of them directly above the A points of the first layer and half directly above the C voids; the voids above the A points are tetrahedral whereas those above the C voids are octahedral. In Figure 2(a') the second-layer centers A' are directly above the C voids, the B' voids are directly above the A points, and the C' voids are directly above the B voids. Both position and orientation of the octohedral voids as well as of the tetrahedral voids therefore differ from Figures 2(a) to 2(a'). Thus, when a third layer is added (Figure 3) there are four possibilities of stacking: ABA, ACA, ABC, and ACB—that is, any sequence in which neither A nor B nor C succeeds itself. The ABA pattern leads to the hexagonal wurtzite structure, and the ABC pattern leads to the cubic zincblende structure. The ACA and ACB, the respective reverse stacking patterns for wurtzite and zincblende, are equivalent to the obverse patterns but often neglected.

Without stacking faults in the cubic structure, the sequence ABCABC would repeat indefinitely. If, however, during growth, any layer shifted position to present one set of voids rather than the other (Figure 2), the subsequent layers would follow the reverse pattern rather than the obverse, or, ABCABC ... would become

	Obverse	Reverse
(2H wurtzite)	ABA	ACA
(3C zincblende)	ABC	ACB

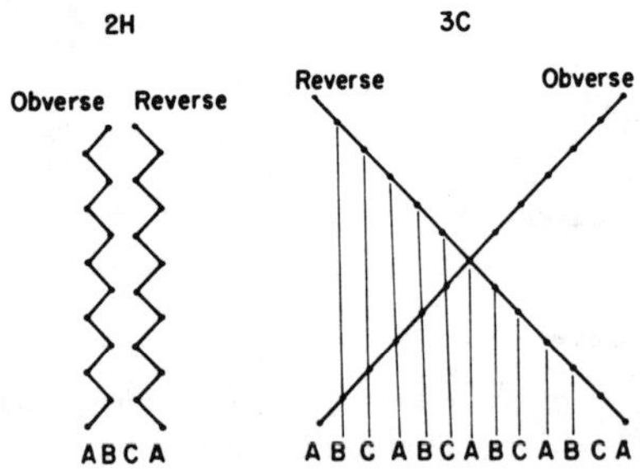

Fig. 3. Three layers of close-packed spheres of constant radius.

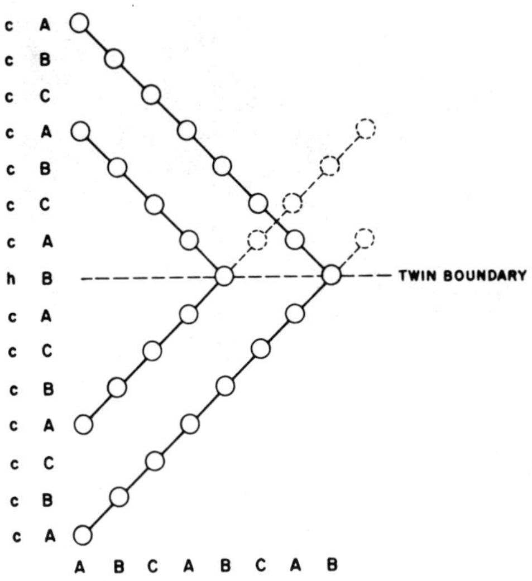

Fig. 4. Twin fault in cubic configuration.

ABCABCACBACB with the fault in the position shown by the arrow. One such fault is equivalent to a twin fault in the cubic configuration (see Figure 4).

The polytypes we discuss here have the ABC sequences and standard nomenclature shown in Figure 5. The H, C, and R in the Ramsdell notation designate hexagonal, cubic, and rhombohedral respectively. The numeral represents the number of layers in the Z direction before the cycle repeats. In the Zhdanov notation, the first figure represents the number of transitions in the obverse direction, the second figure represents those in the reverse direction, and so on. The sum of the figures represents the number of layers in the period. Designations such as $(23)_3$ are simply shorthand expressions for writing 232323, etc. The Hagg and Pauling-

Ramsdell	Zhdanov	ABC sequence	Hagg	P.W.J.* hk notation
2H	11	AB	+−	$(h)_2$
3C	∞	ABC	+++	kkk
4H	22	ABCB	++−−	hkhk = $(hk)_2$
6H	33	ABCACB	+++−−−	hkkhkk = $(hkk)_2$
8H	44	ABCABACB	++++−−−−	hkkkhkkk = $(hkkk)_2$
15R	$(23)_3$	ABCBACABACBCACB	$(++−−−)_3$	$(hkhkk)_3$
21R	$(34)_3$			
33R	$(3332)_3$			
51R	$(333332)_3$			
90R	$[(23)_4 3322]_3$			

*Pauling, Wyckoff, Jagodzinski

Fig. 5. Common polytype notations.

Fig. 6. Model of 2H polytype with both silicon and carbon atoms.

Wyckoff-Jagodzinski notations are slight variations of the Zhdanov scheme. A few of the higher polytypes have been included to indicate the complexity of some of the rarer polytypes and their structural relationships to the simpler ones.

Figure 6 is a photograph of a model of the 2H polytype, with the silicon and carbon atoms respectively represented by the large and small spheres. The next six figures (Figures 7 through 12) are sketched models of some of the more common polytypes, showing 3C, 2H, 4H, 6H, 8H, and 15R respectively. Figure 13 shows the ABC sequences of these polytypes plotted in the $11\bar{2}0$ plane. This type of diagram shows more clearly than the models that the 2H type reverses stacking sequence every layer; the 4H, after every two layers; the 6H, after every three layers; the 8H, after every four layers; and the 15R reverses alternately after first two and then three layers, and hence must go through three cycles to complete a period. The order of frequency of polytype occurrence is: 6H (by far the most common), 15R, 4H, and 8H; all others are quite rare.

So far, we have considered only the geometry of the problem and found polytypes to be ordered sets of C axis stacking faults, in which the faults are most likely to occur after every third layer and somewhat less frequently after two or four layers. Thus, the Zhdanov notation for the more common polytypes is dominated by 3, 2, and 4 sequences, in that order.

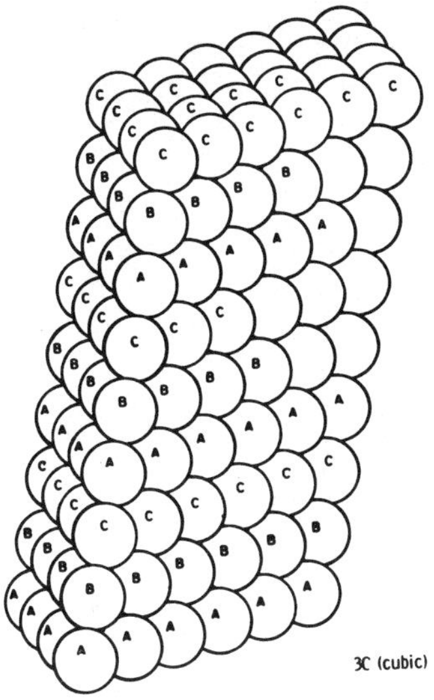

Fig. 7.

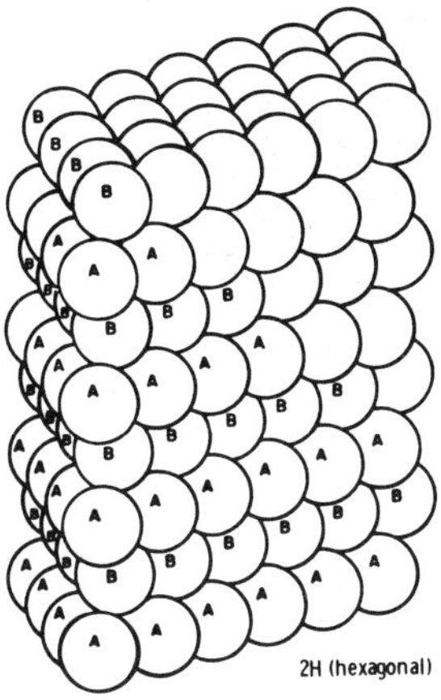

Fig. 8.

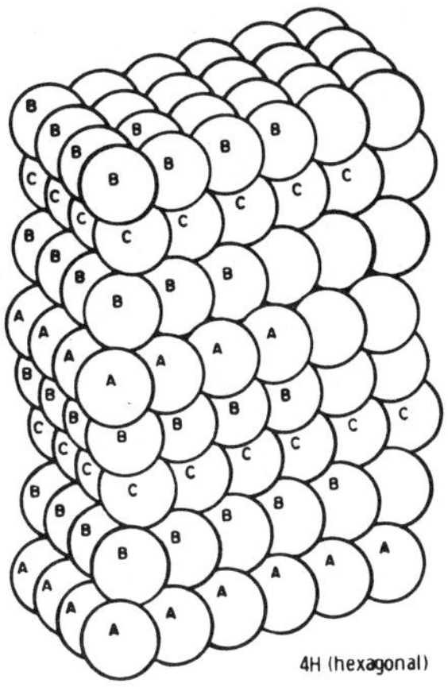

Fig. 9.

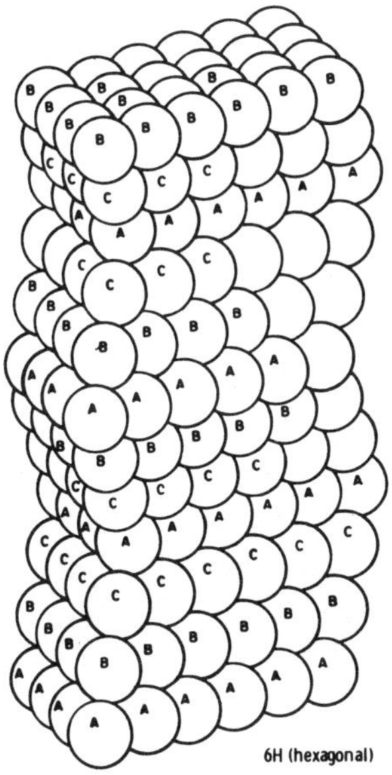

Fig. 10.

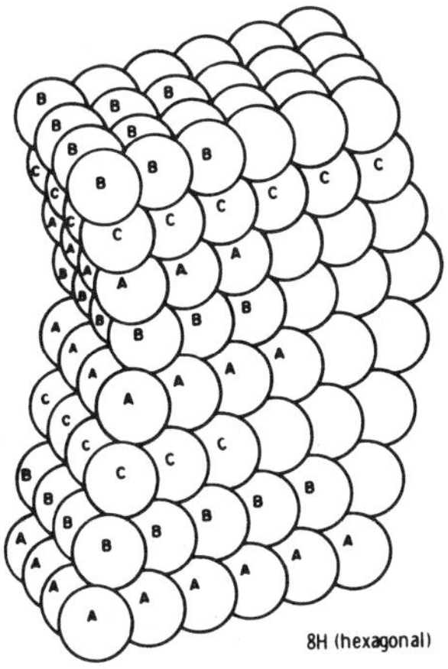

Fig. 11.

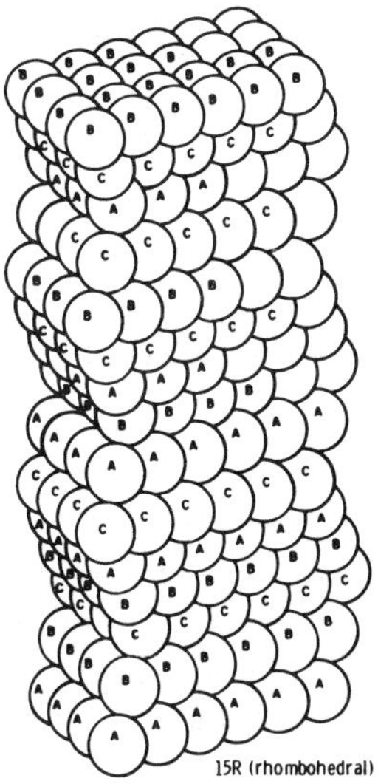

Fig. 12.

Figs. 7-12. Model configurations.

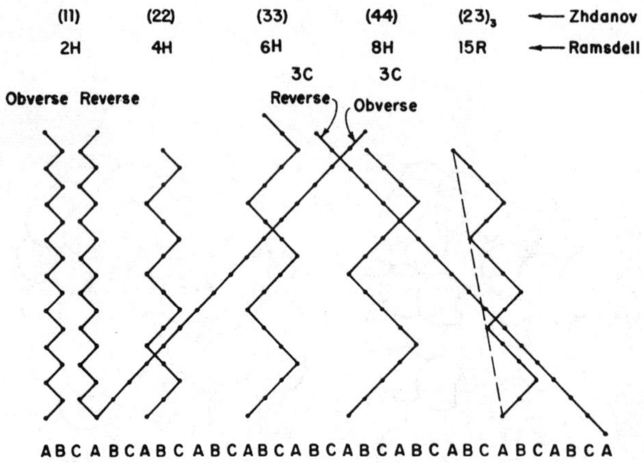

Fig. 13. Position of Atoms in the $11\bar{2}0$ plane.

4. THEORIES ON THE ORIGIN OF POLYTYPES

The real problem for the chemist and physicist is to explain why the stacking faults occur in orderly sequences leading to discrete polytypes with large precisely defined unit cells. In brief, the problem is to explain the origin of polytypes. There are almost as many theories as there are workers in the field, but they are generally based on considerations of temperature and energy, such as proposed by Jagodzinski; based on screw dislocations as the means of propagating polytypes, as proposed by Frank; or based on impurities, as proposed by Lundquist and others. These theories and others are explained in Verma and Krishna's book on polymorphism and polytypism in crystals (1966, Chap. 8).

We fail to see how energy theories can explain the relative stability and frequency of various polytypes. The screw dislocation theories may in certain cases explain the propagation of polytypes but hardly their origin. In general, SiC is prepared under experimental conditions that make it difficult to control its purity, and experimental evidence in support of current impurity theories is contradictory. Verma concludes that there is no satisfactory unified theory to explain the phenomenon of polytypism. A more complicated theory is apparently required—a theory that includes several variables rather than energy, imperfections, or chemical impurities alone. We will make some speculations about such a theory after the companion problem of the beta-to-alpha transformation is developed.

5. METHODS OF PREPARING SILICON CARBIDE

Figure 14 shows the various methods by which silicon carbide is usually prepared. First, neglecting the double-bordered blocks, it is seen that alpha silicon carbide, as generally prepared from the vapor phase by the Lely process or some modification of it (Knippenberg, 1963, pp. 161-274; Hamilton, 1960), has been obtained in the temperature range from about 2300° to 2700° C. Beta silicon carbide, prepared either from silicon melts (Nelson, et al., 1966) in carbon crucibles, or by the hydrogen reduction of organo-silanes (Knippenberg, 1963, pp. 161-274; Jennings et al., 1966), has been obtained at temperatures well below 2000° C. Heat-treated in an inert atmosphere in the range 2100° to 2300° C, beta silicon carbide transformed to alpha silicon carbide (Knippenberg, 1963, p. 242). Hence, the simple conclusion that beta was the low-temperature modification and alpha the high-temperature modification.

Two papers at the Boston Conference on Silicon Carbide, sponsored by AFCRL in 1959, confused this simple picture. Merz and Adamsky (1960, 1959) reported the growth of the 2H modification (wurtzite) for the first time. Their 2H (alpha) crystals were small whiskers intergrown with beta silicon carbide prepared by the hydrogen reduction of methyltrichlorosilane (CH_3SiCl_3) in the temperature range 1300° to

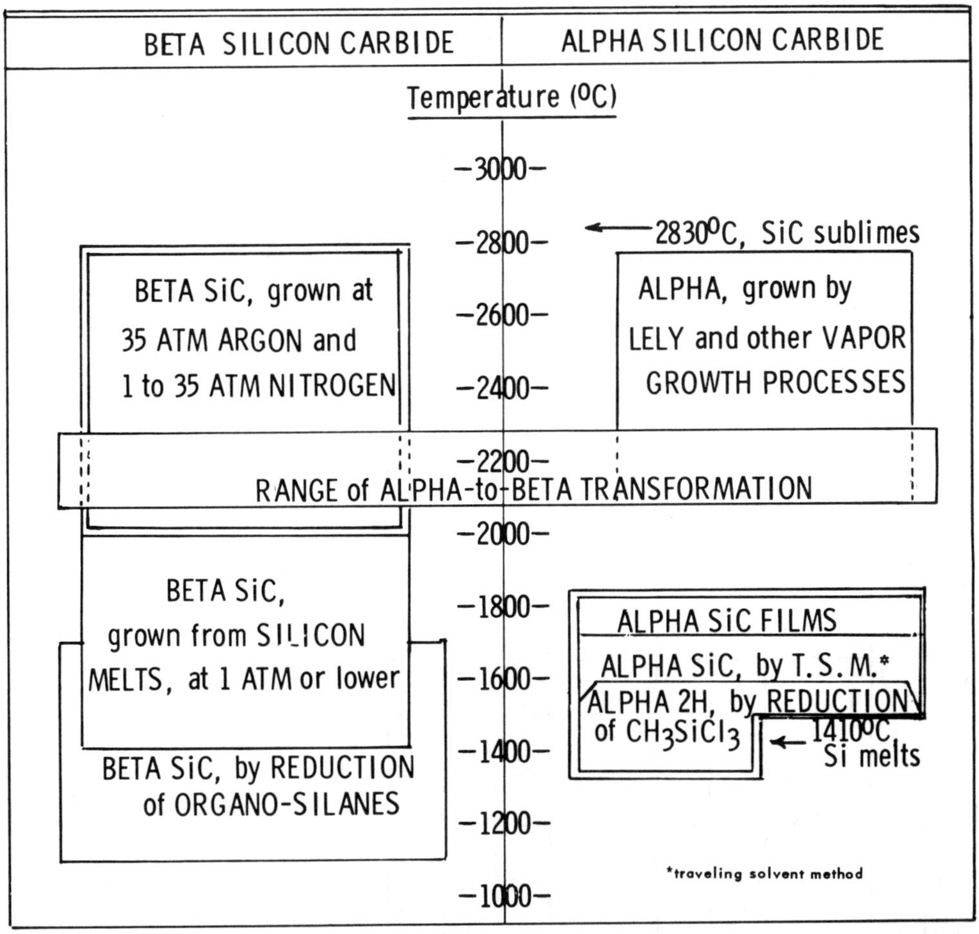

Fig. 14. Temperature ranges for various methods of growing alpha and beta silicon carbide.

1600°C. They concluded that this finding required reevaluation of the theory that it was beta that was the low-temperature form of silicon carbide. Scace and Slack (1959, 1960) reported that the silicon carbide obtained in the course of their solubility studies was beta even at preparation temperatures above 2600°C. Thus, in these cases the alpha form was prepared at low temperatures and the beta at high temperatures.

Figure 14 shows that factors other than temperature determine the form, whether cubic or hexagonal, that will be attained. Experiments conducted at AFCRL (Ryan et al., 1966), reported at the International Conference on Crystal Growth in June 1966, verified Merz and Adamsky's results in preparing 2H SiC whiskers. Further, they extended those results by demonstrating that the growth of the 2H crystals was obtained by means of the vapor-liquid-solid method, hence dominated by impurities, and that through an appropriate choice of impurities, 2H (beta) whiskers could be grown at will. The low-temperature growth of alpha SiC was thus adequately explained.

Both in our laboratory and through contractors (Nelson et al., 1967), we have conducted extensive investigations of beta silicon carbide grown from silicon melts. Practical limitations of this process have encouraged us to consider the possibilities of suppressing the beta-to-alpha transformation and to reevaluate the possibility of growing beta crystals in the higher temperature region. This is the subject of our present investigations.

Figure 15, taken from Elliott's (1966) First Supplement to Hansen's Constitution of Binary Alloys, compares silicon carbide phase diagrams from two sources. The

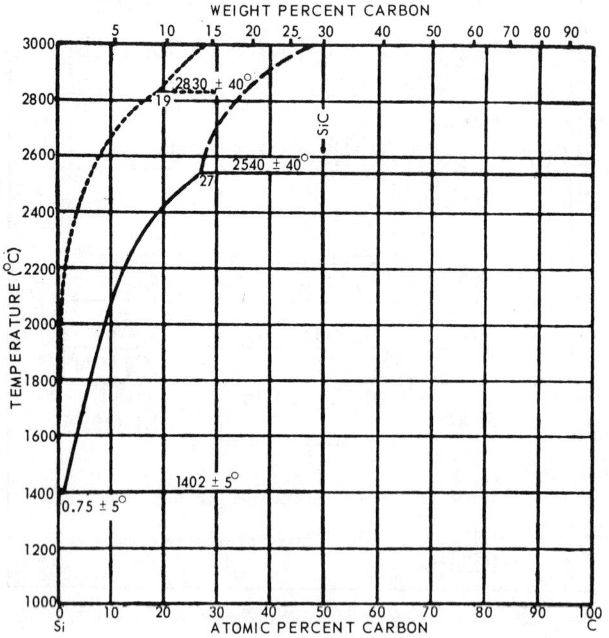

Fig. 15. Silicon carbon phase diagrams (from Elliott, 1966: --- Scace and Slack, 1959; —— Doloff, 1959).

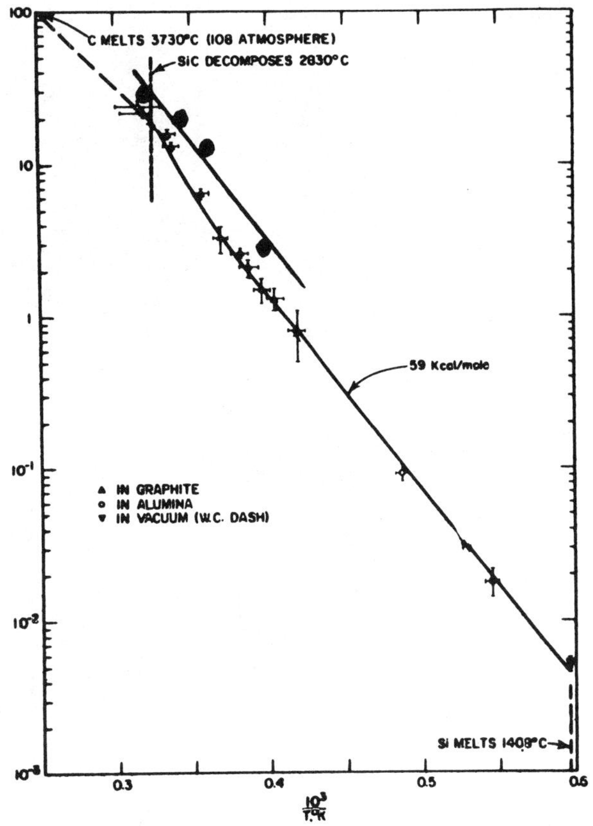

Fig. 16. Solubility (atomic percent) of carbon in liquid silicon as a function of reciprocal temperature (from Scace and Slack, 1960).

large difference between the two is some indication of the difficulty of making measurements of this type at high temperatures. If the solid curve were not unfortunately incorrect, the problem of growing beta silicon carbide from solution would be relatively easy.

The broken curve in Figure 15 is based on Scace and Slack's data (1959, 1960), which they replotted in terms of solubility in the figure we show as Figure 16. Now we can clearly see the problem of growing SiC from silicon solution at 1 atm. If we attempt growth below 2000°K (~1700°C), which is 0. 5 on the $10^3/T$, °K scale, the solubility is less than 0. 1 atomic percent. Low solubility and high supersaturation leads to uncontrolled nucleation and limits the size of the crystals. If we increase the solubility by increasing the temperature or by introducing additives in the melt, the crystals have alpha inclusions.

6. EXPERIMENTS

Experiments were conducted in a modified A. D. Little type-MP furnace that was driven by a 50-kW Lepel rf dual-frequency generator, monitored by a Leeds-

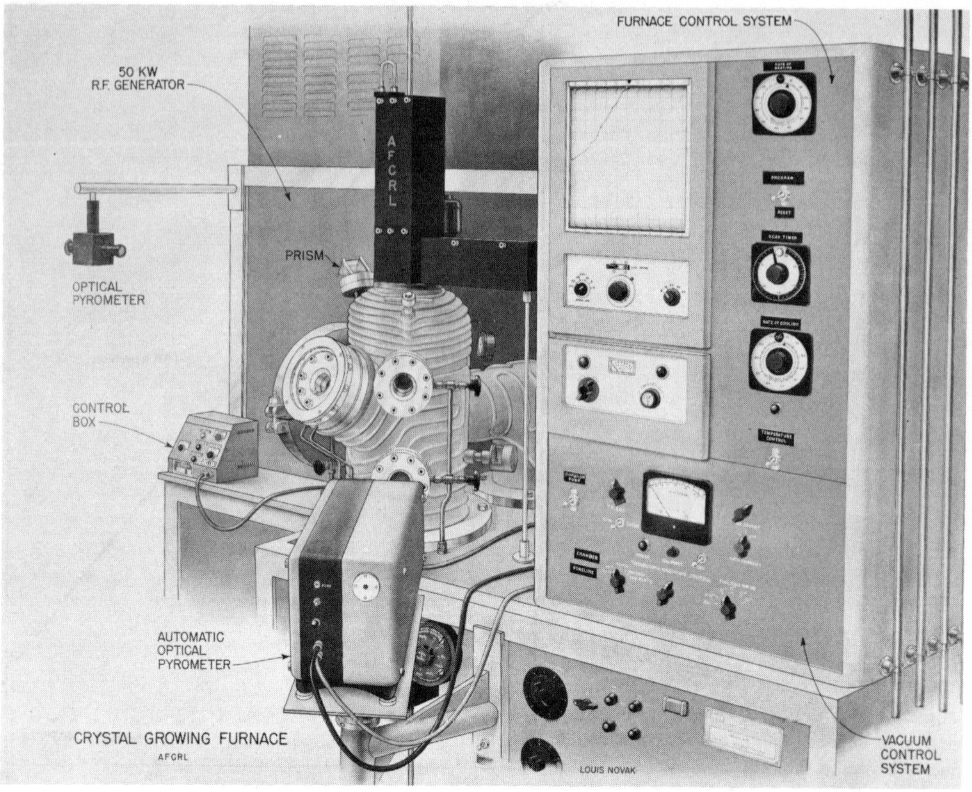

Fig. 17. Modified MP furnace for high-temperature experiments.

Northrup automatic optical pyrometer, and capable of operation from 10^{-5} Torr to 20 atm (Figure 17). A cutaway view of the furnace arranged for Czochralski growth is shown in Figure 18. Details of the test chamber as used in the experiments reported here are shown in Figures 19 to 21.

A typical calibration curve is shown in Figure 22. The pyrometer was sighted into the crucible chamber through a small hole drilled through the top of the crucible. The crucible chamber radiated like a black body. The pyrometer readings plotted against the induction-heater plate current were easily reproducible to ±20°C. An independent check on this relative calibration was made with materials of known melting points. High-temperature reactions and eutectics that formed with the carbon of the crucible presented problems. Among the materials successfully used were sapphire (mp 2050° C) and a specially prepared sintered molybdenum carbide (mp 2695°C). The melting points were in excellent agreement with the calibration curves. We judge our readings to be good to ±30° C and at least better than ±50° C.

The carbon susceptors were outgassed for several hours at about 2500° C and 10^{-5} Torr, after which they were mass spectrographically pure. The crucible

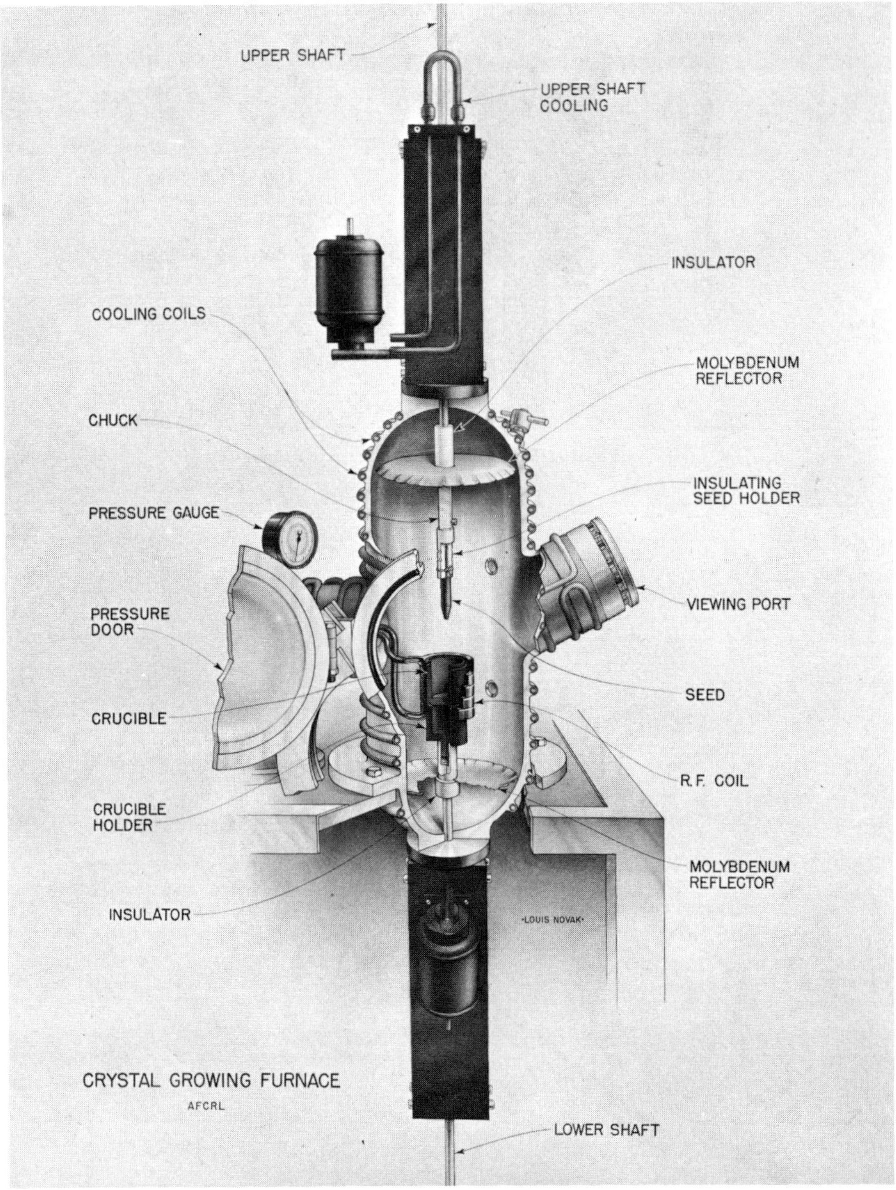

Fig. 18. Cutaway view of furnace.

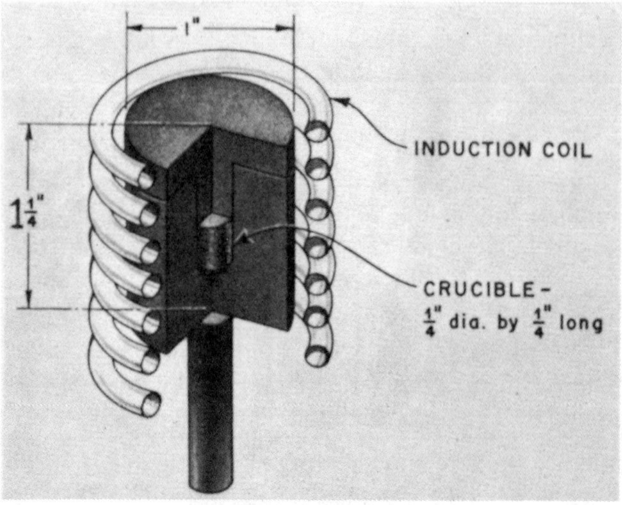

Fig. 19. Susceptor and crucible arrangement.

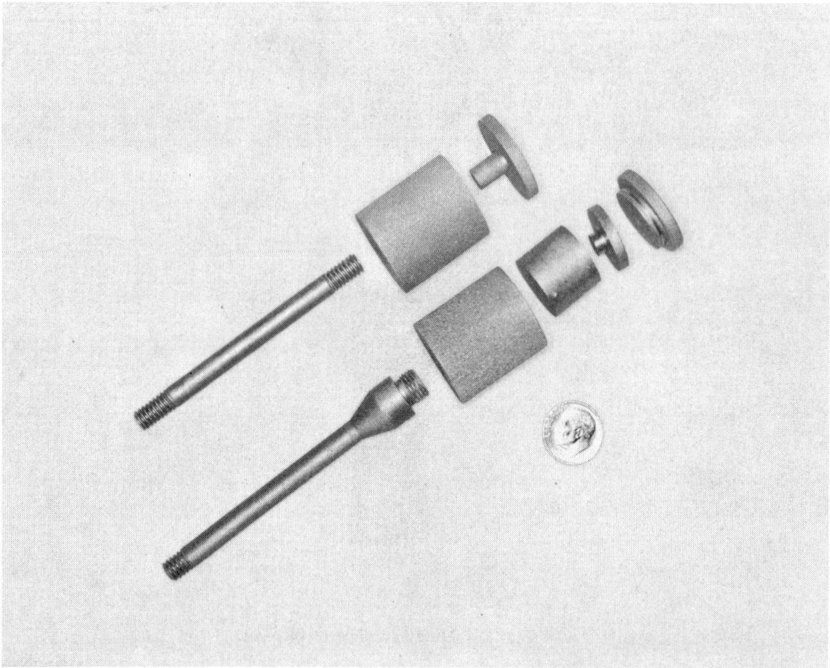

Fig. 20. Disassembled view of susceptor and crucible.

Fig. 21. Disassembled view of crucible.

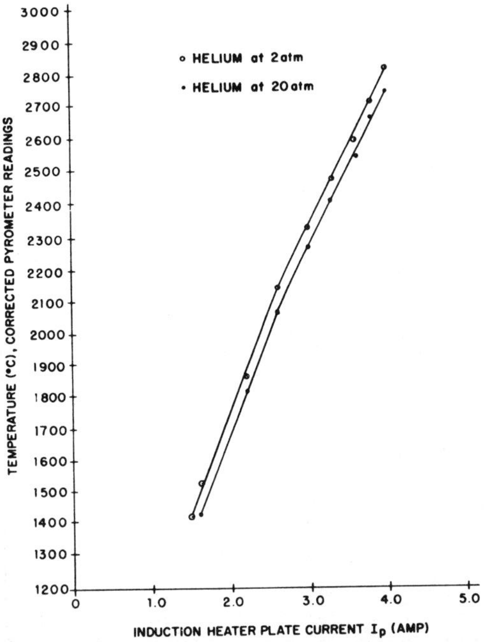

Fig. 22. Calibration of MP furnace in helium atmosphere.

charge was silicon in which the impurities averaged no more than 1 part/10^9.

The solubility points plotted in Figure 16 correspond to 2250°, 2450°, 2650°, and 2850° C. Although our own three sets of measurements are in good agreement with each other, isothermal conditions did not exist in the test chamber and we consider our apparent solubility values to be substantially too high. Hence, we are quite confident that the Scace and Slack phase diagram (in Figure 15) is much nearer correct than Doloff's, which shows even higher solubility than ours.

Table 1 lists the data on 16 experiments in which we used pure silicon as the crucible charge. The first 11 experiments provided the solubility data. More important, however, they showed that even at 2850° C the silicon carbide yield is primarily beta silicon carbide and that the beta-to-alpha transformation is suppressed at 20 atm of argon.

Runs 79 and 80 showed that the beta form is not due to the short time-temperature cycles used in the solubility runs. Runs 79 and 98B showed that the beta-to-alpha transformation is suppressed by 20-atm pressure, the pressure that was used to prevent dissociation of SiC by evaporation of silicon. This finding is important because it indicates that the transformation is associated with physical or chemical defects in the alpha structure and that it can be prevented by practical procedures such as the application of small excess pressures of inert gas.

Table 1. Summary of 16 Experiments With Pure Silicon Charge in Crucible

Exper. No.	Temp. (°C)	Time (min)	Gas	Press. (atm)	Beta only	Beta and trace α	Alpha and trace β	Alpha only	Solubility (atomic %)
72	2850	2	argon	20		x			30
94A	2850	5	argon	20		x			31
74D	2650	10	argon	20		x			22
85A	2650	5	argon	20		x			19
94B	2650	5	argon	20	x				18
74C	2450	10	argon	20		x			10
85B	2450	5	argon	20		x			-
94C	2450	5	argon	20	x				15
74B	2250	10	argon	20		x			3.2
85C	2250	5	argon	20		x			2.2
94D	2250	5	argon	20	x				-
79	2450	60	helium	20		x			
80	2450	60	argon	20	x				
98B	2450	3	helium	2			x		
97	2250	3	argon	20	x				
98A	2250	3	argon	2		x			

Addiamano and Staikoff (1965) have succeeded in suppressing beta-to-alpha conversion up to 2500° C by using 1 atm of nitrogen, or 1 atm of argon that includes at least 400 Torr of nitrogen. They found that the beta-to-alpha transformation was irreversible in the nitrogen atmospheres used (apparently to about 1 atm).

Slack and Scace (1965) have reported the formation of beta silicon carbide from 6H alpha material in 30 atm of nitrogen at a temperature of about 2450° C. They implied that this was a vapor phase conversion.

Quite recently, Kieffer Gugel, Ettmayer, and Schmidt (1966) have reported that silicon carbide can be converted from beta to alpha reversibly and repeatedly. They converted alpha to beta by treatment for 30 min at 2500° C in nitrogen at from 2 to 30 atm pressure and then converted the beta to alpha by reheating the material for 30 min at 2500° C in 1 atm of argon. Since these experiments were carried out with polycrystalline microgranular silicon carbide, it is not clear whether or not the direct and/or reverse transformation was via the vapor state.

In a series of experiments with argon, helium, and nitrogen atmospheres in the temperature range from 2200°to 2800° C, we used the same type of crucible as in our solubility experiments. As the charge, however, instead of pure silicon we used beta silicon carbide. This material was made by the silicon melt process at about 1550° C under conditions of extreme purity in the furnace already described. It showed no trace of alpha inclusions on x-ray measurements, and only very small traces of impurities by mass spectrographic and neutron activation analyses. Seventeen of these experiments are summarized in Table 2.

Runs 95 and 96 show unambiguously that it was the excess pressure rather than the excess silicon that suppressed the beta-alpha transformation. In runs 78, 87, 88, 89, and 91, helium was used as the inert gas. Helium acts essentially like argon, but it seems that there may be more impurities in the helium, leading to more alpha inclusions in the SiC. A comparison of the effect of nitrogen with that of helium strongly indicates that nitrogen suppresses the beta-alpha transformation not because of atom size alone but because it enters substitutionally into the carbon sublattice.

Anomalies of nitrogen in SiC have been noted or studied by many people, including Hamilton (1960), Carroll (1960), Kroko (1966), Griffiths (1966), Slack and Scace (1965), Addiamano (1966), Knippenberg (1963, pp 161-274, and Kieffer et al. (1966), but no clear picture has emerged. Many people have speculated that nitrogen, which is a donor impurity, tends to stabilize the beta form; whereas boron, which is an acceptor, tends to stabilize the alpha. Such speculations are intuitively unsatisfying because they do not give a chemical or physical picture of the role of either the nitrogen or the boron. Since the doping level will rarely exceed 10^{19} atoms cm^{-3}, the conduction electron-to-atom ratio seems too small to influence the polytype structure. In Runs 92 and 93, a single alpha crystal was implanted in the granular beta charge. The single crystals were recovered after the run. The granular sand remained beta and the single crystals remained alpha, indicating that the reverse transformation from alpha to beta probably takes place only through the vapor state. In Runs 101, 102, and 103, beta crystals were inserted into beta sand in the crucible. The results indicate that the direct transformation from beta to alpha probably takes place by a diffusion process, as suggested by Knippenberg (1963, pp 161-274), as well as via the vapor state.

7. CONCLUSIONS

The clearer picture emerging from these experiments is roughly illustrated in Figure 23. A relatively small excess pressure favors the growth of beta silicon carbide al all temperatures. This indicates that the alpha material is a defect structure. We speculate the the alpha is deficient in the carbon sublattice. It may be that at temperatures somewhat below those at which silicon evaporates, nitrogen

Table 2. Summary of 17 Experiments With Beta Silicon Carbide Charge in Crucible

Exper. No.	Temp. (^oC)	Time (min)	Gas	Press. (atm)	Beta only	Beta and trace α	Alpha and trace β	Alpha only	Notes
71	2850	60	argon	20		decom-posed			
75	2550	120	argon	20		x			
96	2550	30	argon	20		x			
95	2550	30	argon	2			x		
76	2450	60	argon	20		x			
77	2250	60	argon	20		x			
87	2550	30	helium	20		xx			(1)
88	2550	30	helium	2			x		
89	2350	30	helium	20		xx			
91	2350	30	helium	2			x		
78	2250	60	helium	20		xx			
92	2550	30	nitrogen	20	x, sand			x, SC	(2)
93	2550	30	nitrogen	2	x, sand			x, SC	(2)
100	2550	30	nitrogen	20				x	(3)
101	2400	30	nitrogen	20	x				(4)
103	2400	30	nitrogen	20	x, sand x, SC				(4)
102	2400	30	argon	2			x		(4)

Notes: (1) xx = beta with larger trace α
(2) alpha single crystal (SC) inserted into beta sand
(3) alpha single crystal (SC) inserted into alpha grains
(4) beta single crystal (SC) inserted into beta sand

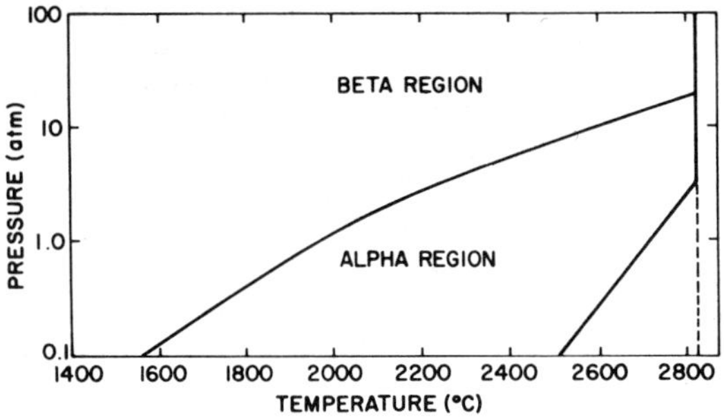

Fig. 23. Estimated temperature–pressure ranges for formation of alpha and beta silicon carbide.

outdiffuses from the carbon sublattice and leaves excess vacancies in it. Then boron, which normally substitutes for silicon in cubic silicon carbide (Gorin and Pletyushkin, 1964), may migrate to the carbon sublattice (Woodbury and Ludwig, 1961), introducing strain and precipitation of boron, which acts as the nucleus for the generation of screw dislocations, which in turn determine the polytype. Impurities, defects, temperatures, and ambient gas would in such a picture all influence toward polytypism.

We are quite confident that beta silicon carbide is the more perfect form and that all the alpha modifications are associated with chemical and physical defects. Nitrogen and boron seem to be the principal impurities whose presence or absence not only enhances or inhibits the transformation from beta to alpha silicon carbide, but determines the particular polytype. Our investigations to clarify some of these points are continuing.

Acknowledgments

We wish to acknowledge the contributions of many members of the Solid State Sciences Laboratory— in particular, the chemical, mass spectrographic, and neutron activation analyses performed by the Properties and Analysis Branch, and the x-ray measurements by G. Sheehan and J. Bruce.

We are indebted to C. O. Dugger for suggesting and preparing the molybdenum carbide samples used in the furnace calibration.

References

Addamiano, A. (1966) Preparation and photoluminescence of silicon carbide phosphors doped with group III-a elements and/or nitrogen. J. Electrochem. Soc. 113(No. 2):134-136, Feb. 1966.

Addamiano, A., and Staikoff, L. (1965) Stabilization of cubic silicon carbide, J. Phys. Chem. Solids 26:669-672.

Carroll, P. (1960) Resistivity of granular silicon carbide—impurity and crystal structure dependence, in O'Connor, J. R., and Smiltens, J., Eds., Silicon Carbide—a High Temperature Semiconductor (Proc. Conf. on Silicon Carbide, sponsored by the U. S. Air Force Cambridge Research Center), Pergamon Press, N. Y. (521 pp), pp. 341-346.

Elliott, R. P. (1966) Constitution of Binary Alloys, First Supplement, McGraw-Hill, N. Y., p. 228.

Gorin, S. N., and Pletyushkin, A. A. (1964) The sector structure of beta-silicon carbide crystals, Soviet Physics—Doklady 9(No. 1):10-13, July 1964.

Griffiths, L. (1966) Defect structure and polytypism in silicon carbide, J. Phys. Chem. Solids 27:257-266.

Hamilton, D. R. (1960) Preparation and properties of pure silicon carbide, in O'Connor, J. R., and Smiltens, J., Eds., Silicon Carbide—a High Temperature Semiconductor (Proc. Conf. on Silicon Carbide, sponsored by the U. S. Air Force Cambridge Research Center), Pergamon Press, N. Y. (521pp), pp. 43-52.

Jagodzinski, H., and Arnold, H. (1960) Anomalous silicon carbide structures, in O'Connor, J. R., and Smiltens, J., Eds., Silicon Carbide—a High Temperature Semiconductor (Proc. Conf. on Silicon Carbide, sponsored by the U. S. Air Force Cambridge Research Center), Pergamon Press, N. Y. (521pp), pp. 136-146.

Jennings, V. J., Sommer, A., and Chang, H. C. (1966) The epitaxial growth of silicon carbide, J. Electrochem. Soc. 113(No. 7):728.

Kieffer, Von R., Gugel, E., Ettmayer, P., and Schmidt, A. (1966) A contribution to the phase stability of silicon carbide (in German), Ber. Deut. Keram. Ges. 43(H10):621-623.

Knippenberg, W. R. (1963) Growth phenomena in silicon carbide, Phillips Research Reports 18(No. 3):161-274, 242, 270, June 1963.

Kroko, L. J., and Milnes, A. G. (1966) Diffusion of nitrogen into silicon carbide doped with aluminum, Solid State Electronics 9:1125-1134.

Merz, K. M. (1960) Crystal, whisker, and microcrystalline forms of silicon carbide, in O'Connor, J. R., and Smiltens, J., Eds., Silicon Carbide—a High Temperature Semicondoctor (Proc. Conf. on Silicon Carbide, sponsored by the U. S. Air Force Cambridge Research Center), Pergamon Press, N. Y. (521pp), pp. 73-83.

Merz, K. M., and Adamsky, R. F. (1959) Synthesis and crystallography of the wurtzite form of silicon carbide, Z. Krist. 111:350-361.

Nelson, W. E., Halden, F. A., and Rosengreen, A. (1966) Growth and properties of beta SiC single crystals, J. Appl. Phys. 37(No. 1):333-336; (1967) The Growth and Characterization of Beta Silicon Carbon Single Crystals, Final report under Contract No. AF19(628)-4190, Stanford Research Institute, Feb 1967.

Neuberger, M. (1965) Silicon Carbide Data Sheets (An annotated bibliography dated June 1965), report under Contract No. AF33(615)-2460 with Air Force Materiel Laboratory, Electronic Properties Information Center, Hughes Aircraft Company, Culver City, Calif.

O'Connor, J. J. (1963) Silicon carbide, Chap. 6 in Gilman, J. J., Ed., The Art and Science of Growing Crystals, Wiley, N. Y., pp. 93-118.

O'Connor, J. R., and Smiltens, J., Eds. (1960) Silicon Carbide—a High Temperature Semiconductor (Proc. Conf. on Silicon Carbide, sponsored by the U. S. Air Force Cambridge Research Center), Pergamon Press, N. Y. (521pp).

Parthé, E. (1964) Crystal Chemistry of Tetrahedral Structures, Gordon and Breach Science Publishers, N. Y., Appendix A, pp. 107-117.

Ryan, C. E., Berman, I., Marshall, R. C., Considine, D. P., and Hawley, J. J. (1966) Some Factors Affecting the Growth of Beta Silicon Carbide, AFCRL-66-641, September 1966.

Scace, R. I., and Slack, G. A. (1959), Solubility of carbon in silicon and germanium, J. Chem. Phys. 30 (No. 6): 1551-1555, June 1959; (1960), The Si-C and Ge-C phase diagrams, in O'Connor, JR., and Smiltens, J., Eds., Silicon Carbide—a High Temperature Semiconductor (Proc. Conf. on Silicon Carbide, sponsored by the U.S. Air Force Cambridge Research Center), Pergamon Press, N. Y. (521pp), pp. 24-30.

Slack, G. A., and Scace, R. I. (1965), Nitrogen incorporation in silicon carbide, J. Chem. Phys. 42: 805-806.

Verma, A. J., and Krishna, P. (1966) Polymorphism and Polytypism in Crystals, Wiley, N. Y. (341pp); Chap. 4: Polytypism and the description of polytypic structures, pp. 61-91 (detailed review of the geometry of polytypes); Chap. 8: Theories of polytypism, pp. 233-294.

Woodbury, H. H., and Ludwig, G. W. (1961) Electron spin resonance in SiC, Phys. Rev. 124 (No. 4): 1083-1089, Nov. 15, 1961.

PHASE RELATIONSHIPS AND DEFECT STRUCTURE IN DIMOLYBDENUM CARBIDE

F. W. Vahldiek and S. A. Mersol
Air Force Materials Laboratory
Wright-Patterson Air Force Base, Ohio

SUMMARY

Phase relationships and stoichiometry variations relative to hexagonal Mo_2C were determined and correlated with the dislocation etch pits and spiral growth epitaxy found on the Mo_2C single crystals studied, using a combination of chemical, x-ray, emission spectrographic, electron diffraction, extraction, electron transmission, and electron microprobe analyses. A veining substructure, superposed by a discontinuous lamellar slip substructure, was observed on all planes of the Mo_2C crystals and determined to be off-stoichiometric $Mo_2C_{0.97}$. The veining is postulated to be due to solidification and precipitation during the crystal growth process. A substructure transition from veins to polygon walls to bands, associated with the formation of γ'-MoC and Mo, with a concomitant increase in dislocation density from 5×10^6 to 1×10^8 dislocations/cm^2, was observed after annealing in vacuum at 1800°, 2000°, and 2150°C, respectively. The γ'-MoC formed was found to correspond to the formula $MoC_{0.81}$. After annealing at 2200°C or above, band formation only occurred, with the bands being primarily Mo. The phase relationships found were confirmed by microhardness measurements. It is postulated that the vein platelets on as-grown crystals are oriented with respect to the hexagonal Mo_2C matrix according to the relationship

$$\{2\bar{1}\bar{1}0\}_M \parallel \{2\bar{1}\bar{1}0\}_P \quad \text{and} \quad (0001)_M \parallel (0001)_P$$

The formation of growth spirals is proposed to be directly related to the reaction $Mo_2C \rightarrow \gamma'\text{-}MoC \rightarrow Mo$. The individual step heights of the growth spirals measured from 1500 to 3000 Å.

From the slip and twin traces produced by room-temperature microhardness indentations on the (0001), $\{2\bar{1}\bar{1}0\}$, and $\{10\bar{1}2\}$ planes of the Mo_2C single crystals studied, it was determined that Mo_2C belongs to the primary $(0001)\langle 2\bar{1}\bar{1}0\rangle$ and the secondary $\{10\bar{1}0\}\langle 2\bar{1}\bar{1}0\rangle$ slip system, and to the $\{10\bar{1}2\}[0001]$ twinning system, respectively. Anisotropy in electrical resistivity and elastic modulus was found and correlated with microhardness anisotropy and Mo_2C crystal structure. An average Debye Θ equal to 892°K was found for Mo_2C, using the mean sound velocities. A covalent and a metallic bonding mechanism is postulated for the Mo-C system.

INTRODUCTION

Veining substructure, similar to the one found on as-grown Mo_2C single crystals studied in the present work, was first reported in the literature on alpha iron[1,2]. It was later found on soft steel crystals[3], rock salt[4], and on such metals as cadmium, zinc, magnesium, aluminum, and nickel[5], as well as on alloys such as in over-aged copper-1.05%-beryllium alloy[6]. Veining was previously observed in the microstructure of single-crystal dimolybdenum carbide[7]. The veining substructure has been interpreted in the literature to be associated with polygonization and consequently with subgrain boundary formation[8,9]. Several authors[10,8,11] have published extensive review articles on the polygonization process. The present study also indicates a polygonization mechanism operative in Mo_2C after annealing, while the original veining substructure is due to a solidification and precipitation process.

Dislocations observed on as-grown crystals are commonly revealed by the etch pit technique, whereby small pits are developed at the emergence of dislocations. The general conclusion is that most of these dislocations did not originate during solidification, but were produced after solidification by thermal stresses that were sufficiently large to cause plastic flow[12]. Dislocation densities ranging from 10^5 to 10^8 dislocations/cm^2 have been reported for Verneuil-grown crystals[13]. Dislocations often mark impurity precipitate sites, where the "impurity" can also be a stoichiometric excess or deficiency of one of the constituents of the material studied[14]. Subsequent annealing of as-grown crystals generally changes the dislocation density and configuration, by means of polygonization and dispersion processes[15,16], with the concomitant formation of Cottrell atmospheres[17]. Growth spirals and spiral etch pits have by now been observed on a number of crystals, and several excellent treatises of them exist in the literature[18,19,20]. Excellent spiral etch pits have previously been reported by the authors for WSi_2 single crystals[21].

A considerable amount of work has been done in the past on the crystal structure[22,23,24] of polycrystalline Mo_2C. These workers found a hexagonal L'3 type of structure for Mo_2C. Butorina and Pinsker[25] found a hexagonal C6 type of structure similar to CdI_2, except the metal and the non-metal had been interchanged. Recently, Parthé and Sadagopan[26] reported an orthorhombic structure based on neutron diffraction studies of polycrystalline material. More recent work by Adelson and Austin[27] essentially confirmed Parthé's and Sadagopan's data. Single-crystal x-ray studies by Buerger precession and cylindrical-rotation techniques[28] were used in the present study and compared with dislocation etch pits and polygon walls relative to their plane orientation.

Room-temperature microhardness indentations are useful for studying microhardness anisotropy, slip, and twinning in brittle materials. Slip has previously been produced in this manner in magnesia[29], alumina[30], titanium carbide[31,32], titania[33], and tungsten disilicide[34] single crystals. Recently, the authors[35] reported slip of the $\{10\bar{1}0\}\langle 11\bar{2}0\rangle$ type produced by high pressure and room-temperature microhardness indentations on hexagonal TiB_2 ($c/a = 1.066$) single and polycrystals. This slip system was also reported by French and Thomas[36], Takahashi and Freise[37], and L. Pons[38] for hexagonal

WC (c/a = 0.976) crystals. These results suggest that prismatic rather than basal slip is favored in hexagonal non-metallic materials having a c/a ratio considerably less than the ideal (1.633). Recent data[39] showed hexagonal Mo_2C to have a c/a = 1.570, which is practically the same as that of the Be metal (c/a = 1.57), which slips primarily on the (0001) plane[40], but also slips on the $\{10\bar{1}0\}$ planes[41]. Zirconium[42] and titanium[43], both with c/a = 1.59, deform mainly by slip on non-basal planes which contain a close-packed direction. This is due to the fact that for these two metals the initial resolved shear stress for slip on the $\{10\bar{1}0\}$ planes is lower than that on the (0001) plane. The prominence of basal rather than prismatic slip in metals of high c/a ratios is shown by cadmium (c/a = 1.89), zinc (c/a = 1.86), and magnesium (c/a = 1.62) which deform mainly by basal slip. However, in case of the latter, by stressing magnesium crystals in tension or compression parallel to the basal plane, slip on $\{10\bar{1}0\}$ planes can also be produced[44].

EXPERIMENTAL

The Mo_2C single crystals investigated were prepared by the Linde Company[45] by a Verneuil-type process using an electric arc. The largest specimens grown were boules 7 mm in diameter by 40 mm in length. Some boules were bi- or tri-crystals, as seen by microstructural, and sometimes even visual, examination. All boules had some incipient microcracks. Some of the boules broke off fairly easily. When they broke off, they tended to do so diagonally, at an angle (somewhere between 10 and 60°, but usually 15°) to the long axis of the boule. The cleavage planes produced were rough and jagged compared to those produced on TiB_2 single crystals[46] which were quite smooth. In contrast to Mo_2C boules, TiB_2 boules cleaved longitudinally, and to a lesser degree, cross-sectionally.

A compositional analysis of Mo_2C boules resulted in a Mo + C content of 99.8%. The crystals were found to have a Mo content of 94.08 w/o and a C content of 5.73 w/o (stoichiometric values are: 94.11 w/o Mo and 5.89 w/o C). This means that the overall specimens were

C-poor, with the formula $Mo_2C_{0.97}$, and therefore according to Rudy et al[47] appear to be α-Mo_2C. Analysis by chemical and emission spectrographic methods showed the crystals to have the following major impurities: 100 ppm each Na, Zr, and Ca; 85 ppm O; 55 ppm Fe; and 10 ppm each Cr and Ta.

Electron microprobe traverses on as-grown crystals showed them to be homogeneous Mo_2C with no additional composition detectable. A Phillips-AMR electron microanalyzer was used for this work. Single crystals were ground and analyzed by powder x-ray diffraction technique using CuK_α radiation. These data agree with the standard x-ray powder patterns found in the literature[23]. Buerger precession and cylindrical-rotation patterns were taken on cleaved sections of the boules and found to be hexagonal Mo_2C with $a_o = 3.0233$ Å, $c_o = 4.7344$ Å and $c/a = 1.5660$. These studies were accomplished with a Siemens x-ray unit equipped with standard x-ray diffraction, precession, and Unicam cameras. Fries and Kempter[48] reported the following values for polycrystalline Mo_2C: $a_o = 3.00292$ Å, $c_o = 4.72895$ Å, and $c/a = 1.575$. More recent x-ray diffraction crystal structure analysis data by Fenter et al[39] on a number of single crystal fragments cleaved from the Mo_2C single crystal boules used in this work resulted in the average values for lattice constants,as follows: $a = 3.010 \pm 0.004$ Å, $c = 4.727 \pm 0.006$ Å, and $c/a = 1.570$.

Average specimen density was determined to be 9.04 gm/cm^3 compared to 9.18 gm/cm^3 obtained by Westgren and Phragmen[22] for polycrystalline Mo_2C. Taking the latter value as theoretical density for Mo_2C, the density of Mo_2C single crystals studied was then 98.47% of theoretical.

Specimens mounted in Lucite were polished on a vibratory polisher consecutively using diamond paste grades ranging from 9-1 μ as well as Linde A powder for times ranging from 15 minutes to 48 hours to insure highly polished and scratch-free surfaces. The detailed polishing procedure has been reported previously[46, 49]. Chemical etching with dilute nitric acid for 1-15 minutes produced the best substructure and etch pits. For other chemical etchants see Table I.

TABLE I

CHEMICAL ETCHANTS FOR Mo_2C

No.	Etchant	Remarks
1*	1 part** 70% HNO_3 1 part Dist. H_2O	General purpose. Fast Acting. Average time: 10-30 sec at R. T. Corrosive
2	1 part 70% HNO_3 3 parts Dist. H_2O	General purpose. Slower Acting. Produces fine Detail. Average time: 5 min. at R. T. Non-corrosive
3	10 gm $K_3Fe(CN)_6$ 10 gm NaOH 100 cc Dist. H_2O	Murakami's Reagent. Slow acting. Average time: 3-7 min. at R. T. depending on surface etched. Slightly corrosive.
4	1 part Conc. HF 1 part Conc. HNO_3 3 parts Lactic Acid	General purpose. Very fast acting. Average time: 1-5 sec. at R. T.
5	4 parts Conc. HF 1 part Conc. HNO_3 3-7 parts Dist. H_2O	General purpose. Fast acting. Average time: 2-10 sec. at R. T.

* Previously reported by Kiffer[45]

** By volume

Dilute nitric acid was also used for thin section polishing. For the transmission work cleaved sections were polished with a mixture of dilute nitric and sulfuric acids.

Carbon replication, extraction, and thin film techniques were used to prepare electron micrographs. For the extraction of veins and bands, specimens were deeply etched with dilute nitric acid to bring out the substructure in relief prior to the stripping procedure. A JEM-6A electron microscope, a Unitron metallograph BU-11, and a Zeiss Ultraphot II metallograph were used for the microscopy studies. Polished and/or etched samples were annealed in a Ta resistance furnace at 1600°, 1800°, 2000°, 2150°, and 2200°C for 2-72 hrs. using Mo_2C setter material, inside a Ta specimen holder, in a vacuum of 1-3 x 10^{-5} mm Hg. The Mo_2C setter material was the same powder from which the Mo_2C single crystals had been made. The same areas of samples were etched before and after annealing, for comparison purposes. X-ray, emission spectrographic, electron microprobe, and electron diffraction analyses were made after each annealing run and the results compared. Electron microprobe traverses were also done across the dislocation etch pits, polygon walls, veining and band-type substructure after annealing up to 2200°C. Temperatures were measured with a calibrated W, W-26% Re thermocouple.

For hardness measurements, a Tukon Microhardness Tester Type FB with Knoop and Vickers indenters was used. Measurements were taken at loads ranging from 25 to 1000 grams; however, the 100-g load was chosen as the standard load. All measurements were taken at room temperature. Indentations of cracking classes 1 and 2 only were considered for hardness determinations[50, 51].

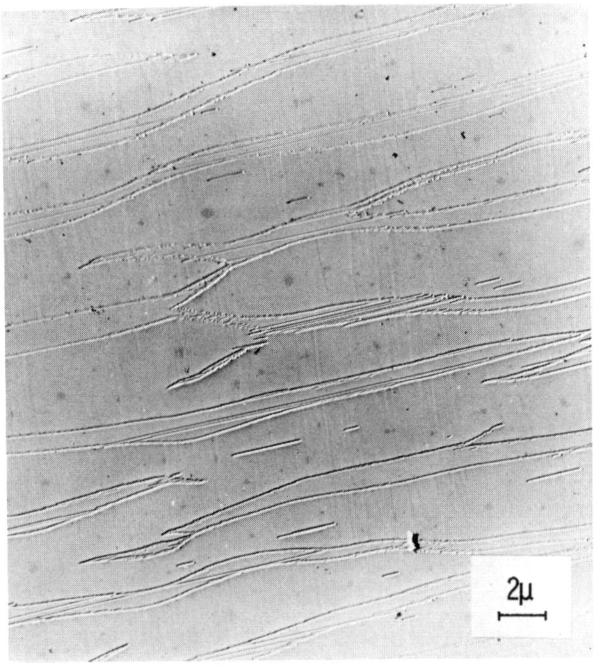

Fig. 1. Veining on the (0$\bar{1}$12) plane of unannealed Mo_2C single crystals, showing slip lines ⊥ to vein direction. Several microcracks are also seen.

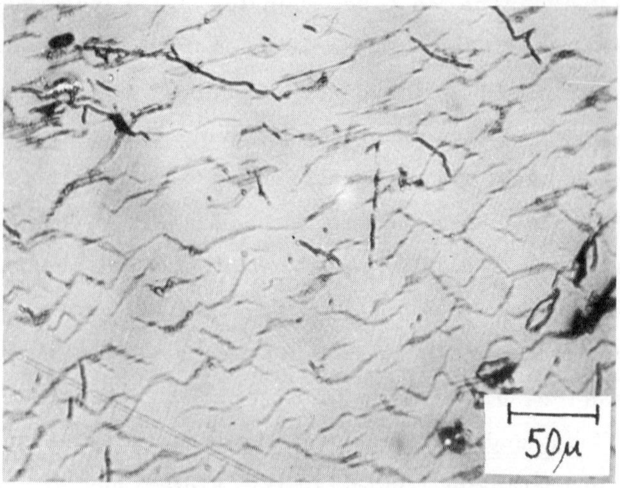

Fig. 2. Veining on the (0001) plane of unannealed Mo_2C single crystals.

Fig. 3. The honeycomb substructure within the matrix on the (0001) plane of unannealed Mo_2C single crystals.

(There are six cracking classes, ranging from "class 1" for a perfect indentation with no cracking to "class 6" where the cracking is severe and the outline of the indentation is virtually destroyed by the cracking. The Knoop numbers based on indentations of classes 1 and 2 are not affected by cracking class. Most of the indentations taken on Mo_2C at 100-g loads were of cracking classes 1 and 2.) Electrical resistivity measurements at room temperature were taken with a Rubicon bridge, using gold contacts. A Sperry ultrasonic attenuation comparator at a frequency of 5 megacycles was used to measure some of the elastic modulus values. The mean sound velocity method[52] was used to calculate the Debye temperatures.

Substructure and Phase Relationships

Optical and electron microscopy revealed that all surfaces of as-grown Mo_2C single crystals studied showed a characteristic veining substructure upon etching, a detailed discussion of which has been published previously[7, 53]. The $\{2\bar{1}\bar{1}0\}$ and similar planes (see Fig. 1) were found to have the straightest veins and the highest density of veins per given length and/or per given area. It was found that the veining density decreases and the individual vein width increases as one proceeds from the $\{2\bar{1}\bar{1}0\}$ planes to the (0001) plane[7]. The veining on the (0001) planes assumes the configuration of malformed hexagons (Fig. 2), with a honeycomb substructure being seen within the matrix in-between the veins, as shown by Fig. 3. The veining substructure on the $(10\bar{1}2)$ plane near the center of the boule is shown in Fig. 4. Going clear to the edges of the boule, a gradual broadening of the veins was observed. This broadening is associated with a slight loss of C near the outer edges of the boules, as determined by electron microprobe analyses.

Electron microscopy showed that each vein consists of two boundaries, and that each boundary is composed of mutually parallel "lines" or platelets oriented parallel to the veining direction. Within the matrix between any two given veins there are imbedded discontinuous platelets, which are similar to the above-mentioned "lines"

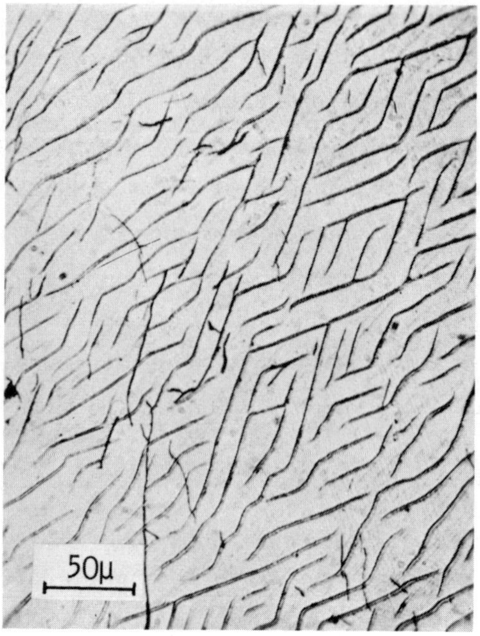

Fig. 4. Veining on the (10$\bar{1}$2) plane of unannealed Mo_2C single crystals, showing discontinuous slip lines.

within the vein boundaries themselves and are amazingly comparable to the well-known Widmanstätten-type precipitate, such as was previously observed by the authors[49, 54] on TiB_2 single crystals. These microstructural features are amply illustrated by Figs. 5 and 6, with the latter also showing a tilt boundary. Occasionally, a transition from the Widmanstätten platelets to the veining substructure was observed. This means that under certain crystal growth conditions the Widmanstätten substructure is produced, and under other conditions, the veining substructure is produced. The interesting connection between the Widmanstätten precipitate and the veining substructure well warrants further study.

By a comparison of the cylindrical-rotation patterns and the corresponding veining substructure on the same samples, it was determined that the resultant direction of the veins on the (2$\bar{1}\bar{1}$0) plane is parallel to the primary growth or c-axis (i. e. the [0001] direction). The veins on the (0001) plane were found to be oriented

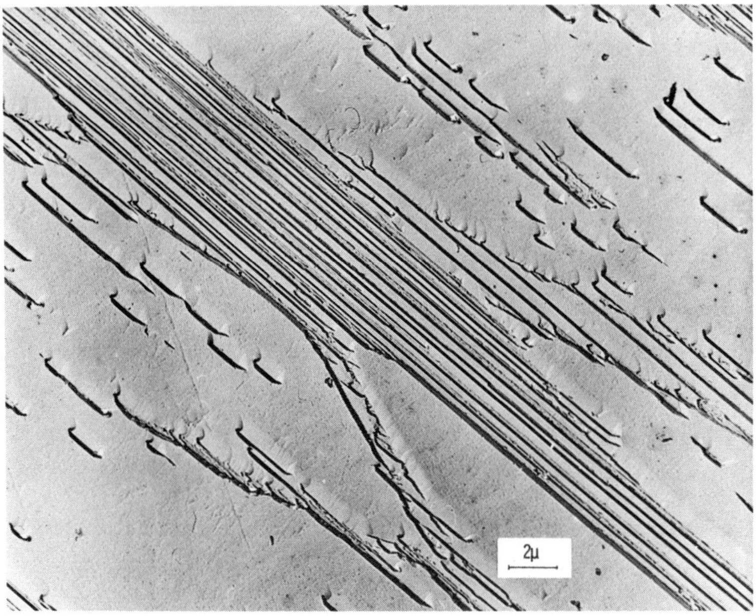

Fig. 5. Veining substructure and Widmanstätten-type precipitate platelets on $(2\bar{1}\bar{1}0)$ plane of an unannealed Mo_2C single crystal.

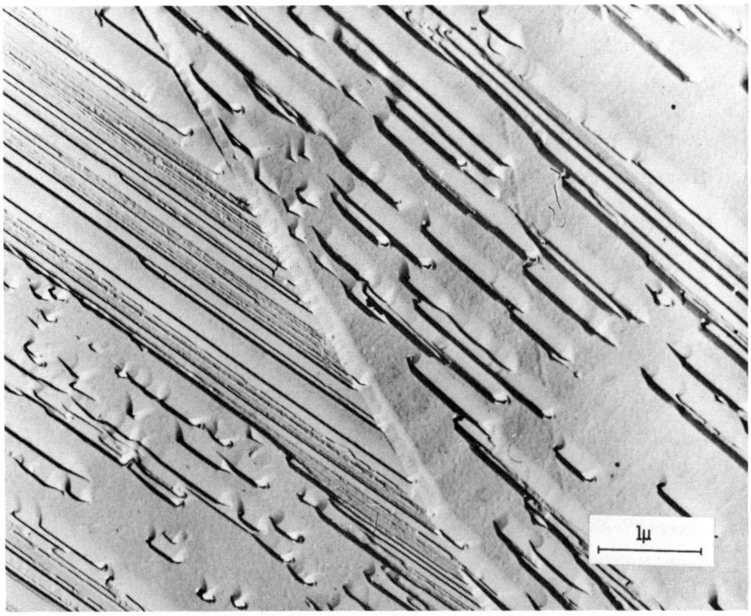

Fig. 6. Veining and tilt subboundary separating two near $(2\bar{1}\bar{1}0)$ planes of an unannealed Mo_2C single crystal.

in the three directions corresponding to the three a-axes of the Mo_2C crystal lattice, indicating secondary crystal growth in the $\langle 2\bar{1}\bar{1}0 \rangle$ directions.

A close examination of the substructure of Mo_2C single crystals revealed the presence of discontinuous slip lines on all planes of as-grown crystals. Slip lines were never observed within individual veins on any crystals studied. Where veins are oriented primarily in one direction, slip lines are at 90° to the vein direction (see Fig. 1). Where veins go in two different directions, the slip lines are perpendicular to the resultant of the two vein directional component vectors (see Fig. 4). These slip lines are an indication of a layer or lamellae substructure in Mo_2C single crystals superposed on the veining substructure. Electron transmission of the as-grown crystals on the $(2\bar{1}\bar{1}0)$ plane showed the veining substructure in fine detail, as seen in Fig. 7.

From all the available data it was determined that the veins are more or less discontinuous boundaries of subgrains or crystallites, which take the form of irregular truncated double cones or hexahedral dipyramids, elongated parallel to the [0001] direction. A cut through the center sections of these "subgrains" and parallel to the (0001) plane results in the appearance of the above mentioned more or less malformed hexagons. Recently, Hopkins and Kraft[55] found a "fault network" substructure in $CuAl_2$-Al eutectic alloy which is analogous to the veining substructure found in Mo_2C crystals. Occasional incipient cracks and microcracks found in Mo_2C single crystals (see Figs. 1 and 4) are thought to be due to rapid cooling (see also Rudy et al[47], p. 55).

Upon annealing at temperatures of 1600°, 1800°, and 2000°C, the number and width of the veins decreased somewhat and the distance between two individual veins increased with increasing temperature, with a concomitant straightening of the veins. This indicates that during annealing runs at increasing temperatures, increasing polygonization of the veining substructure takes place, which

Fig. 7. Electron transmission micrograph of $(2\bar{1}\bar{1}0)$ plane of unannealed Mo_2C single crystal, showing vein boundaries (composed of platelets) in center section of a vein and discontinuous slip lines, ⊥ to the vein platelets.

then is the predominant process at these temperatures. This is in agreement with existent polygonization and subboundary formation theories[8, 9, 11].

The veining substructure on the $(10\bar{1}2)$ plane of a Mo_2C crystal after annealing at 1800°C for 2 hrs. is seen in Fig. 8, where a number of the veins combined during annealing to form a series of discontinuous polygon walls (see center of Fig. 8). These polygon walls are composed of dislocation etch pits, perpendicular and corresponding to each one of which there is a slip line. The spacings between the slip lines are equal and in the order of 0. 2 μ (compare with lamellar spacings on unannealed samples).

A transition of the veining to the band-type substructure (see Fig. 9) with the intermediate polygon walls formation, was observed after annealing at 2000° to 2150°C for 2 to 72 hrs. Long-term annealing resulted in a more widespread and uniform polygonization on the basal

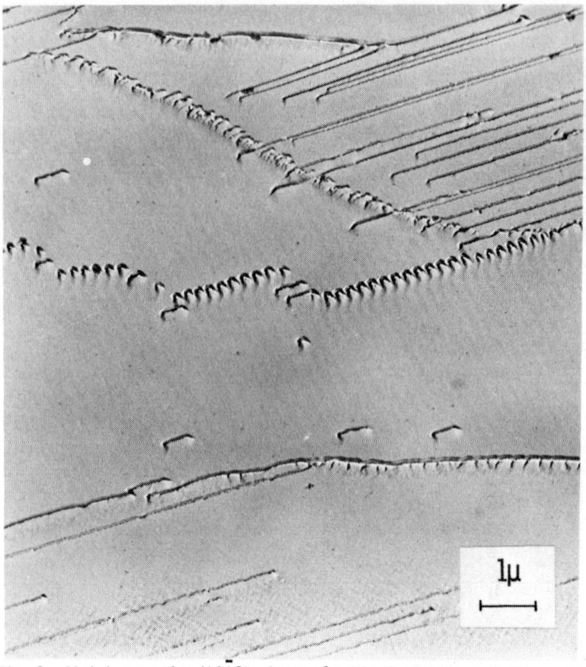

Fig. 8. Veining on the (10$\bar{1}$2) plane of a Mo_2C single crystal annealed at 1800°C.

plane (as seen by Fig. 10) and increasing formation of the bands on the prism planes of annealed crystals, with the bands oriented parallel to the [0001] direction. After annealing at 2200°C for up to 4 hrs, predominantly the band-type substructure was observed, as seen by Fig. 11, with the polygonization process rapidly coming to completion. Similar band-type substructure, with the bands oriented parallel to the growth direction, has recently been found in Zr single crystals with Cd additions grown from the melt [56].

From a combination of chemical, electron microprobe, x-ray, and electron diffraction analyses [7, 39, 53] it was determined that the matrix of the Mo_2C single crystals studied is indeed hexagonal Mo_2C, which, however, is slightly off-stoichiometric, with the approximate formula Mo_2C_{1-x} (with x = 0.01). The veining substructure

Fig. 9. Transition from veining to polygon walls to bands on Mo_2C single crystal, annealed at 2000°C.

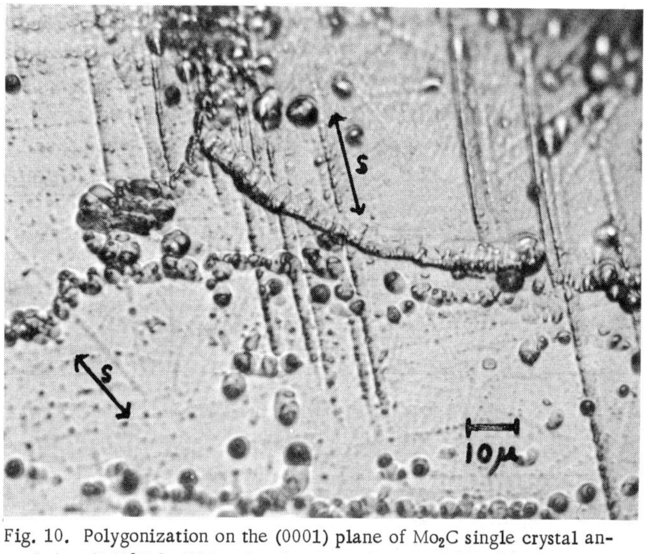

Fig. 10. Polygonization on the (0001) plane of Mo_2C single crystal annealed at 2150°C for 72 h, also showing sub-etch pits produced along slip (s) traces.

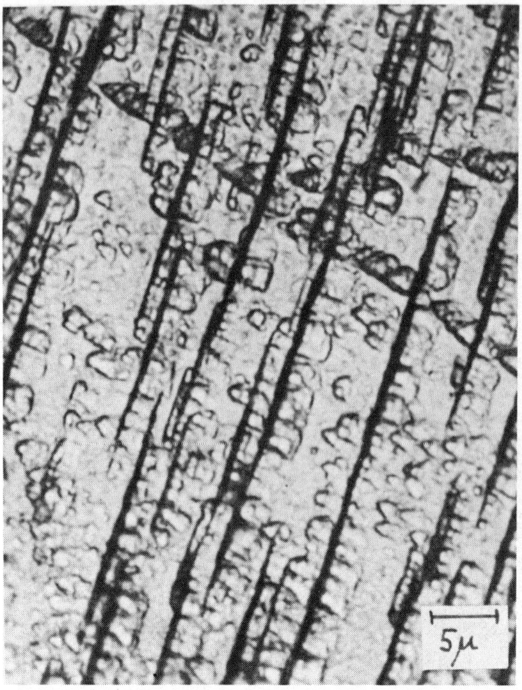

Fig. 11. Pile-up of etch pits along the bands on the $(2\bar{1}\bar{1}0)$ plane of Mo_2C single crystal annealed at 2200°C.

of the as-grown Mo_2C crystals studied was determined to be off-stoichiometric $Mo_2C_{0.97\pm x}$ (where x = 0.01-0.02), using a combination of electron microprobe analyses[7] and electron diffraction of extracted veins[7,57]. Annealing at 1600° to 2000°C resulted in the presence of both Mo_2C and MoC within the veins, as shown by Fig. 12. The accurate composition of the veins is difficult to determine by the electron probe because of the small size of the individual veins. However, a carbon content (see Table II) of approximately 9 w/o was found for the veins after annealing at 2000°C. Both Mo and C concentrations in Table II were obtained by the point count method, which for C determination has a relative error[58] of ~10%. The electron diffraction data showed MoC to be γ'-MoC (as reported by Kuo and Hägg[59]) of the MoC_{1-x} type (see Rudy et al[47]). On the basis of these data, the MoC present within the veins was calculated to correspond to the formula $MoC_{0.81}$.

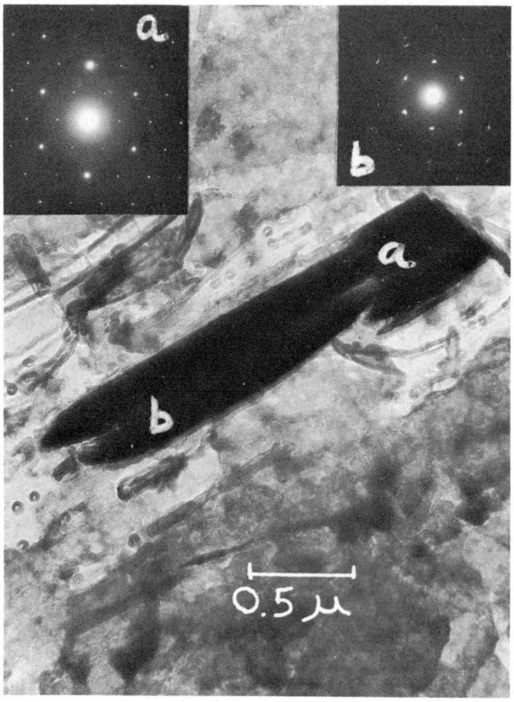

Fig. 12. Two extracted veins of Mo_2C single crystal annealed at 1800°C. Respective electron diffractions are inserted: (a) upper vein (Mo_2C), (b) lower vein (γ'-MoC).

TABLE II

CONCENTRATION DISTRIBUTION (w/o)

Mo_2C Crystal	Mo	C
Matrix	94.1	6.0
Matrix	94.0	5.9
Matrix	93.8	6.0
Matrix	94.2	6.0
Matrix	94.6	6.0
Matrix	94.1	6.0
Vein	90.8	8.9
Vein	90.9	9.0
Vein	91.0	8.9

When corrected for matrix excitation, the carbon concentration of the veins is equal to 11 w/o.

It must be mentioned that occasional Mo within the Mo_2C matrix[7] (in-between the veins) was found after annealing at 2000°C, and that the bands found after annealing at 2200°C consisted of $>$98% Mo present in the form of subgrains, as shown by electron microprobe traverses and electron diffraction of extracted bands[57]. The other $<$2% of the bands was accounted for by γ'-MoC and Ta contamination ($<$1%). It should be mentioned that the matrix of crystals annealed up to 2200°C remained hexagonal Mo_2C, however with the composition $Mo_2C_{0.93}$ after annealing at 2200°C, as shown by a combination of chemical and electron microprobe analyses.

It has been recently reported by Rudy et al[47] that the γ'-MoC phase is probably an oxygen stabilized carbide phase. The formation of this phase in the veins and the bands after annealing from 1600° to above 2200°C can be postulated by a slight oxidation of Mo_2C observed in an increase of oxygen content from 0.05 w/o in unannealed samples to 0.2 w/o in samples annealed at 2200°C in a vacuum of $1\text{-}3 \times 10^{-5}$ mm Hg. A loss of up to 0.4 w/o C was determined by chemical analysis of annealed crystals. By electron diffraction no free Mo was detected in the veins of specimens annealed below 2000°C. However, this does not preclude its presence in very minute quantities at these lower temperatures. The formation of a Mo-Mo_2C eutectic at 2200°C is well-known and has been reconfirmed[47]. The presence of a Mo-γ'-MoC-Mo_2C eutectic at temperatures lower than 2200°C seems reasonable.

The phase relationships in the Mo-C system[60,61,47] and oxidation kinetics of Mo and Mo_2C[62,63] are not completely understood. However, based on the analytical data, the overall reaction can be written as:

$$Mo_2C \xrightarrow{[O_2]} MoC + Mo \qquad [1]$$

The maximum impurity pickup during annealing for 2-24 hrs. at 1600° to 2000°C and for up to 72 hrs. at 2150°C was 100 ppm Ta.

Table III Powder X-Ray Diffraction of Ground Single Crystal Mo_2C

Mo_2C (unannealed)			Mo_2C (annealed at 2000°C)		
d Å	hkl	I/I_o	d Å	hkl	I/I_o
2.601	100	20	2.578	100	25
2.373	002	20	2.347	002	20
2.276	101	100	2.262	101	100
1.753	102	35	1.740	102	30
1.506	110	35	1.495	110	35
1.351	103	40	1.343	103	40
1.305	200	5	1.295	200	5
1.272	112	20	1.262	112	25
1.257	201	10	1.249	201	10
1.188	004	10	1.176	004	10
1.142	202	5	1.135	202	5
1.082	104	5	1.074	104	5
1.010	203	5	1.002	203	5

Chemical analysis showed that during annealing up to 24 hours at 2200°C a total of 0.1 w/o Ta was picked up. Electron microprobe analyses showed that the Ta concentrated primarily within the band substructure after annealing at 2200°C. Results also showed that the Ta enhanced the Mo-Mo_2C formation at 2200°C. Powder x-ray diffraction data before and after annealing at 2000°C are shown in Table III.

Epitaxy of Dislocation Etch Pits

Etched as-grown and annealed Mo_2C single crystals showed hexahedral and spiral etch pits on the (0001) planes (see Fig. 10), triangular etch pits on the $\{10\bar{1}2\}$ planes (see Fig. 13), and hexahedral, rhombohedral, and occasional spiral etch pits on the $\{2\bar{1}\bar{1}0\}$ planes, the latter often shown in the form of a "shingles" substructure as can be seen from Fig. 14. The "shingles" substructure is formed by slip lines, which are normal to the [0001]

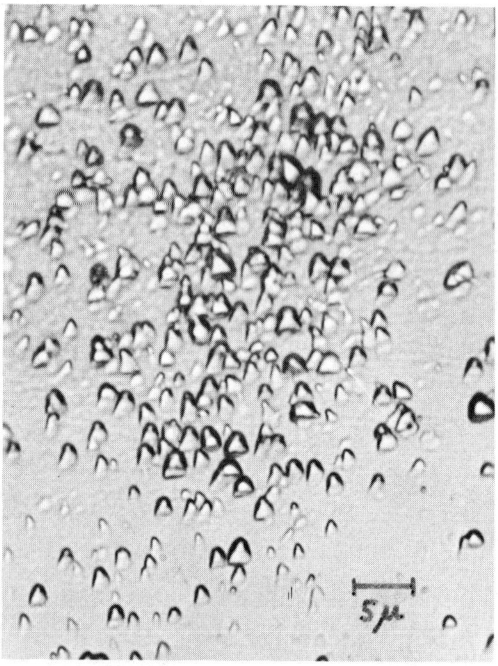

Fig. 13. Forest of triangular etch pits on the $(10\bar{1}2)$ plane of annealed Mo_2C single crystal.

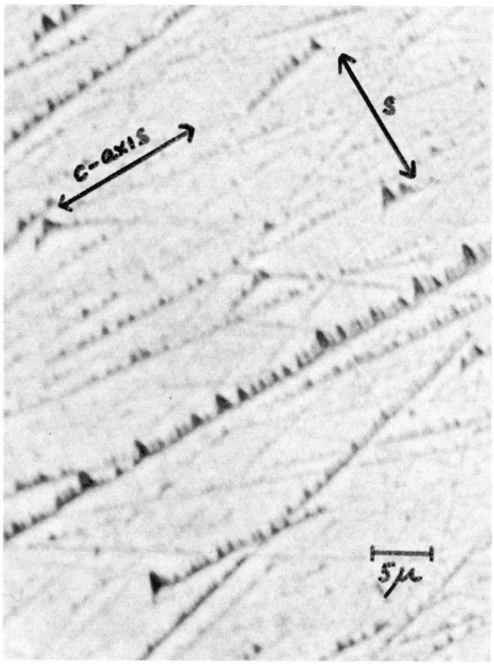

Fig. 14. Shingles substructure on the $(2\bar{1}\bar{1}0)$ plane of annealed Mo_2C single crystal, showing slip (s) and arrays of primarily edge dislocations

direction, and which are intersected by rows of etch pits, determined primarily to be arrays of edge dislocations more or less aligned with the [0001] direction. Since it is known[14] that polygon walls are forming perpendicular to the slip plane, the (0001) plane is then one slip plane for the crystals studied. The Burgers vector for an edge dislocation is defined as being perpendicular to the dislocation line, which in this case is the $[2\bar{1}\bar{1}0]$ direction. The $\langle 2\bar{1}\bar{1}0 \rangle$ directions are then also the slip directions for Mo_2C. It was shown by Amelinckx[16] and Cottrell[11] that after a crystal is sufficiently well annealed, impurities will concentrate along Taylor dislocations, and to a much lesser degree or not at all along pure Burgers dislocations. This seems to be the case here, since most of the Ta impurity was found to be concentrated within or along the bands of crystals annealed at 2200°C and higher, and the bands are oriented parallel to the [0001] direction. It is also apparent that during annealing to 2000°

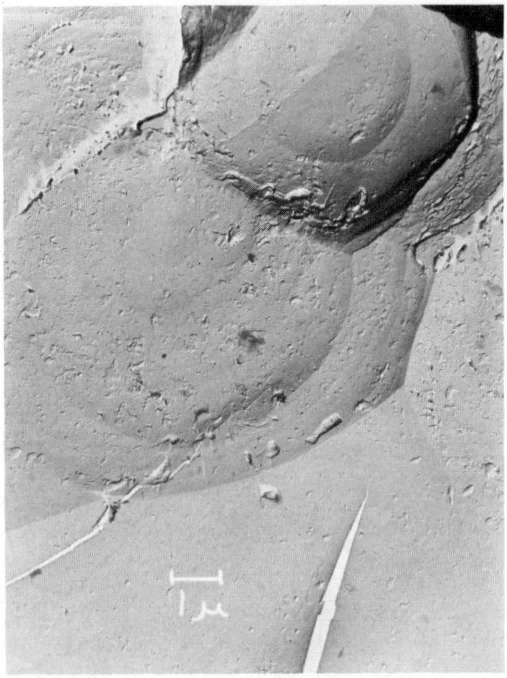

Fig. 15. Interacting growth spirals going from hexagonal into spherical symmetry on (0001) plane of an Mo_2C crystal annealed at 2000°C.

Fig. 16. Multiple elliptical interacting spirals on (0001) plane of Mo_2C crystal annealed at 2000°C.

and 2200°C, the dislocations formed along the [0001] direction provide vacancy sites for the γ'-MoC and Mo metal particles to settle upon[16].

The best growth spirals were observed on the (0001) planes of Mo_2C single crystals, being spherical elliptical, or hexahedral in shape. A transition from the hexagonal symmetry growth spiral into spherical symmetry is shown in Fig. 15. It can be seen from this figure that a subboundary is going right through the spiral etch pit. Well-developed elliptical spirals with many steps, and topped with a ridge of interconnected individual etch pits, is shown in Fig. 16. A similar growth pattern, but with a flattened top and fewer steps is shown in Fig. 17. Interactions of dipole spirals (Fig. 18 and 19) were also observed. Considering that spiral etch pits are sites for screw dislocations, and that in general helical dislocations have their axis parallel to a possible Burgers vector, and that the best developed spirals should be observed on planes perpendicular to a possible

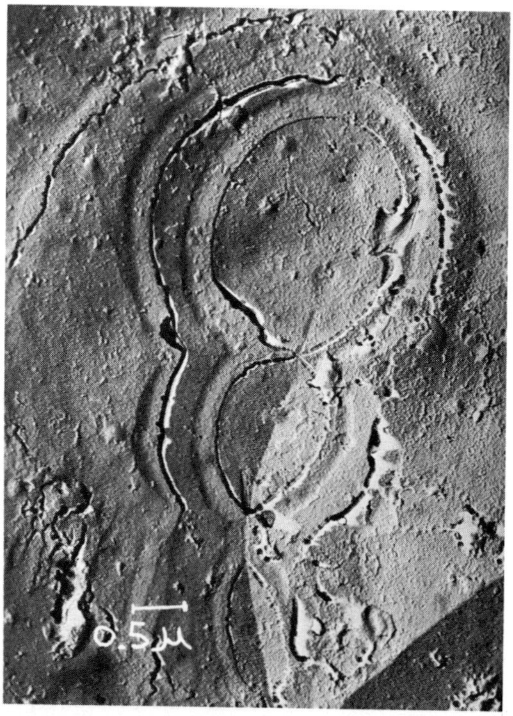

Fig. 17. Flat-bottomed elliptical growth spirals of Mo_2C annealed at 2000°C (0001).

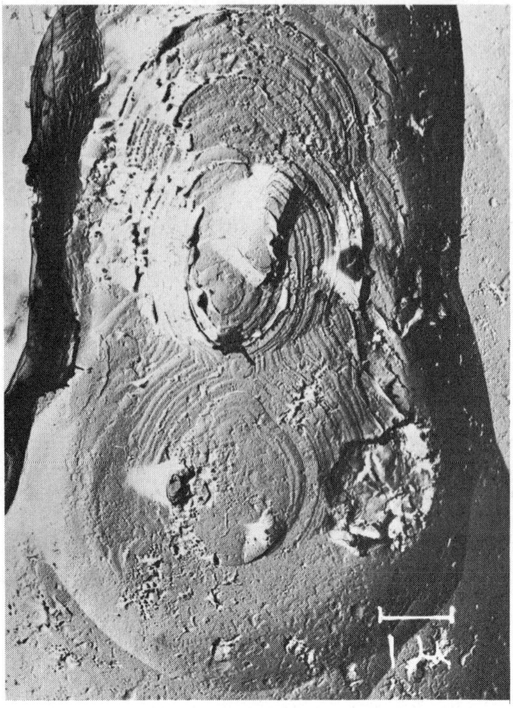

Fig. 18. Dipole growth spirals with well defined steps on the (0001) plane of Mo_2C annealed at 2000°C.

Burgers vector - which in the present case are the (0001) planes - then the Burgers vector for screw dislocations in Mo_2C crystals must be parallel to the [0001] direction. It must be kept in mind, however, that most of the growth spirals found on Mo_2C single crystals studied have both screw and edge components, and that therefore, both Burgers vectors apply. The [0001] direction, as previously pointed out, coincides with the primary growth axis for Mo_2C single crystals which is the growth axis of the helicoidal growth spirals. On the (0001) plane, etch pits and sub-etch pits were found to form polygon walls after annealing, with the angles between them being approximately 60° to 120° during polygonization; in other words, the rows of etch pits on (0001) planes tend to align themselves parallel to the three $\langle 10\bar{1}0 \rangle$, which indicates that the $\{10\bar{1}0\}$ planes are also the slip planes for Mo_2C, since polygon walls are generally perpendicular to a possible slip plane. Occasional spiral etch pits observed on the $\{2\bar{1}\bar{1}0\}$ planes indicate that these crystals can grow also in the $\langle 2\bar{1}\bar{1}0 \rangle$ directions; this, however, is strictly secondary growth. The finding of the various shapes of etch pits is perfectly compatible with Mo_2C hexagonal structure. Figure 10 also shows that dislocations forming polygon walls etch somewhat more easily than dislocations produced by slip (compare the size of etch pits), which is in agreement with similar data previously published by Amelinckx[16].

In the determination of the individual step heights as well as the total height of spiral etch pits (see Figs. 15 to 19), with respect to Mo_2C crystals annealed at 2000°C, a wide range of values was found. The calculations were based on two types of measurements. The first method consisted in trigonometrically measuring the angle of inclination of the overall conically-shaped spiral with respect to the surrounding specimen surface, which in this case, for simplicity reasons was the (0001) plane. From this, the angle and the step height of the total cone can be determined trigonometrically. The

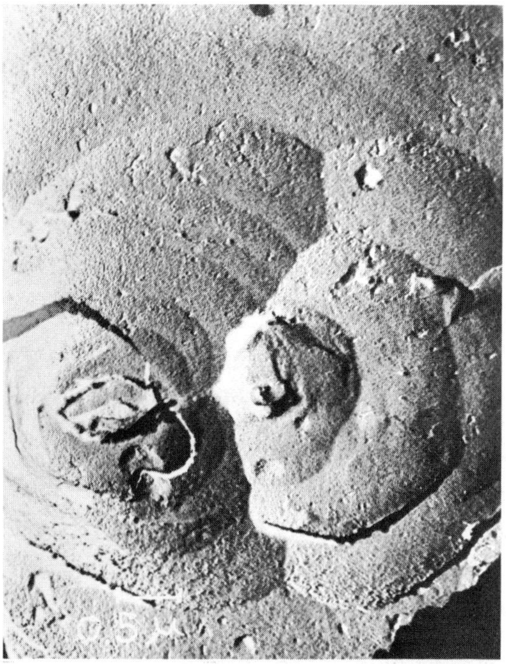

Fig. 19. Interaction of two dipole spirals on the (0001) plane of Mo_2C crystal annealed at 2000°C.

individual step height is then simply the total step height divided by the number of steps measured. This technique was then checked against an ideal situation, namely, assuming a 1:1 ratio in width: height consideration, and in a few cases reasonable agreement between the two techniques was actually found. However, due to the overall complexity of the phase relationships in the annealed Mo_2C crystals, the trigonometric method was preferred and a range of step heights was measured, with the individual steps found to be from 1500 to 3000 Å in height. Horn[65], Freise[66], and Thomas and Roscoe[67] reported spiral step height for natural graphite ranging from 300 to 2000 Å, while Fitzer and Schlesinger[68] reported an average step height of 1000 Å for pyrolytic carbon.

The determination of the spiral height is further complicated by the etching or polishing in the preparation of the specimens, especially in the Mo_2C crystals, because of impurities and large phase changes discussed above, which have slightly different etching characteristics during different etching procedures. It should be mentioned that on the basis of theoretical considerations relative to the helical dislocation growth[19, 20, 69, 70], it is rather difficult to determine from the complex theoretical considerations the actual calculations based on experimental measurements.

During the initial growth of the crystal, nucleation sites for imperfections were introduced in the [0001] direction being the primary growth direction of the crystal, which in turn were brought out at 1800°-2200°C with the formation of γ'-MoC from the initial Mo_2C. Well developed spirals and secondary polygonization reaction then are dependent on the reaction of compositional changes from Mo_2C to γ'-MoC to Mo. The speed and type of formation (from single etch pits to spirals to multiple spirals to polygon walls) of spiral etch pits (growth spirals) is thus directly proportional to the reaction $Mo_2C \rightarrow \gamma'\text{-MoC} \rightarrow Mo$. Fully established spirals were observed at times up to 8 hrs.

Electron microprobe data across dislocation etch pits indicated no change in Mo content compared to the matrix and veining substructure on as-grown crystals. A substantial decrease in Mo content, however, across etch pits and polygon walls of samples annealed at 2000° and 2200°C was noted[57]. This decrease in Mo varied, with a maximum observed across well-developed polygon walls having about 89 w/o Mo, and corresponding to γ'-MoC. Impurities, such as Ta, Zr, and Ca, were not detected (in $>$ 0.1%) scanning across etch pits and polygon walls.

The average dislocation density on as-grown Mo_2C crystals was calculated to be 5×10^6 dislocations/cm^2. Annealing to 2200°C increased the average dislocation density to 1×10^8 dislocations/cm^2. These values are in good agreement with dislocation densities reported for other crystals grown by the Verneuil method[13]. The increase of dislocation etch pits during annealing is postulated to be due to the reaction [1] above, and is directly related to the γ'-MoC and Mo formation, with a concomitant dispersion of impurities such as Ta, Zr, and Ca throughout the Mo_2C matrix.

Microhardness, Slip and Twinning

Knoop microhardness measurements were taken on the matrix of approximately twelve selected polished and/or etched as-grown and annealed Mo_2C single crystals, and the results were compared. Table IV shows the results on polished-only samples at 100-g loads. (In this Table, "Indenter orientation" refers to the orientation of the long diagonal of the Knoop indenter with respect to the given crystal axis.) In addition, Vickers hardness measurements were taken on similar planes on the same samples at 100-g loads for comparison purposes. It was determined that at this load the Vickers hardness values were 5 to 6% higher than the corresponding Knoop hardness values for any given plane. In general, etching reduced the hardness by approximately 100 units in any given orientation. This was due to a small amount of oxidation of the surface associated with the observed formation of amorphous carbon upon etching with dilute nitric acid[7]. Microhardness over broadened veins decreased by some 110 units in any given orientation. This decrease in hardness is associated with a decrease in C content in broadened veins. Similar decrease in

TABLE IV

KNOOP MICROHARDNESS OF Mo_2C SINGLE CRYSTALS
WITH RESPECT TO INDENTER ORIENTATION

(in kg/mm^2 at 100-g loads)

Plane	Indenter Orientation	Unannealed *	Annealed **			
			1800°C, 2 hrs	2000°C, 2 hrs	2150°C, 72 hrs	2200°C, 4 hrs
(0001)	∥ to $\underline{a}_1$ axis	1540±50	1510±50	1360±50	1780±10	1430±50
	At 45 deg to $\underline{a}_1$ axis	1580±30	1570±30	1430±30	1820±10	1470±50
	⊥ to $\underline{a}_1$ axis	1560±40	1540±40	1400±40	1800±10	1450±50
	Avg.	1560±40	1540±40	1430±40	1800±10	1450±50
$(2\bar{1}\bar{1}0)$	∥ to $\underline{c}$ axis	1230±100	1130±100	1060±100	1550±30	1330±100
	At 45 deg to $\underline{c}$ axis	1680±30	1610±30	1570±30	1970±10	1860±100
	⊥ to $\underline{c}$ axis	1560±50	1520±50	1480±50	1850±20	1730±100
	Avg.	1490±60	1420±60	1370±60	1790±20	1640±100
$(10\bar{1}2)$	∥ to $\underline{a}_1$ axis	1630±30	1600±30	1500±30	1920±20	1810±60
	At 45 deg to $\underline{a}_1$ axis	1420±50	1350±50	1260±50	1690±20	1610±60
	⊥ to $\underline{a}_1$ axis	1480±40	1430±40	1320±40	1820±20	1770±60
	Avg.	1510±40	1480±40	1360±40	1810±20	1730±60

*Each value represents 75 measurements

**Each value represents 50 measurements

hardness with a decrease in C content has previously been found on carbides such as TiC[31] and TaC[72]. This decrease in hardness seems to be associated with the decrease of strong Me-C bonds and a decrease in Peierls stress as the C content decreases.

The least hardness anisotropy was observed on the (0001) plane in as-grown and annealed specimens. This plane was also the hardest for as-grown crystals. The hardness decreased with increasing annealing temperatures up to 2000°C for the three planes studied but then increased after annealing at 2200°C. The $(10\bar{1}2)$ plane had the highest hardness after annealing at 2200°C. The highest and lowest microhardness values at any given temperature were exhibited on the $(2\bar{1}\bar{1}0)$ plane with the indenter orientation at 45° to the c-axis, and parallel to the c-axis, respectively. (Note that the direction of the veins and the bands was previously found to be aligned with the c-axis[7]. This hardness increase with annealing temperatures $>$ 2000°C can be explained on the basis of the observed formation of Mo and γ'-MoC with a concomitant dispersion of minute particles of these two phases within the Mo_2C matrix. Also contributing to the increase in hardness after annealing at 2200°C is a significant increase (up to 0.1%) of Ta at these temperatures. The very small amount of carbon loss upon annealing, which by itself would cause a decrease in hardness, was more than offset by the abovementioned dispersion hardening. The highest overall hardness values for all planes and all orientations were obtained after long-term (72 hrs) annealing at 2150°C (see Table IV). This means that there is a true precipitation hardening reaction going on, which confirms Hall's[73] suggestion.

Preliminary data on 90.7% dense and 99+% pure hot pressed MoC yielded a KHN_{100} = 1040 kg/mm^2 and a KHN_{500} = 730 kg/mm^2 (cf. Table V). A VHN_{100} = 1100 kg/mm^2 was also found (cf. Table VI). The density of 7.97 g/cm^3 found for the tested MoC sample was compared with the reported[74] pycnometric density of 8.78 g/cm^3 for polycrystalline MoC. The MoC hardness found by us agrees with the

TABLE V

AVERAGE RANDOM HARDNESS VS. LOAD FOR MOLYBDENUM CARBIDES

	98.47% dense, 99.8% pure Single Crystal Mo_2C (basal plane)		95.55% dense, 99.6% pure Hot Pressed Mo_2C	90.7% dense, 99+% pure Hot Pressed MoC
LOAD (gm)	KHN (kg/mm^2)	VHN (kg/mm^2)	KHN (kg/mm^2)	KHN (kg/mm^2)
25	1640	1480	1680	1030
50	1600	1850	1650	1020
100	1560	1640	1540	1040
200	1450	1610	1420	940
300	1420	1500	1360	730
400	1370	1480	1330	750
500	1360	1450	1380	730
600	1350	1490	1310	770
700	1350	1460	1320	720
800	1360	1480	1310	770
900	1370	1440	1320	770
1000	1340	1450	1300	720

Note 1. Density cited is the bulk density of tested samples.

Note 2. The average deviation for $KHN_{25\text{-}1000}$ and $VHN_{400\text{-}1000}$ values of Mo_2C single crystals is approximately ±50 units; that for $VHN_{25\text{-}400}$ of Mo_2C single crystals and for all hot pressed Mo_2C and MoC values is approximately ±150 units.

Note 3. Polycrystalline Mo_2C was hot pressed from powder obtained from Cerac, Butler, Wis.; polycrystalline MoC was hot pressed from powder obtained from Semi-Elements, Saxonburg, Penna.

reported hardness for MoC being between 7 and 8 on Moh's scale[75]. (Note that 7 Mohs ≃ 750 KHN and 8 Mohs ≃ 1250 KHN)[76].

The hardness, at 100-g loads, taken on the Mo within the bands of the samples annealed at 2200°C ranged from 215 to 230 kg/mm^2, with the average value being 223 kg/mm^2. Samsonov[77] reports a value of 190 kg/mm^2 for Mo metal, at 100-g loads, and Lyman[78] reports a $DPH_{10\ kg}$ hardness of 180 kg/mm^2 for unalloyed arc-cast Mo, and a hardness range of 200-220 kg/mm^2 for recrystallized Mo, respectively. We have found the hardness of 99.99% pure and highly dense single crystal Mo to be KHN_{100} = 220 kg/mm^2 and VHN_{100} = 172 kg/mm^2. Since the bands had some γ'-MoC and up to 0.1% Ta present in addition to Mo metal, their hardness was usually higher than the average value of 223 kg/mm^2 as found on pure Mo bands. Depending on the amount of γ'-MoC and Ta present, the hardness ranged approximately from 230 to 800 kg/mm^2, at 100-g loads. Hardness was also taken across Mo-Mo_2C and Mo-γ'MoC-Mo_2C eutectoids formed after annealing at 2200°C, and found to vary between 334 and 800 kg/mm^2, with the average KHN_{100} value being 605 kg/mm^2.

For an ideal case, on the (0001) plane of hexagonal crystals, the following relationship between the indentation orientation, the angles at which it cuts the three a-axes, and the respective hardness, is established:

Indentation Oriented	Cuts a_2 at	Cuts a_3 at	Hardness Should Be
⊥ to a_1	30°	30°	Lowest
45° to a_1	75°	15°	Highest
∥ to a_1	60°	60°	Intermediate

There is experimental evidence that this relationship holds true for Mo_2C (cf. Table IV). It is to be noted that the hardness variations on the (0001) plane of hexagonal crystals are indeed very small. For the $\{2\bar{1}\bar{1}0\}$ planes of Mo_2C crystals, the following has been established experimentally:

Indentation Oriented		Hardness is
$\perp$ to c	i.e. $\parallel$ to slip lamellae	Intermediate
45° to c	i.e. 45° to slip lamellae	Highest
$\parallel$ to c	i.e. $\perp$ to slip lamellae	Lowest

It is to be noted that although a material theoretically should have only one room temperature hardness value independent of load, experimentally this is not so. Invariably at low loads, both Knoop and Vickers hardness values increase. The question why low loads produce a higher hardness has not been resolved. The reason most often quoted is that this increase is due to the error of the observer and the instrument[50, 79-82], and that it is not inherent in the material itself[83].

From the applications point of view, the bulk hardness value (i.e. value which does not change with load) is more truly the hardness of a given material. Thus, it follows that when comparative measurements of the microhardness of various materials are made, the microhardness numbers should be determined at various loads and compared in the region where the variation of microhardness with load is of no practical importance. For this purpose, Knoop and Vickers hardness were taken on the (0001) plane of Mo_2C single crystals and on polycrystalline hot pressed Mo_2C and MoC samples at loads ranging from 25 to 1000 grams. The data are listed in Table V. The (0001) plane of single crystals was chosen for this study because it shows the least anisotropy. It is seen from Table V that hardness for Mo_2C reaches a "constant" value starting with a load of approximately 400-g. This value remains essentially the same (within the experimental error) up to a load of 1000-g, and can thus be considered the representative hardness of bulk

Mo_2C being equal to $KHN_{400-1000} = 1360\pm50\ kg/mm^2$ for single-crystal Mo_2C and $KHN_{400-1000} = 1320\pm50\ kg/mm^2$ for polycrystalline Mo_2C. The bulk hardness of MoC is equal to $KHN_{300-1000} = 750\pm50\ kg/mm^2$. Similar leveling off of hardness at loads of approximately 400-g was previously observed by the authors on TiB_2 single crystals[49] and WSi_2 single crystals[84]. Samsonov et al[85] point out that for carbides such as TiC, ZrC, TaC, WC, and Mo_2C as well as for refractory borides, such as TiB_2, NbB_2, and ZrB_2, this occurs at a load of 120-g, using a diamond pyramid indenter. In view of the results of the present work, Samsonov's stated load is too low. In any case, however, comparable brittle materials, such as refractory carbides, borides, and silicides, of similar purity and % theoretical density, may be compared in hardness at any one stated given load.

It has been stated above that the direction of the veins on the $\{2\bar{1}\bar{1}0\}$ planes was parallel to the [0001] direction, as determined by x-ray data and microstructural observations.[86] Since, as a rule, no lamellar slip lines were found on (0001) planes of the crystals, and since slip lines are evidence of slip planes intersecting a surface, and the faint lamellar slip lines are perpendicular to the veins, then the ((0001)) planes must be the primary slip planes. At annealing temperatures of 1800° to 2000°C, the veins were found to form polygon walls oriented parallel to the [0001] direction and composed of dislocation etch pits, perpendicular and corresponding to each one of which there was a slip line (see Fig. 8). According to Amelinckx and Strumane[87], polygon walls are perpendicular to the slip plane, which in this case is the (0001) plane. Since the Mo_2C samples were found to cleave best parallel to the lamellae, the (0001) plane is then also the primary cleavage plane for Mo_2C single crystals.

By analogy with a previous work[88], from dislocation etch pit studies, and from single crystal x-ray data, it was determined that the pertinent Burgers vector is parallel to the $\langle 2\bar{1}\bar{1}0\rangle$ directions, which then are also the primary slip directions. Thus, the (0001) $\langle 2\bar{1}\bar{1}0 \rangle$ slip is the primary slip system for Mo_2C, which is consistent with that found for h.c.p. metals such as Be and Mg, which have the c/a ratios equal to 1.57 and 1.62, respectively (as compared to 1.570 for Mo_2C). This primary slip was also produced mechanically on these crystals by microhardness indentations.

Slip traces belonging to a secondary slip system - occasionally also found on non-indented crystals[53] - were produced on all planes studied after the application of mechanical stress (i.e. when indenting the surface). On the (0001) plane three sets of secondary slip traces (cross-slip) were observed, being at 60° or 120° to each other (Fig. 20), and corresponding to the three a-axes of hexagonal Mo_2C. On the $(2\bar{1}\bar{1}0)$ plane (fig. 21) slip traces of both the primary and the secondary slip were produced by indenting, with the former being parallel to the vein direction (i.e. parallel to the c-axis). By reconstructing the secondary slip system from slip traces found on (0001) and $\{2\bar{1}\bar{1}0\}$ planes, it was determined that the $\{10\bar{1}0\}$ planes are the secondary slip planes, with $\langle 2\bar{1}\bar{1}0 \rangle$ directions being the slip directions. Since $\{10\bar{1}0\}$ planes are planes of symmetry for the lattice, mechanical twinning on these planes is not likely[89].

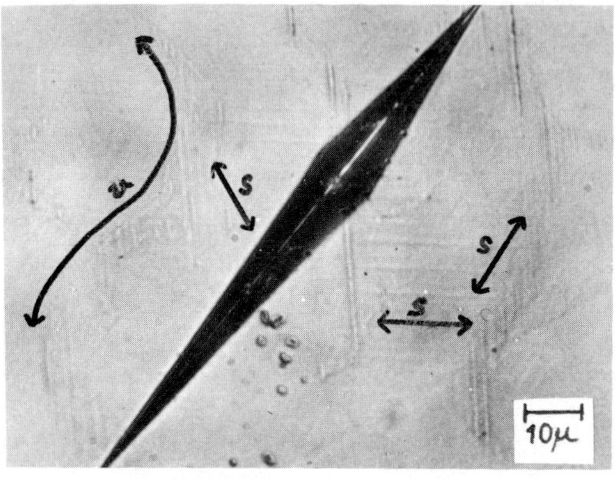

Fig. 20. Secondary slip (s) produced by Knoop hardness indentation at a 1000-g load on the (0001) plane of polished-only unannealed Mo_2C crystal. Note the angles between slip traces being 60°. A faint "weaving" veining (v) pattern is also discernible.

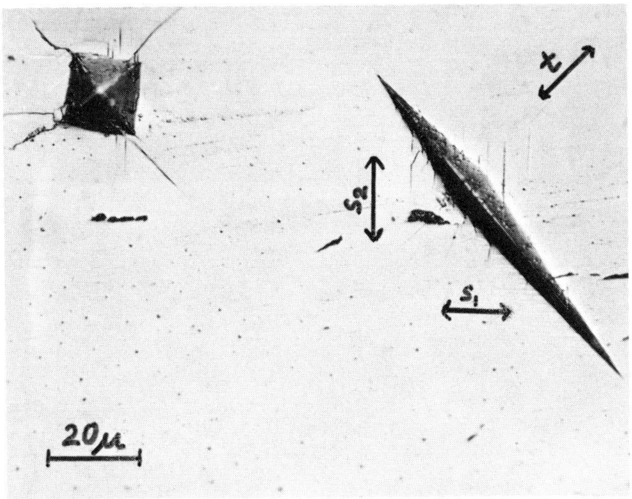

Fig. 21. Slip and twin traces on $(2\bar{1}\bar{1}0)$ plane of polished-only Mo_2C crystal annealed at 2000°C, produced by Knoop and Vickers hardness indentations at a 500-g load.

Mechanical twinning produced by microhardness indentations was observed on all planes studied. By measuring the angle between the primary slip and the twin traces on the $(2\bar{1}\bar{1}0)$ plane (see Fig. 21), it was determined that this angle was 45°. Comparing and correlating this with the twin traces found on the $(10\bar{1}2)$

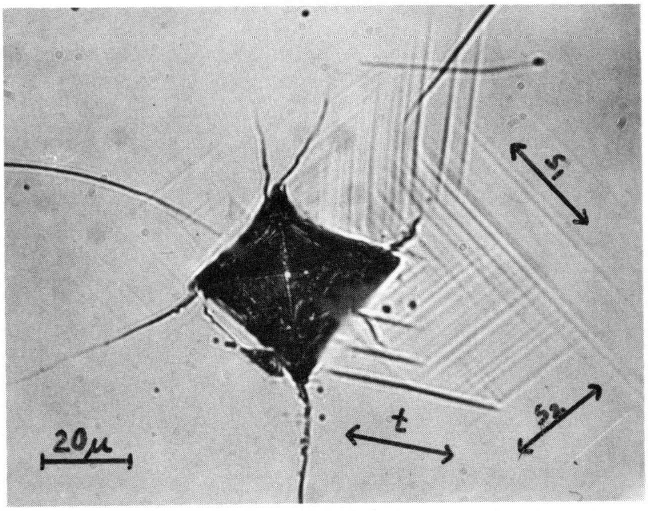

Fig. 22. Primary slip (s_1), secondary slip (s_2), and twinning (t) produced by a Vickers indentation at a 1000-g load on a $(10\bar{1}2)$ plane of polished-only unannealed Mo_2C single crystal.

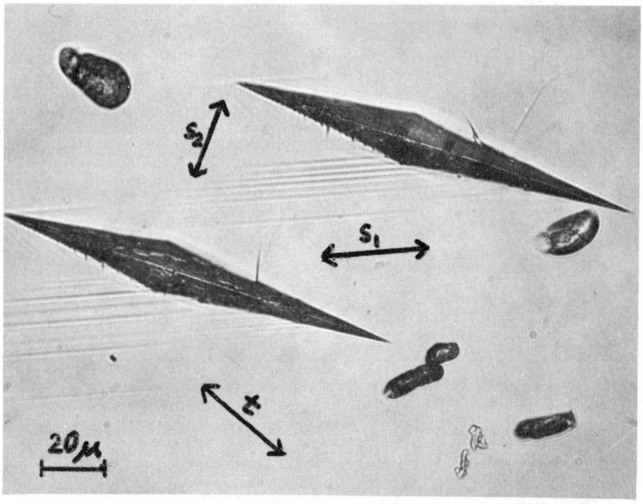

Fig. 23. Primary slip (s_1), secondary slip (s_2), and twinning (t) produced by two Knoop indentations at 1000-g load on a $(10\bar{1}2)$ plane of polished-only unannealed Mo_2C crystal.

plane (Figs. 22 and 23), the $\{10\bar{1}2\}$ planes were determined to be the twin planes. The [0001] was determined to be the twinning direction, which agrees with that for all twins reported on planes of the $\{10\bar{1}n\}$ type, the direction of twinning shear being normal to $\langle 2\bar{1}\bar{1}0\rangle$ according to Rosenbaum[90]. The determination of the $\{10\bar{1}2\}$ [0001] twin system on Mo_2C crystals (c/a = 1.570) is

Fig. 24. High resolution reflection electron diffraction of indented specimen of an unannealed Mo_2C single crystal, showing twinning on the $(10\bar{1}2)$ plane.

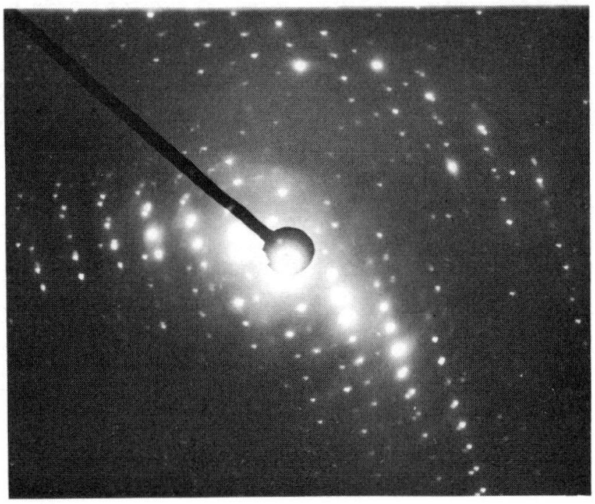

Fig. 25. High resolution transmission electron diffraction of indented thin section of an unannealed Mo_2C single crystal, showing twinning on $(10\bar{1}2)$ plane.

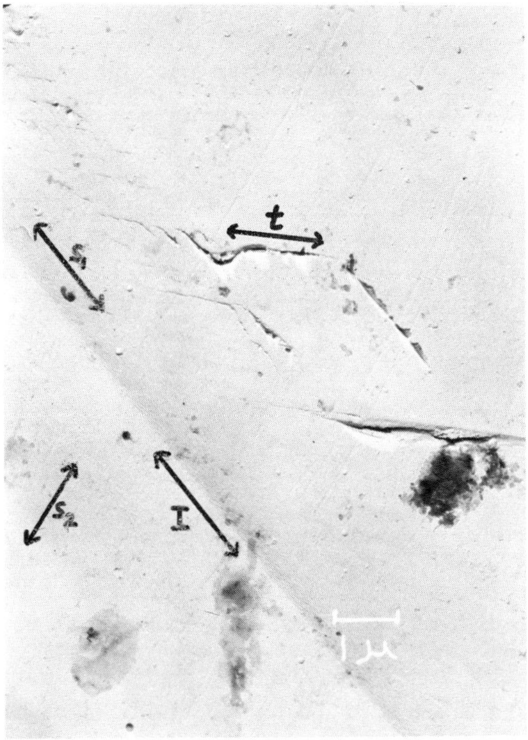

Fig. 26. Electron micrograph showing cross-slip (s_1 and s_2) and twinning (t) in greater detail. (I = direction of long diagonal of Knoop hardness of indentation.)

consistent with that commonly found on h.c.p. metals with $c/a < 1.633$, such as zirconium[91]. A combination high resolution reflection and transmission electron diffraction (Figs. 24 and 25) were used to identify twinning produced by microhardness indentations on edges of thin sections of Mo_2C single crystals and to positively differentiate it from slip. Electron micrographs of areas surrounding microhardness indentations were also taken and are illustrated by Fig. 26. Bulging out of the indentations after annealing was noted[53]. This is apparently due to internal stresses being released during the annealing process.

Both cross-slip and twin traces were frequently found on as-grown crystals as produced by impurity inclusions and scratches[86]. These traces compare well with those produced by hardness indentations on the same surfaces with respect to their orientation. This leads one to conclude that prismatic slip, and twinning, can be produced in Mo_2C single crystals only by stresses induced mechanically or by second phase particles, meaning that prismatic slip and twinning in Mo_2C require a higher critical resolved shear stress than basal slip.

Electrical Resistivity, Elastic Modulus, Debye Temperature, and Bonding

Anisotropy was also noted in the electrical resistivity of Mo_2C single crystals. The resistivity value along the a-axes was higher ($331 \mu\Omega cm$) than that along the c-axis ($279 \mu\Omega cm$). These measurements were taken at room temperature on polished crystal boules. At first mercury contacts were tried but it was found that mercury formed a thin film on the polished surfaces of the crystals, and therefore these resistivity measurements were not considered reliable. However, good reproducibility of ±1% was obtained with gold contacts. Note that the hardness along the a-axes is also higher than that along the c-axis. (See Table IV.) Samsonov[92] (see Table VI) previously reported a value of $133 \mu\Omega cm$ for electrical resistivity of polycrystalline Mo_2C. A resistivity value of $\sim 41 \mu\Omega cm$ was found along the c-axis for Mo_2C single crystals annealed at 2200°C for 4 hrs. (Compare with data of Table VI.)

A novel method for determining elastic modulus values from microhardness indentations was developed.

It is generally known that elastic recovery occurs during hardness measurements, when the indenter is withdrawn[50, 107]. Based on this finding, the modulus of elasticity can be derived from Knoop hardness indentations by assuming that the elastic stress

$$\sigma_E = \frac{P}{A} \qquad [2]$$

where P is the applied load in kg and A the projected area of the Knoop impression arising from quasi-elastic deformation. The "elastic" strain, ϵ_E can be simplified as

$$\epsilon_E = \frac{\Delta L}{L + \Delta L} \qquad [3]$$

Here ΔL is the reduction of the long diagonal of the indentation caused by elastic recovery (i.e. the difference between the unrecovered and recovered long diagonal), and L is the measured long diagonal of the indentation after recovery. Experimentally, this means measuring the difference in the long diagonal (ΔL), optically appearing as a fading-out silhouette surrounding the recovered indentation (Compare with Fig. 23). Analogous results are obtainable using Vickers indentations (see Fig. 22). It should be pointed out that only indentations of cracking classes 1 and 2 can be used for measurement of L and ΔL. The elastic modulus then can be approximated as

$$E \simeq \frac{\sigma_E}{\epsilon_E} \qquad [4]$$

This experimental method is considered accurate to within 10%. Recently, a similar technique was proposed by Kaufman and Clougherty[93],

Anisotropy in elastic modulus values with respect to crystallographic planes and orientations within a given plane was noted,

TABLE VI

SELECTED PROPERTIES OF MOLYBDENUM METAL AND MOLYBDENUM CARBIDES

Property	Mo	Ref	MoC	Ref	Mo_2C	Ref	Mo_2C *	Ref
Molybdenum content, wt %	100	97	88.9	75	94.11	101	94.11	101
Carbon content, wt %	-	-	11.1	75	5.89	101	5.89	101
			γ hex (B_h)	99				
Structure	bcc	97	γ' hex (B_i)	100	cph (L'_3)	71	cph (L'_3)	39
Lattice parameters (Å): a	3.1405	97	γ 2.898	71	3.00292	48	3.010±0.004	39
			γ' 2.932	71				
c	-	-	γ 2.809	71	4.72895	48	4.727±0.006	39
			γ' 4.724	71				
c/a	-	-	γ 0.972	71	1.575	48	1.554	39
			γ' 1.611	71				
Melting point (°C)	2610	97	2695	97	2475	47	-	-
Density at 20°C: theoretical (g/cm^3)	10.22	97	9.15	71	9.18	71	9.18	71
			8.78	74				
experimental	10.2	98	8.4	71	9.18	22	9.04	7
			7.97	⊠				
Hardness: Mohs	4-5	78	7-8	75	7-9	75	8-9	⊠
$DPH_{10\ kg}$	180-220	78	-	-	-	-	-	-
					1479	102		
Vickers, 100-g	190	77	1100	⊠	1630	⊠	1610	⊠
(kg/mm^2)	172 *	⊠			1800	103		
Knoop, 100-g	220 *	⊠	1040	⊠	1540	⊠	1520	86

TABLE VI (CONT'D)

SELECTED PROPERTIES OF MOLYBDENUM METAL AND MOLYBDENUM CARBIDES

Property	Mo	Ref	MoC	Ref	Mo_2C	Ref	Mo_2C *	Ref
Rockwell	90-100B	97	-	-	74A 88A	104 105	-	-
Young's modulus: 10^6 psi	47	97	28.59	75	31.4	100	-	
: kg/mm^2	-	-	-	-	54,400	77	57,800	86
Th. expansion coeff. ($10^{-6}/°C$)	5.43	97	5.95 41	75 ⊠	5.48, 6.15	106	-	-
Elec. resistivity (μΩcm) (27°C)	5.78	97	-	-	133	92	331; 279	86
Oxidation resistance Upper limit (°C)	650	97	500-800	75	500-800	75	-	-
Debye temperature (°K)	-	-	-	-	-	-	892	86

Legend: * Single Crystal Data

⊠ Data Resulting from Present Work

Note: This table does not list all the values for any given property reported in literature.

corresponding to respective microhardness anisotropy. Using the above described method, several elastic modulus values were determined. For unannealed Mo_2C crystals, virtually no elastic modulus anisotropy was noted on the (0001) plane, with the average value being $E = 53,900\ kg/mm^2$. For the $(2\bar{1}\bar{1}0)$ plane, an average value of $E = 39,300\ kg/mm^2$ was obtained. A value of $E = 30,800\ kg/mm^2$ was derived for the $(10\bar{1}2)$ plane from hardness indentations oriented perpendicular to the a_1 axis, and is therefore lower than would be expected for the average value for this plane. Any other elastic modulus values can be derived - by comparison with indentations at loads of 500-g and 1000-g at given orientations - from microhardness values listed in Table IV.

For comparison purposes, the modulus of elasticity along the long axis ($\simeq$ c axis) of the boule was obtained by the ultrasonic method and determined to be $29,400\ kg/mm^2$. This value is comparable with that determinable from hardness data for the parallel to the c-axis orientation for the $(2\bar{1}\bar{1}0)$ plane. Köster and Rauscher[94] reported a value of $22,100\ kg/mm^2$ for polycrystalline Mo_2C using a sonic method. Along the short axis of the boule, a value of $57,800\ kg/mm^2$ was determined by the authors using the same ultrasonic method - a value in fair agreement with that calculated from hardness indentations on the (0001) plane. Samsonov[77] reported a value of $54,400\ kg/mm^2$ for the elastic modulus of polycrystalline Mo_2C.

It is known[52] that the mean sound velocity, $\bar{v}_m$, of a crystal can be used to calculate the Debye temperature. A $\bar{v}_m = 6.99 \times 10^5$ cm was found for Mo_2C crystals studied, a $\Theta_{\bar{v}_m}$ value of 892°K was determined for Mo_2C. A mean shear velocity, $\bar{v}_s = 8.62 \times 10^5$ cm, and a mean longitudinal velocity, $\bar{v}_l = 5.68 \times 10^5$ cm were also determined for these crystals. From these values the respective Debye temperature values were calculated and found to be $\Theta_{\bar{v}_s} =$ 1098°K and $\Theta_{\bar{v}_l} = 725$°K.

The bonding mechanism in the transition metal carbides is not well understood. Dempsey[95] recently argued that an indication of the metallic behavior can be seen by large changes from stoichiometry, which is evidenced in the case of the Mo-C system in the formation of MoC from Mo_2C in the substructure. A small change from stoichiometry (ranging from Mo_2C_{1-x} to $Mo_2C_{0.93}$) was also determined in the matrix of the crystals after annealing which is probably due to the covalent nature of the bonds being responsible for the higher hardness of Mo_2C with respect to Mo. On the basis of Dempsey's work, it appears that there are a number of pros and cons relative to the metallic and covalent bonding in transition metal carbides. Dempsey[95] concludes that covalent bonding does not generally play an important role, and that bonding in carbides is similar to that of the transition metals. However, on the basis of the present work, a covalent and a metallic bonding mechanism is postulated for the Mo-C system, the metal-nonmetal bonds being responsible for the hardness, and the metal-metal bonds causing Mo_2C to behave in some ways like a transition metal. A recent Russian work[96] further elaborates on the bonding mechanism in transition metal carbides, in general.

ACKNOWLEDGEMENTS

The authors would like to thank R. E. Ogilvie, S. H. Moll and G. W. Bruno of Advanced Metals Research Corporation, Burlington, Massachusetts for furnishing the electron microprobe data. Special thanks are due to Clayborn A. Garthwait III and Robert E. Leasure, at the time of this research with the University of Dayton Research Institute, Dayton, Ohio, for general laboratory assistance. The authors are also indebted to R. E. Pence, Monsanto Research Corporation, Dayton, Ohio, for his help in electron microscopy. Special thanks are due to Dr. C. T. Lynch for his valuable suggestions and discussions, and to Mrs. Jean Gwinn for her patient help in the preparation of the manuscript.

REFERENCES

1 O.W. Storey, Trans. Am. Electrochem. Soc., 25 (1914) 489.

2 N.P. Goss, Trans. AIME, 145 (1941) 272.

3 A. Hultgren and B. Herrlander, Trans. AIME, 172 (1947) 493.

4 S. Amelinckx, Acta Met., 2 (1954) 848.

5 L. Northcott, J. Inst. Metals, 59 (1936) 225.

6 M.G. Corson, The Iron Age, 148 (1941) 56.

7 S.A. Mersol, F.W. Vahldiek, and C.T. Lynch, J. Less Common Metals, 10 (1966) 373.

8 A. Guinier, Imperfections in Nearly Perfect Crystals, edited by W. Shockley, J.H. Hollomon, R. Maurer and F. Seitz, John Wiley and Sons, Inc., New York, 1952, p. 402.

9 J.J. Gilman, Acta Met., 3 (1955) 277.

10 R.W. Cahn, J. Inst. Metals, 76 (1949) 121.

11 A.H. Cottrell, Dislocations and Plastic Flow in Crystals, Clarendon Press, Oxford, 1953, p. 180.

12 W.G. Johnston, GE Research Laboratory Report 61-RL-2649M, April 1961, Research Information Section, The Knolls, Schenectady, New York, p. 38.

13 C. Kittel, Introduction to Solid State Physics, John Wiley and Sons, Inc., New York, 1953, p. 565.

14 S. Amelinckx, The direct observation of dislocations in Advances in Materials Research in the NATO Nations, The MacMillan Co., New York, 1963, pp. 117-169.

15 N.F. Mott, Phil. Mag., 43 (1952) 1151.

16 S. Amelinckx, Acta Met., 2 (1954) 848.

17 A.H. Cottrell, Theoretical Structural Metallurgy, St Martin's Press, New York, 1962.

18 J. Friedel, Dislocations, Addison-Wesley Publishing Company, Inc., Reading, Massachusetts, 1964, p. 12.

19 S. Amelinckx, The Direct Observation of Dislocations, Academic Press, New York, 1964.

20 A.R. Verma, Crystal Growth and Dislocations, Butterworths Scientific Publications, London, 1953.

21 F.W. Vahldiek, S.A. Mersol, and C.T. Lynch, Japan. J. Appl. Phys., 4 (1965) 269.

22 A. Westgren and G. Phragmen, Z. anorg. allg. Chem., 156 (1926) 27.

23 H. Tutiya, Sci. Rep. Inst. Phys. Chem. Res., Tokyo, 19 (1932) 384.

24 K. Kuo and G. Hägg, Nature, 170 (1952) 245.

25 L.N. Butorina and Z.G. Pinsker, Soviet Physics, Crystallography 5 (1960) 560.

26 E. Parthé and V. Sadagopan, Acta Crysta., 16 (1963) 202.

27 E. Adelson and E.A. Austin, Battelle Memorial Institute, Columbus, Ohio, 43210, Private Communication.

28 F.W. Vahldiek and S.A. Mersol, Unpublished Data.

29 A.S. Argon and E. Orowan, Phil. Mag., 9 (1964) 1003.

30 H. Palmour, III, J.J. DuPlessis, and W.W. Kriegel, J. Am. Ceram. Soc., 44 (1961) 400.

31 W.S. Williams and R.G. Lye, Research to determine the mechanisms controlling the brittle-ductile behavior of refractory cubic carbides, Technical Documentary Report ML-TDR-64-25, Part II, April 1965, USAF Contract AF 33(657)-10109.

32 F.W. Vahldiek, J. Less Common Metals, 12, No. 6 (1967) 429.

33 S.A. Mersol and F.W. Vahldiek, Unpublished Data.

34 S.A. Mersol, F.W. Vahldiek, and C.T. Lynch, Trans. Met. Soc. AIME, 233 (1965) 1658.

35 F.W. Vahldiek, S.A. Mersol, and C.T. Lynch, Science, 149, No. 3685 (1965) 747.

36 D.N. French and D.A. Thomas, Trans. Met. Soc. AIME, 233 (1965) 950.

37 T. Takahashi and E.J. Freise, Phil. Mag., 12 (1965) 1.

38 L. Pons, presented at the International Symposium on Anisotropy in Single-Crystal Refractory Compounds, Dayton, Ohio, 13-15 June 1967.

39 J.R. Fenter, F.W. Vahldiek, and C.T. Lynch, Crystal structure analysis of dimolybdenum carbide, AFML-TR-66-38, January 1966.

40 M. Herman and G.E. Spangler, J. Inst. Metals, Monograph Rept. Series, 28 (1963) 75.

41 H.T. Lee and R.M. Brick, Trans. Am. Soc. Metals, 48 (1956) 1003.

42 E.J. Rapperport and C.S. Hartley, Trans. Met. Soc. AIME, 218 (1960) 869.

43 E.A. Anderson, D.C. Jillson, and S.R. Dunbar, Trans. Met. Soc. AIME, 157 (1953) 1191.

44 R.E. Reed-Hill and W.D. Robertson, Trans. Met. Soc. AIME, 209 (1958) 496.

45 A.D. Kiffer, Research investigation to determine the optimum conditions for growing single crystals of selected borides, silicides and carbides, WADD Technical Report 60-52, Contract AF 33(616)-6326, April 1960, Office of Technical Services, U.S. Department of Commerce, Washington, D.C., 20025.

46. C.T. Lynch, F.W. Vahldiek, S.A. Mersol, and C.R. Underwood, Investigation of single-crystal and polycrystalline titanium diboride: metallographic procedures and findings, ASD Technical Report 61-350, November 1961, Office of Technical Services, U.S. Department of Commerce, Washington, D.C., 20025, AD 271965.

47 E. Rudy, S. Windisch and Y.A. Chang, Ternary phase equilibria in transition metal-boron-carbon-silicon systems, part I. Related Binary Systems, Vol. I. Mo-C System, U.S.A.F. Report AFML-TR-65-2, Part I, Vol. I, Contract AF 33(615)-1249, March 1965.

48 R.J. Fries and C.P. Kempter, Anal. Chem., 32 (1960) 1898.

49 S.A. Mersol, C.T. Lynch, and F.W. Vahldiek, Investigation of single-crystal and polycrystalline titanium diboride: microstructural features and microhardness, ML Technical Documentary Report 64-32, Office of Technical Services, U.S. Department of Commerce, Washington, D.C., 20025, April 1964.

50 N.W. Thibault and H.L. Nyquist, Trans. Am. Soc. Metals, 38 (1947) 271.

51 C.T. Lynch, S.A. Mersol, and F.W. Vahldiek, Trans. Met. Soc. AIME, 233 (1965) 631.

52 O. L. Anderson, J. Phys. Chem. Solids, 24 (1963) 909.

53 F. W. Vahldiek, S. A. Mersol, and C. T. Lynch, Investigation of single-crystal dimolybdenum carbide, Technical Report AFML-TR-66-288, August 1966.

54 C. T. Lynch, S. A. Mersol, and F. W. Vahldiek, J. Less-Common Metals, 10 (1966) 206.

55 R. H. Hopkins and R. W. Kraft, Trans. Met. Soc. AIME, 233 (1965) 1526.

56 I. K. Zasimchuk and D. E. Ovsienko, Dokl. Akad. Nauk SSSR, 168, No. 1 (1966) 80.

57 F. W. Vahldiek, S. A. Mersol, and C. T. Lynch, Japan. J. Appl. Phys., 5, No. 8 (1966) 663.

58 G. W. Bruno, Advanced Metals Research Corporation, Burlington, Massachusetts. Private Communication.

59 K. Kuo and G. Hägg, Nature, 170 (1952) 245.

60 L. C. Browning and P. H. Emmett, J. Am. Chem. Soc., 74 (1952) 4773.

61 M. Gleiser and J. Chipman, J. Phys. Chem., 66 (1962) 1539.

62 J. B. Berkowitz-Mattuck, Technical Documentary Report No. ASD-TDR-62-203, Part II, March 1963, Office of Technical Services, U. S. Department of Commerce, Washington, D. C., 20025.

63 R. W. Bartlett, Technical Documentary Report No. ML-TDR-64-290, September 1964, Office of Technical Services, U. S. Department of Commerce, Washington, D. C., 20025.

64 S. Amelinckx, W. Bontinck, and W. Dekeyser, Phil. Mag., 2 (1957) 1264.

65 F. H. Horn, Nature, 4327 (1952) 581.

66 E. J. Freise, Ph. D. Dissertation, Cambridge University (1962).

67 J. M. Thomas and C. Roscoe, Paper presented at the Second Conference on Industrial Carbon and Graphite, London (1965).

68 E. Fitzer and H. Schlesinger, Carbon, 3 (1965) 247.

69 F. C. Frank, Discussions Faraday Soc., 5 (1949) 48.

70 A.J. Forty, Advan. Phys. (Phil. Mag. Suppl.) 3, (1951) 1.

71 R. Kieffer and F. Benesovsky, Hartstoffe, Springer Verlag, Wien, 1963, pp. 168-169.

72 G.V. Samsonov and V.B. Rukina, Dopovidi Akad. Nauk Ukr. SSR, No. 3 (1957) 247. Translated as F-TS-9994/V.

73 Richard C. Hall, Materials Research Laboratory, Martin Company, Orlando Division, Orlando, Florida, 32805, Private Communication dated 22 November 1966.

74 O.J. Whittemore, Jr., J. Can. Ceram. Soc., 28 (1959) 43.

75 P.T.B. Shaffer, High Temperature Materials, No. 1 Materials Index, Plenum Press, New York, 1964, pp. 97-100.

76 V.E. Lysaght, Indentation Hardness Testing, Reinhold Publishing Corporation, Norwood, Massachusetts, 1949, p. 237.

77 G.V. Samsonov and K.I. Portnoy, Alloys Based on High-Melting Compounds, Gosudarstvenoye Nauchno-Tekhnicheskoye Izdatel'stvo Oborongiz, Moscow, 1961, p. 181. Translated as FTD-TT-62-430.

78 T. Lyman, Editor, Metals Handbook, 8th Edition, 1, ASM, Metals Park, Novelty, Ohio, 1961, p. 1216.

79 H. Bückle, Z. Metallkunde, 45, No. 11 (1954) 623.

80 H. Bückle, Progress in micro-indentation hardness testing, Metallurgical Reviews, 4, No. 13 (1959) 49.

81 H. Bückle, Microhardness testing in powder metallurgy research in Powder Metallurgy (W. Leszyuski, Ed.), Interscience Publishers, New York, 1961, pp. 221-249.

82 L.P. Tarasov and N.W. Thibault, Trans. ASM, 38 (1947) 331.

83 D.P. Moak, D.H. Fisher and R.D. Koester, Feasibility determination and development of an experimental hot-microhardness device satisfactory for use with ceramic materials to temperature approaching 1800 to 2000°C, AFML-TR-66, Contract AF 33(615)-2212, April 1966, Battelle Memorial Institute, Columbus, Ohio, 43210.

84 S.A. Mersol, F.W. Vahldiek, and C.T. Lynch, Unpublished Data.

85 G.V. Samsonov, V.S. Neshpor, and L.M. Khrenova, Fiz. Metal. i Metalloved, 8, No. 4 (1959) 622.

86 F.W. Vahldiek, S.A. Mersol, and C.T. Lynch, Trans. Met. Soc. AIME, 236 (1966) 1490.

87 S. Amelinckx and R. Strumane, Acta Met., 8 (1960) 312.

88 F.W. Vahldiek, S.A. Mersol, and C.T. Lynch, Japan. J. Appl. Phys., 4 (1965) 269.

89 F.W. Cahn, Advances in Physics, 3 (1954) 202.

90 H.S. Rosenbaum, in Deformation Twinning (R.E. Reed-Hill, J.P. Hirth, and H.C. Rogers, Eds.), Gordon and Breach Science Publishers, New York, 1964, p. 73.

91 D.G. Westlake, in Deformation Twinning (R.E. Reed-Hill, J.P. Hirth, and H.C. Rogers, Eds.), Gordon and Breach Science Publishers, New York, 1964, p. 34.

92 G.V. Samsonov, Zh. Tekh. Fiz., 26 (1956) 716.

93 L. Kaufman and E.V. Clougherty, Investigation of boride compounds for very high temperature applications, Technical Documentary Report No. RTD-TDR-63-4096, Part I, Contract AF 33(657)-8635, December 1963, pp. 191-193. Office of Technical Services, U.S. Department of Commerce, Washington, D.C., 20025.

94 W. Köster and R. Rauscher, Z. Metallkunde, 39 (1948) 111.

95 E. Dempsey, Phil. Mag., 8 (1963) 285.

96 Yu. N. Surovoi, L.A. Shvartsman, and V.I. Alekseev, Fiz. metal. i metalloved., 20, No. 2 (1965) 251.

97 R.S. Archer, Molybdenum in Rare Metals Handbook (C.A. Hampel, Ed.), 2nd Edition, Reinhold Publishing Corporation, New York, N.Y., 1961, pp. 283-303.

98 R.C. Weast, S.M. Selby, and C.D. Hodgman, Eds., Handbook of Chemistry and Physics, 45th edition, The Chemical Rubber Co., Cleveland, Ohio, 1964, p. D-28.

99 H. Nowotny, E. Parthe, R. Kieffer, and F. Benesovsky, Monatshefte Chemie, 85 (1954) 255.

100 J.H. Westbrook and E.R. Stover, Carbides for high-temperature applications, GE Research Laboratory Report 60-RL-2565M, November 1960, Research Information Section, The Knolls, Schenectady, New York.

101 M. Hansen and K. Anderko, Constitution of Binary Alloys, 2nd Edition, McGraw-Hill Book Company, New York, 1958, pp. 370-372.

102 G. V. Samsonov, Dokl. Akad. Nauk SSSR, 93, No. 4 (1953) 689.

103 H. Bückle, Metall., 9 (1955) 549, 1067.

104 G. V. Samsonov, High Temperature Materials, No. 2 Properties Index, Translated from Russian, Plenum Press, New York, 1964.

105 R. Kieffer and F. Kölbl, Powder Met. Bull., 4 (1949) 4.

106 O. H. Krikorian, Thermal expansion of high temperature materials, UCRL-6132, September 1960.

107 D. Tabor, The Hardness of Metals, Oxford University Press, Amen House, London, 1951, pp. 85-86 and 146-148.

INFLUENCE OF OCTAHEDRAL SUBCELLS ON THE ELECTRICAL PROPERTIES AND BONDING CHARACTERISTICS OF MOLYBDENUM CARBIDES

L. E. Toth, J. Zbasnik, and Y. Sato
School of Mineral and Metallurgical Engineering
University of Minnesota
Minneapolis, Minnesota

and W. Gardner
Semiconductor Research and Development Laboratories
Texas Instruments
Dallas, Texas

ABSTRACT

Nowotny and coworkers[1] have demonstrated the importance of the octahedral subcell in the crystal structures of several hundred carbides and nitrides with complex crystal structures and compositions. Molybdenum is unique in that it forms more carbides than any other transition metal and that all these carbides have a common subcell. We have measured the heat capacities of αMoC_{1-x}, ηMoC_{1-x}, Mo_2C, and Mo_2BC in the range 1-20°K, as well as the superconducting transition temperatures and superconducting critical fields of these phases and other molybdenum carbides. These measurements demonstrate the importance of the subcell in determining the electrical properties and they also clearly indicate the importance of the molybdenum-carbon interaction in the bonding of the molybdenum carbides.

Introduction

Bonding in transition metal carbides and nitrides is not completely understood. Low temperature heat capacities of several molybdenum carbides have been measured to clarify certain aspects of the bonding and also some superconducting properties. Low temperature heat capacities yield directly the coefficient of electronic specific heat (γ), the Debye temperature (θ_D), the superconducting critical temperature (T_c), and indirectly information about the strength of the electron-phonon coupling and the band structure. By studying these parameters as a function of composition, order, and structure, information about bonding in refractory compounds is obtained.

Molybdenum carbides were chosen for this investigation because of the unique ability of molybdenum to form more distinct carbides than any other transition metal. Known molybdenum carbides include αMoC_{1-x}(NaCl), η MoC_{1-x}(hexagonal) βMo_2C (hexagonal), αMo_2C (orthorhombic), Mo_2BC (orthorhombic), Mo_2GaC (H-phase), Mo_3Al_2C (β-Mn) and Mo_5Si_3C ($D8_8$). All crystal structures of these phases are related by a common subcell in which molybdenum is located at the corners of an octahedron and carbon is at the center of the octahedron in the interstitial site. Nowotny and coworkers[1] have demonstrated the importance of the octahedral subcell as a building block in the construction of these crystal structures. Essentially, different molybdenum carbide structures can be derived from one another by rearrangements of the geometry of octahedra, variations in the C/Mo ratio, and additions or subtractions of nontransition metal atoms in layers or voids between close-packed arrangements of octahedra.

Several theories[2,3] of bonding in carbides have focused on the octahedral subcell and metal-nonmetal bonding. Other theories[4,5,6] have de-emphasized the importance of metal-nonmetal bonding and emphasized instead the interstitial nature of carbon and metal-metal bonding. Rundle[2] has observed that the distance between transition metal atoms (T-T distances) in octahedra is several percent greater than what it is in the pure transition metal. He further notes that monocarbides and mononitrides nearly always have the B1 structure regardless of the radius of the metal atom or the structure of the metal; the atomic packing of metal atoms has therefore changed to accommodate carbon or nitrogen in an octahedral interstitial site not available in the original crystal structure of the parent phase. Rundle has suggested that the increased T-T distances greatly decrease the importance of T-T bonding and that the principle bonding mechanism is between the sp electrons of carbon and the d^2sp^2 electrons of the transition metal, (See Fig. 1). Electrons

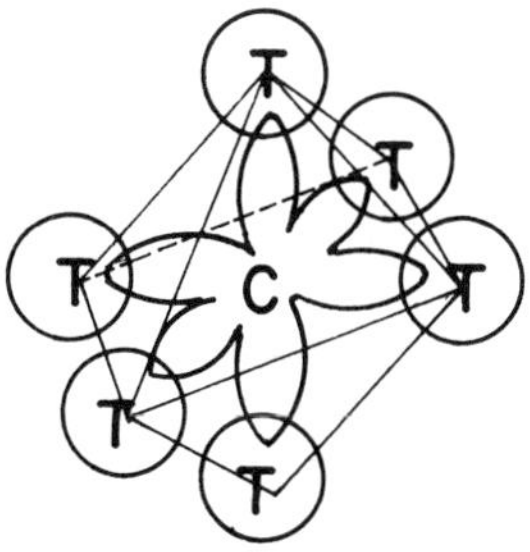

Fig. 1a. Rundle's[2] concept of bonding in carbides emphasizing metal—nonmetal bonding in carbides and de-emphasizing metal—metal bonds. Shown is the octahedral configuration with the bonding sp—p^2 configuration of the central carbon atom.

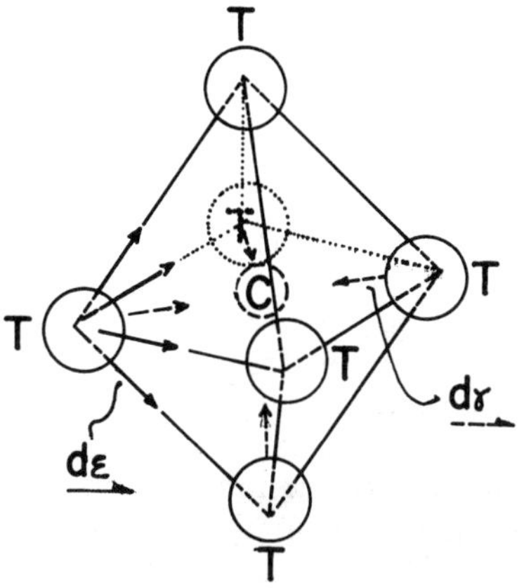

Fig. 1b. Some of the directions of the d_γ and d_ε states in an octahedron. While the d_γ states interact strongly with the 2p states of carbon, the d_ε states can form strong T—T bonds.

are transferred from the metal to the nonmetal atom because of the weaker metal-metal bonds. Kiessling[4], however, emphasized the metallic character of the bonding and proposed a transfer of electrons to the metal atoms from the nonmetal. Bilz[3] attempted to resolve the relative importance of metal-metal versus metal-nonmetal bonds. His LCAO calculation of the band structure considers nine interactions: $(ss\sigma)_T$, $(sd\sigma)_T$ $(dd\sigma)_T$, $(pp\sigma)_X$, $(pp\pi)_X(dd\pi)_T$ $(pd\sigma)_{XT}$, $(ps\sigma)_{XT}$ and $(pd\pi)_{XT}$ where estimations for $(ss\sigma)_T$, $(dd\sigma)_T$ and $(dd\pi)_T$ were taken from values available for pure copper and nickel. Bilz maintains his band structure is applicable to all NaCl structured refractory metals because he used average values for the above two center integrals. Essentially, his band model consists of a narrow d-band (3.5 eV wide) superimposed over a broad s-p conduction band. He finds that the $(pd\sigma)_{XT}$ and $(ps\sigma)_{XT}$ two center integrals lower substantially the energies of the original p band of the interstitial atom so that electrons are transferred from the metal atoms to the nonmetal atoms (2p-bands). The d-band is mainly empty for all refractory compounds from the 4th to the 6th group. Bilz's calculations therefore support Rundle's hypothesis about the direction of electron transfer. Hardness, brittleness and other properties result from the directional character of the XT interactions.

Costa and Conte[5] have criticized Bilz's band model on the following basis:

> "Bilz's model, which suppresses any binding between metal atoms, provides no explanation, in particular, as to why the transition metal carbides are much less stable from group VI onwards. The d band is practically empty in his model for TiC. It should be possible to observe considerable stabilization of the compounds for the metals of groups V, VI and VII similar to that found by filling the bonding half of the d band for the transition metals. In the same line of thought, one cannot

understand why the existing structures always possess metal-metal distances approximately equal to those found in the metallic structure."(6)

In their theoretical model of the band structure, only the metal-metal interactions of the d_ϵ and d_γ type are considered*; metal-nonmetal interactions are neglected except for the effects of the nonmetal potential on the metal d states. The geometry of the d_ϵ and d_γ functions are illustrated in Fig. 1b. According to Costa and Conte the d_ϵ states form a narrow band (approx. 2.5 eV) while the d_γ band is very broad (approx. 10 eV) because of the effect of the nonmetal potential. The contribution of the interstitial atom to the potential in the interaction integrals between two d_γ states is about four times greater than the contribution from the metal atoms. The interaction of d_γ states may also be thought of as $(dp\sigma)_{XT}$ bonding because of the geometry involved. The d_γ states therefore form a broad bonding band. The d_ϵ states with a narrow band give shape to the entire d band. Electrons are transferred from the metalloids to the d band. The Fermi level lies in the d band and electrical properties are explained in terms of the d band, but such properties as hardness and high melting points are explained by the directional character of the d_γ states and their interactions with the metalloids.

Lye(7) has considered bonding in TiC and attempted to modify both the Bilz and Costa-Conte models by utilizing data from optical absorption experiments to adjust the relative eV spacing between bands. His band model differs considerably from that of Bilz and is more general than that of Costa and Conte because of considerations of metal-nonmetal interactions. The essential feature is that both metal-metal and metal-nonmetal interactions are

*The d_ϵ states are d_{xy}, d_{yz}, and d_{zx} and the d_γ states are $d_{x^2-y^2}$ and d_{z^2}.

important to the bonding and that the Fermi level in TiC lies in a band primarily of d character with both the d_ϵ and d_γ contributing to the shape of the band.

Presently there is insufficient experimental information to distinguish between the different theoretical models. Lye maintains that the metallic conductivity in TiC results from a small conduction band derived from the d states and not the 4s states as the Bilz model would suggest. Geballe et.al.[8] maintain that their low temperature heat capacities measurements and superconducting experiments indicate that the Fermi level lies in an sp band. The present experiments indicate that the Fermi level for MoC lies in a band of essentially d character.

Experimental

a. Heat Capacities

Low temperature heat capacities of molybdenum carbides were measured in the range 1.5 - 20°K with a newly built adiabatic calorimeter. The design of the calorimeter is quite standard but because of the newness of the cryostat certain general features will be described as well as the results of the final calibration on a specimen of pure copper.

The design of the calorimeter is similar to those at the University of California, Berkeley, under Professor N. Phillips direction, and we are indebted to Professor Phillips for allowing us so much access to information about the design of calorimeters. Because of porosity in powder metallurgy samples and possible undesirable adsorption and deadsorption of a heat exchange gas, a heat clamp is used to cool the specimen to low temperatures. The temperature sensor is a Cry-O-Cal germanium thermometer, the resistance of which is read with a Leeds and Northrup K6 potentiometer, microvolt D.C.

amplifier, and recorder. The heating circuit consists of a Harrison power supply, Leeds and Northrup K3 potentiometer and detector, and Hewlett-Packard timer. The C_P measurement is by the discontinuous-pulse technique, in which the temperature drift is monitored before and after each pulse. The temperature rise due to each pulse is normally 4% of the absolute temperature.

Typical opening parameters are as follows. The lowest obtainable temperature is 1.06°K. The thermal isolation is better than 10^{-8} watts/deg°K. The temperature control above the λ point is approximately 10^{-3}°K, while below the λ point the control is about 2×10^{-4}°K.

The accuracy of the calorimeter was checked by measuring the heat capacity of 99.999$^+$ pure copper. The present value of γ = 165.0 μ cal/mole°K^2 and θ_D = 345°K compare favorable with Martin's[23] value 166.3 μ cal/mole°K^2 and θ_D = 345.6°K. The greatest difference between the present results and those of Martin occurs between 14°K and 18°K and is about 1%.

Superconducting T_c measurements by inductive techniques were also performed on several molybdenum carbides. These techniques have been described elsewhere.[9]

b. Samples

The molybdenum carbide samples were kindly prepared for this group by Dr. Erwin Rudy of Aerojet General Corporation. The preparation of these phases, many of which decompose unless quenched very rapidly from temperatures above 2000°C, involves equipment and techniques not generally available or known. The techniques of sample preparation and the phase diagram for the Mo-C systems have been described elsewhere by Rudy and coworkers.[10] The notations for the phases in Mo-C follow that established by Rudy and coworkers.[10]

Samples prepared for our group were Mo_2BC (orthorhombic), αMoC_{1-x} (NaCl), and $\eta\ MoC_{1-x}$ (hexagonal). In this lab, samples of Mo_2C (orthorhombic) were prepared by conventional powder metallurgy techniques. All samples were examined by X-rays to confirm the crystal structure.

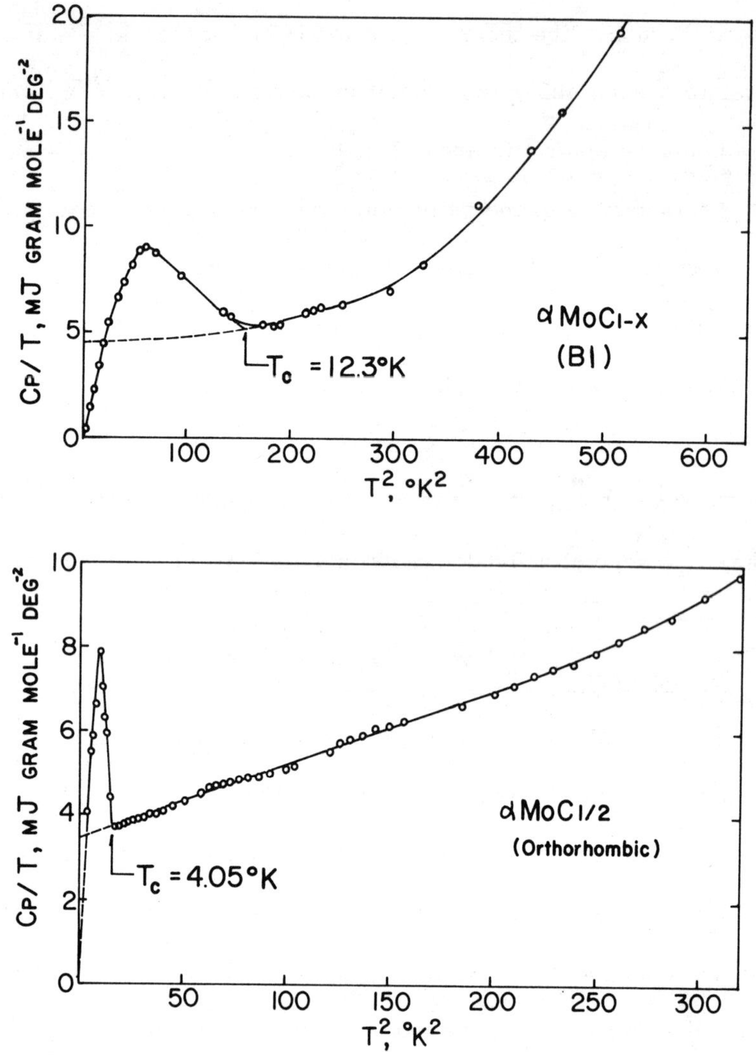

Fig. 2. C_p/T versus T^2.

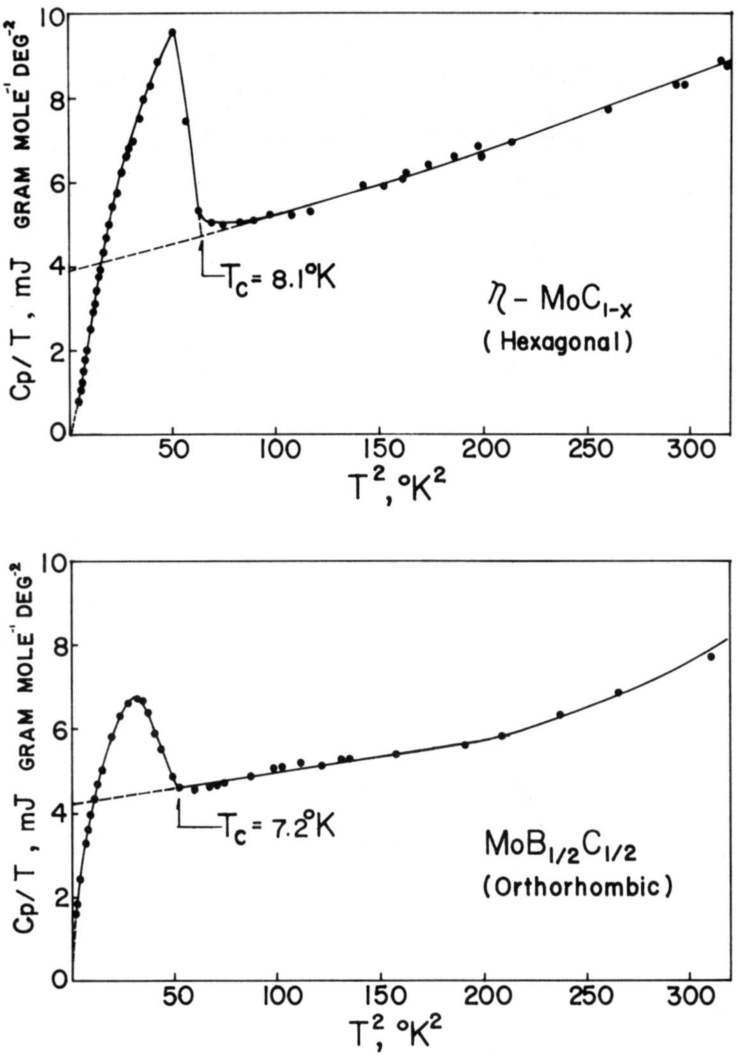

Fig. 3. C_p/T versus T^2.

Experimental Results

Shown in Figures 2 and 3 are the C_P/T vs T^2 curves for $\alpha\,MoC_{1-x}$, $\eta\,MoC_{1-x}$, αMo_2C and Mo_2BC. As shown by the peak in the C_p curves, all of these samples have very high superconducting transition temperatures. In Table I are listed the γ and θ_D values for these phases. The γ values and the C_p/T vs T^2 curves are calculated on the basis of one atom Mo per formula, that is, αMoC_{1-x}, ηMoC_{1-x}, $MoC_{1/2}$, $MoB_{1/2}C_{1/2}$. The data are reported in this manner so that direct comparisons can be made with pure Mo.

For αMoC_{1-x} accurate determinations of θ_D and γ are difficult because of the long extrapolation from the normal region. The αMoC_{1-x} samples were furthermore two phase, containing a small amount of ηMoC_{1-x}. αMoC_{1-x} decomposes extremely rapidly into ηMoC_{1-x} and even though very small samples (~4 gms) were drastically quenched by literally "exploding" them into liquid tin, some decomposition could not be avoided. Fortunately, the thermodynamic properties of the two phases are very similar so that the error due to the second phase was small. Because the normal region of αMoC_{1-x} is no longer a nearly pure T^3 region, the γ and θ_D values were obtained by 3rd law calculations. In zero magnetic field the superconducting transition is second-order and it can be easily shown[13] that

$$S_{es}(O) = S_{en}(O) \text{ at } T_c.$$

where $S_{es}(O)$ is the entropy of the superconducting electrons and $S_{en}(O)$ the entropy of the normal electrons in zero magnetic field. Assuming that the lattice contribution to C_p is the same in both states, we have

$$\int_o^{T_c} \gamma dT = \int_o^{T_c} \frac{C_{p_{es}}}{T} dT.$$

This relationship allows γ and also θ_D to be determined. Third law calculations of γ and θ_D were in agreement with those obtained by the usual graphical extrapolation of the C_p/T vs. T^2 curves for Mo_2BC, Mo_2C and ηMoC_{1-x}. The present calculations also included a T^5 term in the expansion for C_p.

TABLE I

Low Temperature Heat Capacities of Molybdenum Carbides

Phase	Nominal Composition	Heat Treatment	Structure	Lattice Parameter Å	γm joule gm mole^{-1} deg^{-2}	θ_D °K*	T_c °K (on set)
αMoC_{1-x}	$Mo_{0.59}C_{0.41}$	Hot press and quench from 2340°C	B1	a = 4.281	4.40	610	12.3
$\eta\ MoC_{1-x}$	$Mo_{0.61}C_{0.39}$	Hot press and quench from 1920°C	Hexagonal	a = 3.010 c = 14.62	3.79	536	8.05
$\alpha MoC_{1/2}$	$Mo_{0.65}C_{0.35}$	Sinter in argon and heat treat in vac. for 2 hrs at 1900°C and 3 hrs at 1500°C and 10 hrs at 1000° C	Orthor-hombic	a = 4.736 b = 6.024 c = 5.217	3.41	473	4.05
$MoB_{1/2}C_{1/2}$	$MoB_{0.50}C_{0.50}$	Hot press and heat treat at 1900°C for 2 hrs in vac.	Orthor-hombic	a = 3.086 b = 17.35 c = 3.047	4.25	529	7.2
Mo	-	-	BCC	-	1.83[11]	460[11]	0.92[12]

*Based on the formula $\theta_D = \left(\frac{234\ N_o K}{a}\right)^{1/3}$ where a is the value in the expansion $C_p = \alpha T + aT^3 + bT^5$, and C_p is per gm atom of Mo.

Discussion

(1) Band Structures

The γ and θ_D values for molybdenum carbides are more similar to one another than they are to those for pure molybdenum although even here the difference is small. (See Table I.) Other measurements[5,8] show much larger differences between the values for the pure metal (Nb, Ti, Ta) and for the respective carbides or nitrides. Large changes in γ and θ_D compared to the parent metal indicate the importance of metal-nonmetal interactions. The fact that the molybdenum carbides have very similar properties to one another can be easily explained by comparing their crystal structures. Bonding within the octahedra is little affected by structural changes which preserve the unit, and bonding between octahedra should be similar in structures in which the octahedra are close-packed or nearly close-packed, as is the case here. The present results therefore substantiate the importance of the subcell as a structural unit.

Unlike other NaCl structured carbides and nitrides the molybdenum carbides have γ values greater than that of parent element. Geballe and coworkers[8], have found γ values for NbN and NbC substantially lower than that of pure Nb. They have used this information and other related experimental results to argue that the Fermi level lies in a conduction band formed from s-p electrons. The same agreement cannot, however, be applied to the present results; rather, the relatively higher value of γ in molybdenum carbides implies that the Fermi level lies in a band primarily of d character. The γ value for the molybdenum carbides is greater than that of the typically sp elements; i.e. Sn, Pb, and Hg.

An unambiguous interpretation of the present results with any of the previous band models was impossible. The results disagree with the Bilz[3] model, which predicts that the d band be unpopulated for MoC_{1-x}. The model of Geballe et.al.[8] correctly predicts increased d band character for MoC as well as sp band characteristics for NbC, but their interpretation seems to imply that γ should increase with decreasing C/Mo or C/Nb ratios, an increase which is not experimentally observed. Minor modifications of their model could be made to obtain agreement with this experiment. For carbides of the second transition series, however, it appears that a band structure similar to that of Lye[7] for TiC best fits the data. It is necessary, however, to modify his model and assume that the d band split is away from the Fermi level in NbC and ZrC.

The following experimental data are to be explained by the proposed band model for the second transition series:

(1) relatively low γ value for NbC

(2) higher γ value for MoC implying that the Fermi level lies in a d band

(3) a decrease in γ values with decreasing C/T ratio for both NbC and MoC

(4) insensitivity of γ values of molybdenum carbides to changes in structure.

(5) decreased chemical stability of the group six carbides compared to fourth and fifth group carbides

(6) relatively poor mechanical properties indicating strong localized metal-nonmetal interactions[14]

(7) susceptibility measurements by Bittner and coworkers[15,16] and Hall coefficient measurements by Piper[17]

Items 2 and 4 imply that the Fermi level for molybdenum carbides lies in the

d_ϵ band because this part of the band would be the least sensitive to structure changes which leave the metal atoms in close-packed positions (see Fig. 1b) and item 5 implies that the Fermi level lies in the antibonding portion of the d band for MoC. The data in item 7 indicate a low density of states at the Fermi level for ZrC and the formation of a new band at that composition.

The following band model is proposed for the second transition series carbides analogous to that proposed by Lye[7] for TiC. Lowest in energy would be the 2s states of carbon with both the bonding and antibonding portions of the band occupied. Following Costa and Conte's suggestion, next in energy would be the broad 4dγ states, possibly hybridized with the carbon 2p states, and a relatively narrow d_ϵ bonding band. It is necessary to assume that the d_ϵ band is split so that low densities exist at the Fermi levels of NbC and ZrC.[8, 15-17] For these phases the Fermi level would lie in a conduction band, derived presumably from a mixture of 5s, 2p, and 4dγ states. Still higher in energy would the the principal parts of the $4d_\epsilon$ and 4dγ antibonding bands and the remaining 2p states of carbon. We are

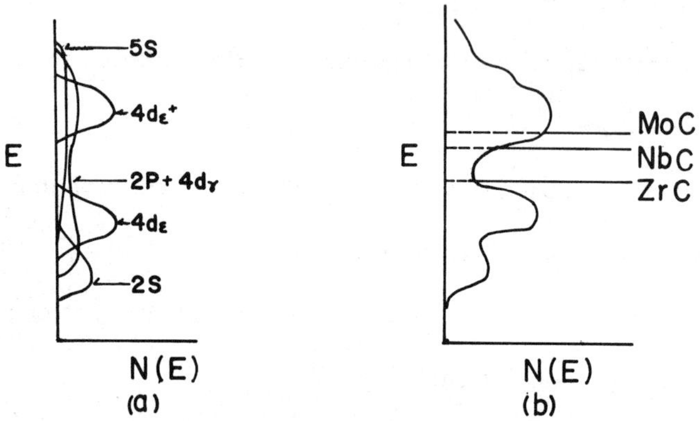

Fig. 4. Qualitative features of proposed band model for 2nd row transition metal carbides. The model is similar to that proposed by Lye[7] for TiC except for a splitting of the d bands. (a) Shows the contributions of different electrons to the band and (b) the Fermi level. Drawings not to scale.

utilizing Lye's controversial suggestion that the 2p band for carbon be elevated in energy. The placement of these states determine the direction of electron transfer between C and the transition metal atom. The 2p band could be elevated in energy by an uncompensated electron transfer from carbon to the metal.[7] The electron configuration in MoC would be approximately: $(4d)^6\ (5s)^{1/2}$ for Mo and $(2s)^2\ (2p)^{1\frac{1}{2}}$ for C, in NbC: $(4d)^5\ (5s)^{1/2}$ for Nb and $(2s)^2\ (2p)^{1\frac{1}{2}}$ for C. Electrons are therefore transferred from the carbon to the d bands of the transition metal in agreement with the theories of Kiessling[4], Dempsey[18], Costa and Conte[5] and Lye[7]. The qualitative features of this hypothetical band are shown in Fig. 4.

The proposed band structure is consistent with the experimental observations listed above. For MoC the Fermi level lies in a predominantly d band while for NbC and ZrC, it lies in a band with a low density of states. Because of the proposed direction of electron transfer in these phases, reduced C/Mo ratios lower the Fermi level and reduce the density of states. Stability of fourth and fifth group carbides results from filling the bonding $4d\gamma$ and $4d_\epsilon$ states, while sixth group carbides are less stable because the $4d_\epsilon$ antibonding states are being occupied.

Mechanical properties are also explained. The observed high values of critical resolved shear stresses necessary to initiate slip result from the directional character of the $4d\gamma$ states of the metal and the 2p states of carbon. This stress obviously becomes less as the C/T ratio decreases.

The proposed band model is really a combination of several previous models. The main difference between this model and that proposed by Lye[7] is the splitting of the d_ϵ band at electron concentrations of 8 and 9 so that low values of N(E) are obtained. The model stresses the importance of both metal-metal and metal-nonmetal interactions. While the band model fits the present results further experiments are necessary to test its validity.

(2) Debye Temperatures

One very puzzling feature of the present experiments was the large discrepancy between the θ_D value for Mo_2C obtained here (473°K) and the θ_D value obtained by Vahldiek and coworkers[14] from elastic constant measurements (892°K). The discrepancy appears to be far broader in scope than just one compound. Other determinations of θ_D by elastic constant measurements on TiC[19] also give very high values, while the θ_D values obtained here are comparable to other values determined by specific heats.[8] Normally θ_D values from elastic constants agree well with θ_D values from specific heats. The present specific heat values of θ_D are in fair agreement with the values calculated by the Lindemann formula.[20] At present no explanation can be offered for the discrepancy.

(3) Superconductivity

Toth[21] has pointed out that the superconducting T_c's of molybdenum carbides are unique in that nearly all members of the family have unusually high T_c's. αMoC_{1-x}(NaCl) has the highest T_c for a binary carbide, η MoC_{1-x} has the second highest known T_c for any phase with a hexagonal structure, Mo_2BC has the second highest T_c for any phase with an orthorhombic structure, Mo_3Al_2C has the highest T_c for a compound with a β-Mn structure and Mo_2GaC is the only known superconductor among the H-phases. It was suggested that the common octahedra subcell played an essential role in fostering these high T_c's. A more general suggestion about octahedral subcells in all carbides and nitrides favoring superconductivity has been prepared by

Sadagopan and Gatos.[22] The low temperature heat capacity experiments now allow a more complete understanding of this interesting phenomenon. The details of this analysis and the critical field measurements will be presented elsewhere.[24] The results show that the molybdenum carbides have unusually high T_c's because of a moderately strong electron-phonon coupling constants and large density of states (compared to other carbides and nitrides) and large θ_D values. The results also show that the molybdenum carbides tested to date have very similar superconducting properties; thus these results support the original hypothesis that the octahedral subcell plays an important part in superconductivity of all these phases as well as in the bonding and the crystal structure.

Acknowledgments

This research was supported in part by a NASA grant (NGR-24-005-065) to the Space Science Center of the University of Minnesota, and by a grant from the Air Force Office of Scientific Research AFOSR-1012-66.

The authors are indebted to Dr. E. Rudy, of Aerojet General Corporation, Sacremento, California, for providing the samples used in this experiment and for providing us with the results of the phase-diagram for Mo-C. The assistance of M. Ishikawa, K. Smith, and D. Stephens in the heat capacity measurements is also gratefully acknowledged. We also thank Professor R. Swalin, University of Minnesota for reading the manuscript and helpful comments.

References

1. H. Nowotny, W. Jeitschko, and F. Benesovsky, Planseeber, Pulvermet., 12, 31 (1964).

2. R. E. Rundle, Acta Cryst. 1, 180 (1948).

3. H. Bilz, Z. Physik 153, 338 (1958).

4. R. Kiessling, Met. Rev. 2, 77 (1957).

5. P. Costa and R.R. Conte, Nucl. Met. AIME 10, 29 (1964).

6. Ibid p. 13-14.

7. R. G. Lye, RIAS Technical Report, 66-8, 1966.

8. T. H. Geballe, B.T. Matthias, J. P. Remeika, A. M. Clogston, V. B. Compton, J. P. Maita and H. J. Williams, Physics Z, 293 (1966).

9. L. E. Toth, C. M. Yen, L. G. Rosner and D.E. Anderson, J. Phys. Chem Solids, 27, 1815 (1966).

10. E. Rudy, St. Windisch, A. J. Stosick and J. R. Hoffman, Trans. AIME 239, 1247 (1967).

11. F. Heiniger, E. Bucher and J. Muller, Phys. Kondens. Materie 5, 243 (1966).

12. B. T. Matthias, T. H. Geballe, and V. B. Compton, Rev. Mod. Phys. 35, 1 (1963).

13. E. A. Lynton, Superconductivity, John Wiley, New York, 1962, p. 14.

14. F. W. Vahldiek, S. A. Mersol, C. T. Lynch, Technical Report AFML-TR-66-28, 1966.

15. H. Bittner and H. Goretzki, Mh. Chem. 91, 616 (1960) and 93, 1000 (1962).

16. H. Bittner, H. Goretzki, F. Benesovsky and H. Nowotny, Mh. Chem 94, 518 (1963).

17. J. Piper, Nucl. Met. AIME 10, 29 (1964).

18. E. Dempsey, Phil Mag, 8, 225 (1963).

19. R. Chang and L. J. Graham, J. Appl. Phys. 37, 3778 (1966).

20. L. Kaufman, Trans AIME 224, 1006 (1964).

21. L. Toth, J. Less-Common Metals, 13, 129 (1967).

22. V. Sadagopan and H. C. Gatos, J. Phys. Chem. Solids 27, 235 (1966).

23. D. Martin, Phys. Rev., 141, 576 (1966).

24. L. Toth and J. Zbasnik, to be published.

SESSION IV

Physical-Chemical Properties

Chairman: Michael Hoch
University of Cincinnati
Cincinnati, Ohio

Co-Chairman: C. T. Lynch
Air Force Materials Laboratory
Wright-Patterson Air Force Base, Ohio

THERMODYNAMIC PROPERTIES OF REFRACTORY TRANSITION METAL COMPOUNDS*

Larry Kaufman and Harold Bernstein

ManLabs, Inc.
Cambridge, Massachusetts

ABSTRACT

The current status of thermodynamic information concerning refractory transition metal compounds is reviewed in order to assess the availability of classical free energy and entropy data for stoichiometric compounds. This assessment is expanded to consider the broader problem of the interaction of thermodynamic information with phase equilibria and vaporization. Applications of models for defining the behavior of compound phases existing over a range of compositions to compute congruent vaporization and melting conditions are included. Moreover, application of computer techniques for predicting complete phase diagrams and thermodynamic properties of binary systems are discussed. Finally, the efficacy of the general approach in predicting ternary phase equilibria and guiding complex experimental investigation of ternary systems, and the application to three component composites are considered.

---

* This work has been supported in part by the Air Force Materials Laboratory under Contract AF 33(657)-9826 and Contract AF 33(615)-2352.

I. INTRODUCTION

Investigation of material systems capable of functioning for extended periods of time at temperatures between 2200° and 3500°K in our laboratory (1-3)** have clearly demonstrated the advantages afforded by complete thermodynamic descriptions. The wide range of environmental conditions over which refractory systems are required to perform has expanded the requisite description of compound properties. Thus, evaluation of the enthalpy and free energy of formation of a stoichiometric compound from absolute zero to its melting point constitutes a beginning, rather than an ending, for describing the high temperature stability of the compound phase (4-9). Since most of our current effort has been directed toward evaluation of transition metal boride and carbide base materials, the present review will begin by considering the availability of classical thermodynamic data for these systems (10-15). Subsequently, attention will be focused on current methods for predicting vaporization phenomena and phase equilibria in refractory transition metal and metal-compound systems (16-19). Application of computer techniques have permitted significant progress in these areas (20-23) which are concerned with the expanded description of thermodynamic stability. This extended description must of necessity deal with compound phases over a range of compositions, in addition to the stoichiometric combination. Moreover, prediction of complete binary and ternary phase diagrams requires information concerning the thermodynamic properties of phases which are normally unstable (19-22). As a consequence, progress in predicting vaporization phenomena and phase equilibria has been based on idealized models for characterizing the extended description in terms of a minimum number of parameters. Evaluation of these parameters from limited observations is then employed to make additional pre-

---

**Underscored numerals refer to references which are contained at the end of this paper.

dictions for comparison with experiment. In spite of the idealized models currently being employed (19-22), the efficacy of this approach in dealing with high temperature phenomena is quite apparent. As a result, it is quite evident that just as classical thermodynamic data (i.e., enthalpies and free energies of formation of stoichiometric compounds) undergo cycles of reassessment, expanded descriptions of compound and unstable phases will continue to be reconsidered and refined in the future. In both cases, the result will be more useful tools for understanding and predicting the behavior of materials at high temperature.

II. REVIEW OF THERMODYNAMIC INFORMATION ON TRANSITION METAL-CARBIDES AND BORIDES

Generation of new experimental data on the specific heats, enthalpies of formation and free energies of formation of the compounds in Group IV, V and VI. Transition Metal-Carbon systems has proceeded at a rapid pace in recent years. In addition, there have been three recent reviews (10-12) of the thermodynamic properties of compounds in the binary systems between carbon and Ti, Zr, Hf, V, Nb, Ta, Cr, Mo and W. The recent review of these systems by K. K. Kelley (supplement to Reference 10) post dates the recent reviews by Schick et al. (12) and Chang (11). Differences between Kelley's "selected values (10)" and those due to Chang (11) are minor. The major differences between these reviews is that the latter study (11) discusses the properties of nonstoichiometric metal carbide phases much more extensively than the reviews of Kelley (10) and Schick (12).

The present thermodynamic description of the transition metal diborides, TiB_2, ZrB_2, HfB_2, NbB_2 and TaB_2 is not as complete as the corresponding carbide systems. This is due to the fact that samples of these materials in quantities and purities suitable for calorimetric measurements have not been available until recently (1-3). The thermodynamic properties of the diborides have been reviewed (5, 6, 12). The values selected by Schick (12) were based in part on estimates of the entropy of the diboride phases at 298°K. Experimental values of the specific heat (5°K-300°K) and entropy at 298°K were employed in the earlier reviews (5, 6). As a consequence, the temperature dependent components of the free energy of formation presented earlier (5, 6) are more reliable than the results given by Schick et al. (12). Data on the enthalpy of formation of the diborides is still incomplete in spite of the fact that suitable samples for combustion and fluorine calorimetry were supplied to laboratories specializing in these measurements several years ago (1-3).

At present, values for the enthalpy of formation of ZrB_2 derived from vapor pressure and fluorine bomb calorimetry are in good agreement. Thus, recently reported measurements of vapor pressures of ZrB_2 (14) are in agreement with the values selected by the earlier reviews (5, 6, 12). Similarly, the results selected for TiB_2 are in good agreement (5, 6, 12). Recently, however, Huber (15) has reported an enthalpy of formation of -25.8 ± 0.3 kcal/g.at (-77.4 ± 0.9 kcal/mole)for TiB_2 obtained by combustion. This result is 3.5 to 4.5 kcal/g.at (or 10.5 to 13.5 kcal/mole) more negative than the selected values (5, 6, 12). The selected values for the free energy and enthalpy of formation of HfB_2 tabulated by Schick et al. (12) are based on an estimated entropy at 298°K which as mentioned earlier, is superseded by the experimental value (5, 6), as pointed out in the Supplementary Note (p. 1-245 of Reference 12) In addition, as indicated in the above mentioned Supplementary Note, Trulson and Goldstein (13) performed a mass spectrographic study of HfB_2 vaporization at temperatures between 2300°K and 2550°K. Coupling their results with free energy functions based on the estimated entropy of HfB_2 (12) yields an enthalpy formation of -20.3 kcal/g.at. Utilization of the observed entropy of HfB_2 (5, 6) corrects this value to -21.3 kcal/g.at. (-63.6 kcal/mole) for the enthalpy of formation of HfB_2 at 298°K. In contrast to this finding, the value tabulated by Schick et al. (12) (Table 91 p. 2-29) is -79.7 kcal/mole while the earlier review estimated -23.0 kcal/g.at (or -69.0 kcal/mole).

Estimates of the enthalpy of formation of NbB_2 and TaB_2 presented in the above mentioned reviews ((5), (6), (12)) differ by 10 to 20 kcal/mole. No calorimetric data are available. Measurements (1) of the vapor pressure of boron performed at 2150°K across the niobium diboride phase field between $NbB_{1.9}$ and $NbB_{2.1}$ and in the $Nb_3B_4 + NbB_{1.9}$ two phase field are in excellent agreement with the values estimated earlier (5, 6). The measured boron activity of 10^{-2} in the $Nb_3B_4 + NbB_{1.9}$ field requires that the free energy of formation of NbB_2 be substantially more negative than -40.0 kcal/mole at 2150°K in contrast to the value tabulated by Schick et al. (12).

In summary then, completely satisfactory thermodynamic data on the diborides is available only for ZrB_2 where specific heat, vapor pressure and fluorine bomb calorimetric measurements of the enthalpy of formation yield consistent results. The situation is less satisfactory for TiB_2 and HfB_2 where fluorine bomb measurements are not available (in spite of the fact that samples were submitted for calorimetry (1)), and where recent combustion results (15) are not consistent with vapor pressure and specific heat measurements. However, in the case of TiB_2 and HfB_2, the current description of thermodynamic properties (5, 6) is quite effective in predicting vapor pressure and vaporization phenomena. A similar situation applied for NbB_2 and TaB_2 (zone refined samples were submitted for calorimetry (1)) where no calorimetric measurements of the enthalpy of formation are available. However, the current description of the thermodynamic properties of NbB_2 predicts the vapor pressure of boron across the ($Nb_3B_4 + NbB_{1.9}$), ($NbB_{1.9}$ to $NbB_{2.1}$) and ($N_bB_{2.1}$ + B) ranges quite satisfactorily at 2150°K.

III. APPLICATION OF THE EXTENDED THERMODYNAMIC DESCRIPTION OF COMPOUND PHASES TO PREDICTION OF VAPORIZATION PROCESSES

The Schottky-Wagner model of compound phases has been applied to describe the compositional dependence of the free energy of the sodium chloride type of transition metal carbide phase (i.e., TiC, ZrC, HfC, NbC and TaC) as well as the corresponding diboride compounds (4-7). The simplest variant of the model was employed, restricting consideration to vacancies on either sublattice. This restriction permitted explicit description of the free energy-composition curves for the relevant compounds in terms of the free energy and enthalpy of formation at stoichiometry. The description proved effective in yielding accurate estimates of phase equilibria, and vaporization phenomena for refractory transition metal monocarbide and diboride compound phases. The formulation was extended to deal with ternary metal-metal-carbides to correctly predict a minimum vaporization rate for $(Ta_{0.41}Hf_{0.10})C_{0.49}$ alloys at 2600°K (9). A compound which vaporizes noncongruently (i.e., where the rate of vaporization of one component exceeds that of the second across the entire single phase field) such as TaC (7), TaB_2 or NbB_2 (5, 6), may exhibit vaporization kinetics which are time dependent.

The time dependence of the vaporization rate for compounds which vaporize congruently and those which vaporize noncongruently have been considered in detail (16, 21). The analysis considers a compound phase $A_{1-x}B_x$ stable in the neighborhood of a stoichiometric composition x_o and obeying the restricted Schottky-Wagner thermodynamic model. The assumptions are made that under Langmuir vaporization conditions: (a) only atomic vaporization is allowed, (b) the atomic vaporization rates obey the Langmuir rate equation, (c) accommodation coefficients are unity, (d) the specimen surface area remains constant, (e) quasi-static equilibrium between solid and gaseous phases holds and (f) mass transport in the solid is not diffusion limited. The atomic mass vaporization rates may now be written:

$$\dot{W}_A[x,T] = -A_i\left(\frac{M_A}{2\pi RT}\right)^{1/2} P_A[x,T] = -A_i\left(\frac{M_A}{2\pi RT}\right)^{1/2} P_A^o[T]\exp\left(\frac{\overline{F}_A - F_A^o}{RT}\right) \quad (1)$$

$$\dot{W}_B[x,T] = -A_i\left(\frac{M_B}{2\pi RT}\right)^{1/2} P_B[x,T] = -A_i\left(\frac{M_B}{2\pi RT}\right)^{1/2} P_B^o[T]\exp\left(\frac{\overline{F}_B - F_B^o}{RT}\right) \quad (2)$$

Here A_i is the area of the vaporizing solid, M_A and M_B are the molecular weights of elements A and B, P_A^o and P_B^o are the vapor pressures of the pure elements at the absolute temperature T, $\bar{F}_A[x, T]$ and $\bar{F}_B[x, T]$ are the partial free energies of A and B in the compound phase $A_{1-x}B_x$, while $F_A^o[T]$ and $F_B^o[T]$ are the free energies of pure A and B in their stable state. The Schottky-Wagner model for nonstoichiometric compound phases yields the following expressions for the partial free energies (4).

$$\bar{F}_A[x,T] = F_A^o[T] - F_{A+} + RT\, ln\left(\frac{1-x}{(1-x_o)y-(1-x)}\right) \tag{3}$$

$$\bar{F}_B[x,T] = F_B^o[T] - F_{B+} + RT ln\left(\frac{x}{x_o y-x}\right) \tag{4}$$

$$\alpha y = (x_o y - x)^{x_o}((1-x_o)y-(1-x))^{1-x_o} \tag{5}$$

$$RT ln\left(\frac{\beta-\alpha}{\alpha}\right) = x_o F_{B+} + (1-x_o)F_{A+} + \Delta F_o \tag{6}$$

$$\beta = x_o^{x_o}(1-x_o)^{1-x_o} \tag{7}$$

In these equations x_o is the stoichiometric composition of the compound, F_{A+} and F_{B+} are the free energies of formation of A vacancies and B vacancies in the lattice, $\Delta F_o[T]$ is the free energy of formation of the stoichiometric compound, and the function $y[x,T]$ derived from the transcendental equation (5) represents the ratio of the total number of lattice sites to the total number of atoms. α is identified as the fraction of empty sites at stoichiometry and is generally small for the refractory metal compounds of interest.

Now consider a specimen of total weight $W = W_A + W_B$ where W_A and W_B are the instantaneous weights of elements A and B, related through the mole ratio $z = \frac{x}{1-x}$ by:

$$W_B = \gamma z W_A = \frac{\gamma z W}{1 + \gamma z} \tag{8}$$

$$\gamma = \frac{M_B}{M_A} \tag{9}$$

Taking the time derivatives of (8) and eliminating the variables W_A and W_B in terms of $\dot{W}_A$ and $\dot{W}_B$ leads to:

$$\gamma \dot{W}_A = \left(\frac{d}{dz} - \frac{\ddot{z}}{\dot{z}} \right) (\dot{W}_B - \gamma z \dot{W}_A) \tag{10}$$

Upon substituting the expressions (1), (2), (3) and (4) into (10), the following differential equation for $z[t]$ results:

$$\ddot{z} + \left[\frac{\omega f_A}{u} - \frac{d\ln u}{dz} \right] (\dot{z})^2 = 0 \tag{11}$$

where

$$f_A = \left[\frac{1+z}{1+z_o} y - 1 \right]^{-1} \tag{12}$$

$$f_B = \left[\frac{1+z}{1+z_o} \frac{z_o}{z} y - 1 \right]^{-1} \tag{13}$$

$$u = \omega z f_A - f_B \tag{14}$$

$$\omega = \gamma^{1/2} \frac{P_A^o}{P_B^o} \exp\left(\frac{F_{B+} - F_{A+}}{RT} \right) \tag{15}$$

The functions f_A and f_B represent the compositional dependence of the chemical activities of A and B. Equation (11) may be integrated directly by some appropriate numerical technique but a formal solution to (11) may be written in the form of an integral equation (16):

$$\int_{z_i}^{z} e^{-g[z, z_i]} \frac{dz}{u[z]} = \frac{t}{\tau} \tag{16}$$

where

$$g[z, z_i] = \omega \int_{z_i}^{z} \frac{f_A}{u} dz \tag{17}$$

$$\tau = \frac{\gamma}{1+\gamma z_i} \frac{W_i}{AP_B^o} \left(\frac{2\pi RT}{M_B} \right)^{1/2} \exp\left[\frac{F_{B+}}{RT} \right] \tag{18}$$

Here, z_i and W_i are the initial composition and total weight at $t = 0$. The instantaneous weight W, vaporization rate $\dot{W}$, and compositional derivative $\dot{z}$ after a vaporization time t are:

$$W = W_i \left(\frac{1 + \gamma z}{1 + \gamma z_i} \right) e^{-g[z,z_i]} \tag{19}$$

$$\dot{W} = \frac{W_i}{\tau(1 + \gamma z_i)} (\omega f_A + \gamma f_B) \tag{20}$$

$$\dot{z} = \frac{u[z]}{\tau} e^{g[z,z_i]} \tag{21}$$

In these equations W, $\dot{W}$ and $\dot{z}$ are implicit functions of time through the variation z[t] found from (16).

Congruent vaporization occurs when the molar rates of vaporization of A and B are in proportion to their respective instantaneous compositions, i.e., at some temperature T it is required that:

$$\frac{1}{\gamma} \left. \frac{\dot{W}_A}{\dot{W}_B} \right|_{z = z_c} = \left. \frac{f_B}{\omega f_A^s} \right|_{z = z_c} = z_c \tag{22}$$

Solving (22) with the aid of (5), (12) and (13) gives the congruent composition $z_c[T]$ as:

$$z_c = \frac{z_o - \alpha (1 + z_o) \omega^{-\frac{(1-x_o)}{x_o}}}{1 - \alpha (1 + z_o) \omega^{-\frac{(1-x_o)}{x_o}}} \tag{23}$$

z_c coincides with the stoichiometric composition when T is such that $\omega z_o = 1$. It follows from (22) that $u[z_c] = 0$ and therefore $\dot{z}[z \to z_c] \to 0$. Consequently, if a congruent composition exists within the range over which the compound phase is stable, the composition will change uniformly with time from its initial value z_i to the terminal composition z_c. The vaporization rate also approaches a terminal value of:

$$\dot{W}_c = \frac{W_i}{\tau} \left(\frac{1 + \gamma z_c}{1 + \gamma z_i} \right) \left(\frac{1 - z_o \omega}{z_o - z_c} \right) \tag{24}$$

A minimum in the vaporization rate is predicted at the composition:

$$z_m = z_o - \frac{(\gamma - z_o\omega)(z_o - z_c)}{\gamma^{x_o}(1 - z_o\omega)} \tag{25}$$

with the value:

$$\dot{W}_m = \frac{W_i}{\mathcal{T}} \left(\frac{\gamma - z_o\omega}{1 + \gamma z_i}\right) \left(\frac{1 + z_m}{z_o - z_m}\right) \tag{26}$$

It can be shown that $0 < z_c < z_m < z_o$ and $\dot{W}_m$ is a true minimum when $\omega < \gamma / z_o$. The integral equation (16) must be solved numerically for general z_o, z_i and T because of the transcendental character of the function $y[z,T]$ defined by Equation (5). For the special case of $z_o = 1$ (e.g., ZrC) however, some analytical results are possible.

In particular when only the nonmetal element B vaporizes, $\omega = 0$, and if $\alpha << 1$ then the integral equation (16) reduces to:

$$ln\left(\frac{z_i}{z}\right) - (z_i - z) = \frac{t}{\mathcal{T}} \tag{27}$$

This situation describes a noncongruently vaporizing system like TaC above temperatures of about 2000°K where the vapor pressure of pure graphite is substantially greater than that of pure Ta. An interesting feature of Equation(27) is that when $z_i = 1$ and $t << \mathcal{T}$, then $\dot{W} \sim t^{-1/2}$. The square root of time dependence suggests a solid phase diffusion limited transport mechanism of mass loss, whereas, in fact, no such mechanisms have been assumed in this analysis. Interpretation of mass loss data in Langmuir vaporization experiments could therefore be erroneously attributed to diffusion processes (9).

Hoch, et al., (24) have investigated the heat of sublimation of carbon in the temperature range 2170° to 2770°K by measuring the partial pressure of C above graphite, TaC and WC. They conclude that only carbon is lost from TaC and that no congruency phenomena occur in that range of temperature, in agreement with the present analysis. In particular at 2500°K, Hoch finds that a TaC specimen of starting composition z_i=0.952 and effective area A_i = 11.4 cm^2

experiences a weight loss $\Delta W = 1.7 \times 10^{-3}$ gms after $t = 6.013 \times 10^{3}$ seconds. Using $2\Delta H_o^{\sigma}[0^oK] = -35.6$ cal/mole, for TaC (10, 25), the expression $F_{C+} = -2\Delta H_o^{\sigma} - RT\, ln$ (refer to Section IV) and a graphite pressure of $P_c^o = 1.77 \times 10^{-7}$ atmospheres at 2500°K (26) then z_c is calculated to be .245 (or $x_c = .197$). This composition is well outside the TaC : TaC + Ta_2C boundary at $z \simeq 0.725$. Since the vaporization time is short in comparison to τ and the total change $z_i - z$ is small, Equation (27) may be expanded and combined with (19) to give:

$$\Delta W \simeq \frac{z_i}{1 - z_i} A_i P_c^o \exp\left[-\frac{F_{C+}}{RT}\right]\left(\frac{M_C}{2\pi RT}\right)^{1/2} t \tag{28}$$

ΔW as calculated from Equation (28) is 1.16×10^{-3} gms, which compares favorably with the measured value of 1.7×10^{-3} gms, considering the possible error in the thermodynamic parameter F_{C+}.

The time required for the TaC phase to become unstable after sufficient loss of carbon may be computed from Equation (27) if z is set equal to the left hand boundary of the TaC phase. Assuming $W_i = 1.0$ gms, $A_i = 1.0\ cm^2$ and $t_i = 1.00$, then at a temperature of 2400°C the time required to reach the TaC : TaC + Ta_2C boundary at $z = 0.705$ is 148 hours with an associated weight loss of 0.0181 gms. At 2900°C, however, this time is rapidly reduced to 25.2 minutes with an associated weight loss of 0.0204 gms. At 2900°C, the boundary has shifted to about $z = 0.667$. At higher temperatures where the TaC phase is in equilibrium with the melt, the present model predicts that the compound will melt almost instantaneously.

The analysis is now applied to the compound phase NbC at 2940°K to illustrate congruent vaporization behavior. Fries (27) investigated the vaporization of NbC at this temperature with a specimen having $W_i = 6.69$ gms, $A_i = 7.01\ cm^2$, $z_i = 0.970$. A congruency mole ratio of 0.785 is calculated from Equation (23) using $\log_{10} P_{Nb}^o = -5.385$, $\log_{10} P_C^o = -4.531$ and $2\Delta H^{\sigma}[0^oK] = -33.6$ kcal/mole (10). Fries's experimentally determined result is $z_c = 0.747 \pm 0.007$, τ from Equation (18) is found to be 137.7 hours and z_m from (25) is 0.923. From these parameters the curves of Figures 1, 2, 3 and 4 have been calculated and com-

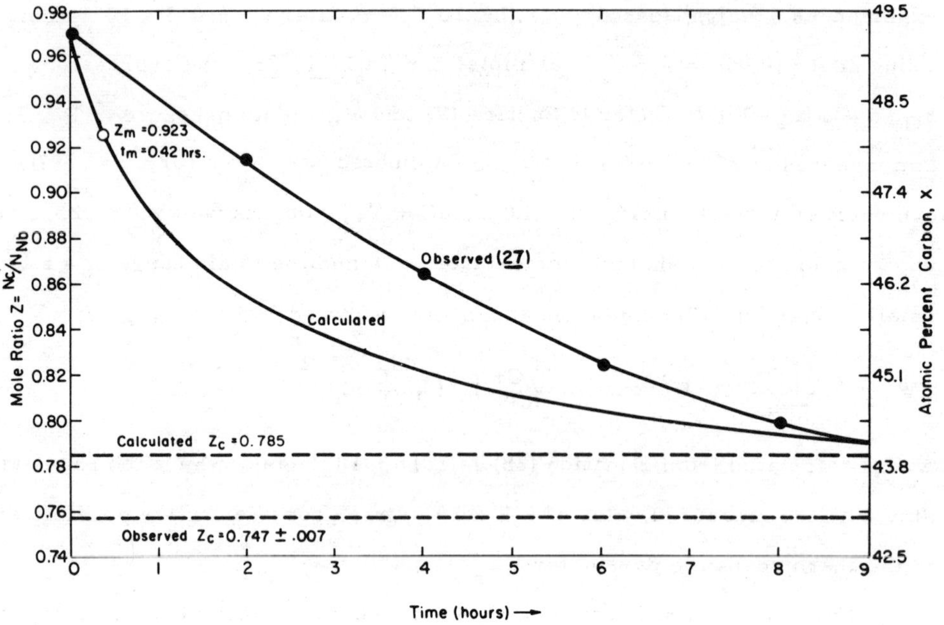

Fig. 1. Comparison of calculated and observed composition of niobium carbide as a function of time at 2940°K.

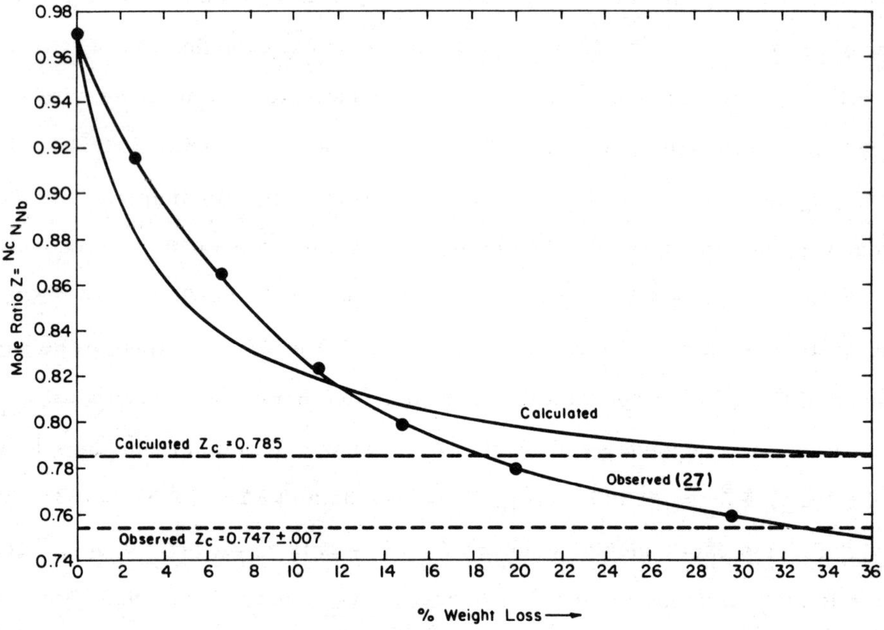

Fig. 2. Comparison of calculated and observed per cent weight loss for niobium carbide as a function of mole ratio at 2940°K.

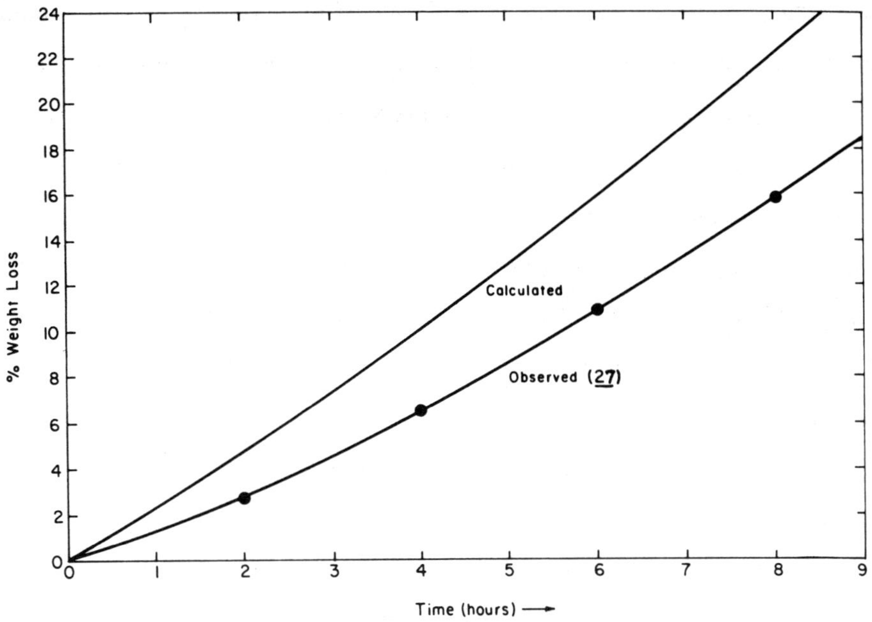

Fig. 3. Comparison of calculated and observed weight loss for niobium carbide as a function of time at 2940°K.

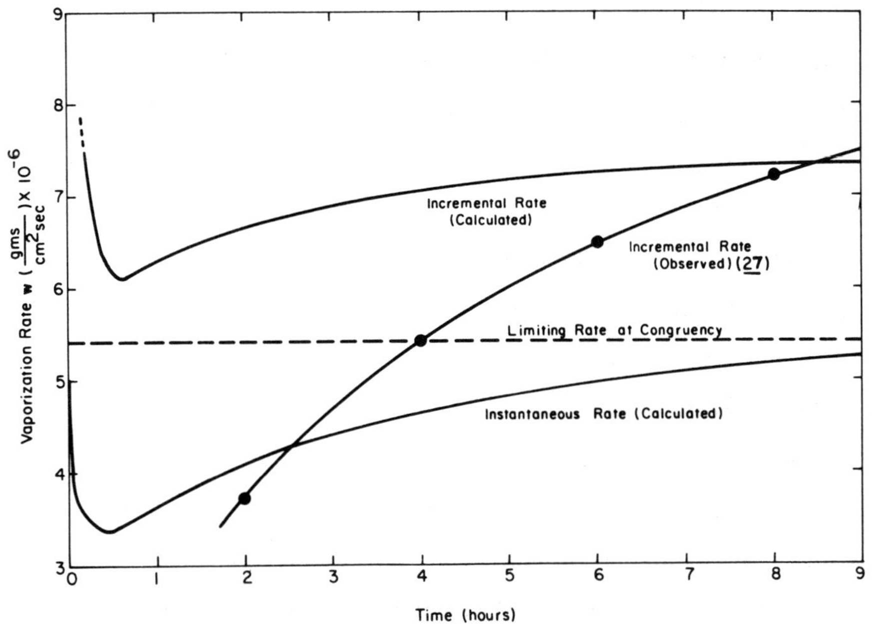

Fig. 4. Comparison of calculated and observed vaporization rates for niobium carbide as a function of time at 2940°K.

pared with experimental results. It is seen that the calculated initial composition change decreases more rapidly with both time and % weight loss than does the observed compositon and that the calculated % weight loss increases more rapidly than the observed % weight loss with time. In Figure 4, the experimental, incremental vaporization rate curve is bracketed by the calculated incremental and instantaneous vaporization rate curves, the incremental rate always being larger than the instantaneous rate. The minimum in the calculated curves of Figure 4 occurs at $t = 0.42$ hours; since no experimental points were reported within the first two hours, it is not known whether a dip in the experimental curve exists.

The differences betweeen the theoretical and observed time dependence can be accounted for, in part, by noting that the surface area of the experimental specimen decreased by 13% after 8 hours of vaporization at 2940°K and that diffusion gradients of carbon were observed. Both of these effects would tend to bring the calculated and observed results into closer agreement, in as much as constant surface area and essentially infinite diffusion parameters were assume in the analysis. The over-all agreement is believed to be sufficiently good to demonstrate the predictive quality of the analytical model.

IV. PREDICTION OF CONGRUENT MELTING COMPOSITIONS

The melting point maxima and in some cases, the entire compound phase field of many congruently melting refractory transition metal compounds are known experimentally to occur on the metal rich side of the stoichiometric composition. By employing a Schottky-Wagner model of the compound phase and a regular solution model of the liquid phase, an analytical description of the temperature-composition-stability features of refractory metal-nonmetal systems has been possible. Figure 5 illustrates the formation of the melting region of a compound phase σ as the temperature is lowered through the melting point (x_m, T_m). Below the melting temperature, the requirement of minimum free energy is equivalent to the familiar common tangent construction between the free energy curves

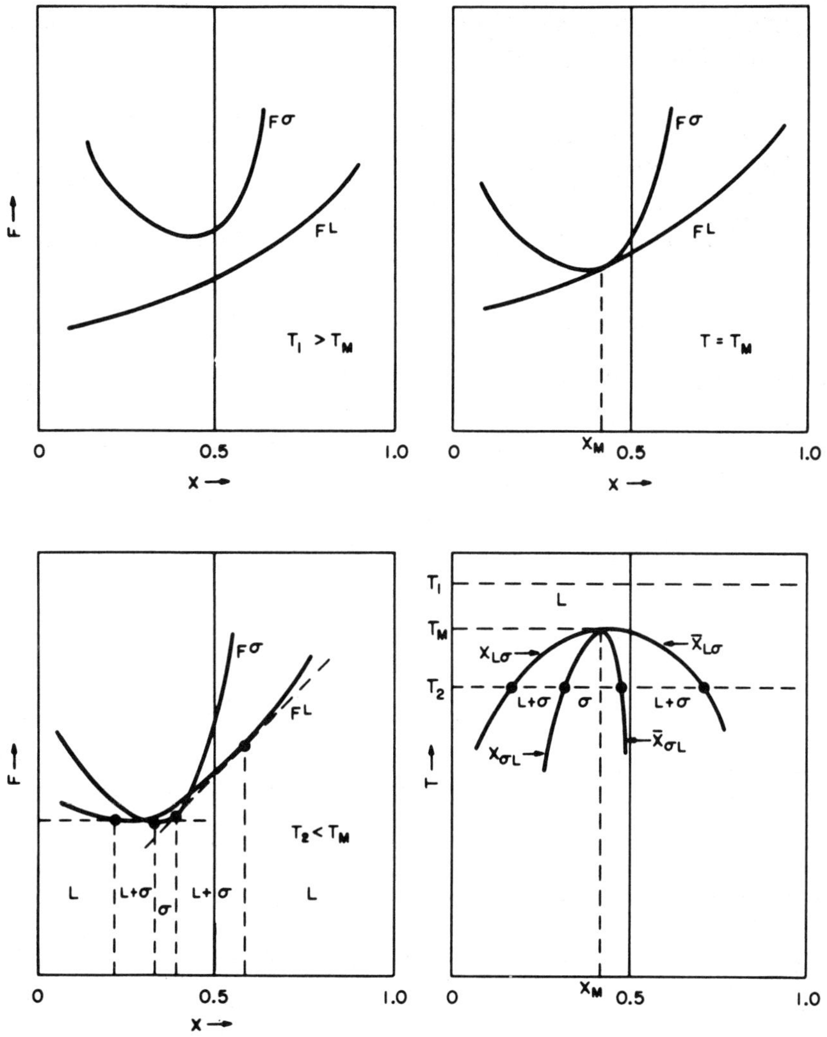

Fig. 5. Illustrating the formation of the melting point maximum of the σ phase.

$F^{\sigma}[x, T]$ and $F^{L}[x, T]$. Equilibration of the partial free energies of the elements A and B across the σ + L field may be stated:

$$\left(F^{\sigma} - x \frac{\partial F^{\sigma}}{\partial x}\right)_{x_{\sigma L}} = \left(F^{L} - x \frac{\partial F^{L}}{\partial x}\right)_{x_{L\sigma}} \tag{29}$$

$$\left(F^{\sigma} + (1-x) \frac{\partial F^{\sigma}}{2x}\right)_{x_{\sigma L}} = \left(F^{L} + (1-x) \frac{\partial F^{L}}{\partial x}\right)_{x_{L\sigma}} \tag{30}$$

The regular solution model of the liquid phase gives an integral free energy expression:

$$F^{L}[x, T] = F_A^L (1-x) + F_B^L x + RT\,[x\,ln\,x + (1-x)\,ln\,(1-x)] + Lx\,(1-x) \tag{31}$$

where the interaction parameter L is a measure of the average interaction between A-A, B-B and A-B atom pairs. The Schottky-Wagner model of a compound phase of stoichiometric composition x_o yields:

$$F^{\sigma}[x, T] = (F_A^{\sigma} - F_{A+})(1-x) + (F_B^{\sigma} - F_{B+})x + RT\left[x\,ln\left(\frac{x}{x_o y - x}\right) + (1-x)ln\left(\frac{1-x}{(1-x_o)y-(1-x)}\right)\right] \tag{32}$$

The function $y[x, \alpha]$ has been defined in equation (5). Performing the operations indicated in equations (29) and (30) on the free energy functions (31) and (32) leads to the following equilibria equations for $x_{L\sigma}[T]$ and $x_{\sigma L}[T]$

$$(1 + U)^{1-x_o} (1 + V)^{x_o} = 1 + \exp\left[\frac{F_{A+}(1-x_o)+F_{B+}x_o+\Delta F_o^{\sigma}}{RT}\right] \tag{33}$$

where

$$U[x_{L\sigma}, T] = (1 - x_{L\sigma}) \exp\left[\frac{Lx_{L\sigma}^2 + \Delta F_A^{o\rightarrow L} + F_{A+}}{RT}\right] \tag{34}$$

$$V[x_{L\sigma}, T] = x_{L\sigma} \exp\left[\frac{L(1-x_{L\sigma})^2 + \Delta F_B^{\sigma\rightarrow L} + F_{B+}}{RT}\right] \tag{35}$$

$$x_{\sigma L} = \left(\frac{1-x_o}{x_o}\ \frac{1+V}{1+U}\ \frac{U}{V} + 1\right)^{-1} \tag{36}$$

$\Delta F_A^{o\to L}$ and $\Delta F_B^{o\to L}$ are the difference in free energy of pure A and B between the liquid state and the stable state at temperature T. Equation (33) is solved by numerical techniques to give two solutions for x_L at any $T < T_m$. Then (36) is employed to calculate the corresponding values of x_σ. In the limit that $x_{\sigma L}[T] = x_o$ or that the σ phase shrinks to a line compound, the liquid phase limit curve $T[x_{L\sigma}]$ is computed from:

$$\frac{T}{T_m} = \frac{(ln\,\beta - \frac{\Delta S}{R})\,[x_o\,(1-2x_{L\sigma}) + x_{L\sigma}^2] - \frac{\Delta H}{RT_m}\,(x_o - x_{L\sigma})^2}{x_o(1-x_o)\,[ln\, x_{L\sigma}^{x_o}\,(1-x_{L\sigma})^{1-x_o} - \frac{\Delta S}{R}]} \tag{37}$$

where

$$\Delta F_A^{o\to L} = \Delta S_A^{o\to L}\,(\bar{T}_A - T) \tag{38}$$

$$\Delta F_B^{o\to L} = \Delta S_B^{o\to L}\,(\bar{T}_B - T) \tag{39}$$

$$\Delta F_o^{\sigma} = -\Delta H_o^{\sigma} + T\,\Delta S_o^{\sigma} \tag{40}$$

$$\Delta H = x_o\,\Delta S_B^{o\to L}\,\bar{T}_A + (1-x_o)\,\Delta S_A^{o\to L}\,\bar{T}_B + \Delta H_o^{\sigma} \tag{41}$$

$$\Delta S = x_o\,\Delta S_B^{o\to L} + (1-x_o)\,\Delta S_A^{o\to L} + \Delta S_o^{\sigma} \tag{42}$$

Here $\bar{T}_A$ and $\bar{T}_B$ are the melting temperatures of pure A and B, $\Delta S_A^{o\to L}$ and $\Delta S_B^{o\to L}$ are their entropies of fusion per gram atom. ΔH_o^{σ} and ΔS_o^{σ} are the magnitudes of the enthalpy and entropy of formation per gram atom of the stoichiometric σ compound.

The composition of the melting point maximum is found from Equations (33) and (36) by setting $x_{L\sigma} = x_{\sigma L} = x_m$ at $T = T_m$. The resulting equation for x_m as a function of T_m is:

$$RT_m\,ln(\beta - \alpha)\,y_m - \left(\frac{x_o(1-2x_m)+x_m^2}{(1-x_m)^2}\right)(RT_m\,ln\,(x_o y_m - x_m) + \Delta F_B^{o\to L} + F_{B+})$$

$$+ x_o \Delta F_B^{o\to L} + (1-x_o)\,\Delta F_A^{o\to L} - \Delta F_o^{\sigma} = 0 \tag{43}$$

In general $x_m \neq x_o$ unless the following condition is obeyed:

$$x_o(1-x_o)(F_{A+}-F_{B+})-x_o(1-2x_o)\Delta F_o^\sigma+(1-x_o)^2\Delta F_A^{o\to L}-x_o^2\Delta F_B^{o\to L}$$

$$+RT_m[(1-x_o)^2 \ln(1-x_o)-x_o^2 \ln x_o] = 0 \tag{44}$$

For the case $x_o = 1/2$ condition (44) reduces to $F_{A+} \simeq F_{B+}$, which describes a symmetrical F^σ vs. x curve. Because of the large difference in atom size for compounds like ZrC, the energy to form Zr vacancies is several times larger than the energy to form C vacancies. Thus $F_{A+} > F_{B+}$ and both the minimum of the F^σ vs. x curve and the composition x_m shift to the left of $x_o = .50$ as indicated in Figure 5. Once x_m is determined from Equation (43) with a knowledge of T_m and the free energy parameters, the interaction parameters L is computed by:

$$L = \frac{\Delta F_o^\sigma - (x_o\Delta F_B^{o\to L} + (1-x_o)\Delta F_A^{o\to L} + RT_m \ln(\beta-\alpha)\, y_m)}{x_o(1-2x_m)+x_m^2} \tag{45}$$

L is necessarily large and negative for refractory compound phases with large heats of formation.

The energy parameters F_{A+} and F_{B+} could be evaluated if the vapor pressures of the elements were known experimentally at specific compositions, or if congruent vaporization data were available. In lieu of such data for the carbides ZrC, HfC, TiC, NbC and TaC the assumptions are made that(4):

$$RT \ln \alpha \simeq -\left|2\,\Delta H_o^\sigma\right| \tag{46}$$

$$F_{C+} \simeq -RT \ln 2\alpha = \left|2\,\Delta H_o^\sigma\right| - RT \ln 2 \tag{47}$$

The latter assumption reflects the fact that the phase boundary $x_{\sigma\gamma}$ of the carbide phase in equilibrium with graphite at low temperatures lies very close to the stoichiometric composition 0.50. If F_{C+} is in fact smaller in magnitude than the value assumed in Equation (47), the boundary $x_{\sigma\gamma}$ shifts to the left of the stoichiometric composition, as observed.

Table 1 summarizes the comparison of computed (Equation 43) and observed melting point maxima for the carbides. The free energy of formation data and experimental compositions are taken from the work of Chang (11). The agreement is good for HfC, NbC and TaC, but TiC and ZrC show lower values of x_m than are calculated by about 10%.

Equations (33) and (36) have been used to calculate the liquid-solid equilibria for TiB_2, ZrB_2, HfB_2, NbB_2 and TaB_2 phases as well as the carbide phases (21). More detailed computations of the phase diagrams of TaC and NbC appear in (17).

V. THERMODYNAMIC PROPERTIES OF UNSTABLE PHASES

Consideration of the free energy of compound phases over a wide range of composition is only one aspect of the extended description required for predicting high temperature behavior. In addition, it has been recognized that the properties of unstable phases play an important role in controlling phase equilibria (28). It is only in recent years, that quantitative estimates and descriptions of the relative stability of phases which do not form at one atmosphere have been made (29 - 31). The utilization of very high pressures, rapid cooling from the melt, and epitaxial vapor deposition methods have provided experimental means for forming

TABLE 1

COMPARISON OF CALCULATED AND OBSERVED CONGRUENT MELTING COMPOSITIONS OF REFRACTORY CARBIDE COMPOUNDS

Carbide	$2\Delta F_o^\sigma$ (cal/mole)	T_m (°K)	L(kcal/gm atom)	(calc) x_m	(obs) x_m
TiC	-48, 490+5.77T	3340	-47.38	0.494	0.44
ZrC	-51, 220+4.34T	3710	-53.37	0.490	0.45
HfC	-82,330+11.11T	4200	-48.63	0.492	0.485
NbC	-38,380+2.65T	3770	-39.42	0.477	0.46
TaC	-64,490-1.795TlogT +8.79T	4260	-43.72	0.475	0.47

these unstable or metastable structures so that some of their properties could be observed. Hexagonal close packed iron formed at high pressure (30), sodium chloride (B1) type WC and MoC formed by splat cooling (32), and fcc vanadium (33), molybdenum and tantalum (34) formed by epitaxial vapor deposition and fcc tungsten (35), tantalum, molybdenum, rhenium, hafnium and zirconium formed by sputtering (36) are recent examples.

Rudy and co-workers have employed observed phase relations in ternary transition metal carbide systems to compute the thermodynamic properties of unstable phases (19, 20, 37, 38). The results are then employed to expand and refine the experimentally observed phase relations. Examples of these findings include the free energy difference between the σ, sodium chloride (B1), and η, hexagonal (WC), structure for tantalum carbide (where the former structure is stable) and for tungsten and molybdenum carbide where the latter structures are stable. Equations 48-50 are typical results (18, 20)

$$\Delta F_{TaC}^{\sigma\rightarrow\eta} = +6000 \text{ cal/g. at.} \tag{48}$$

$$\Delta F_{WC}^{\sigma\rightarrow\eta} = -2600 + 0.7\,T \text{ cal/g. at.} \tag{49}$$

and

$$\Delta F_{Mo_2C_3}^{\sigma\rightarrow\eta} = -1500 + 0.9T \text{ cal/g. at.} \tag{50}$$

In addition, the relative stability of Zr_2C has been evaluated (37) by determining the free energy of disproportionation of this phase into the monocarbide and metal solid solution to be -3,300 cal/g. at. of Zr. Similar techniques are being applied to evaluate the stability of Ti_2C and Hf_2C (37, 38) which are also unstable relative to the monocarbide plus metal. In dealing with the thermodynamic properties of the Ta-W-C ternary diagram, Rudy (19, 20) has described the metal sublattice of the monocarbide phase as a regular solution of Ta and W atoms with an interaction parameter of +6500 cal/g. at. Thus by evaluating experimentally established phase relations in ternary systems with regards to the stability of the participating phases; data has been extracted describing the unstable phases. Investigation of a sufficiently large number of systems can, in principle, "provide a reasonably complete mapping of the thermodynamic characteristics of all crystal types which are of relevance for a particular group of systems." (page 45 of Reference 19).

Along similar lines, (21, 23, 31) the relative stability of the bcc, β, hcp, ϵ, and fcc, α, forms of the transition metals has been estimated as shown in Figures 6 and 7 and employed in performing ideal (31) and regular (21-23) solution calculations of binary phase diagrams between transition metals. Computer techniques have been developed for the regular solution phase diagram computations (21, 22)

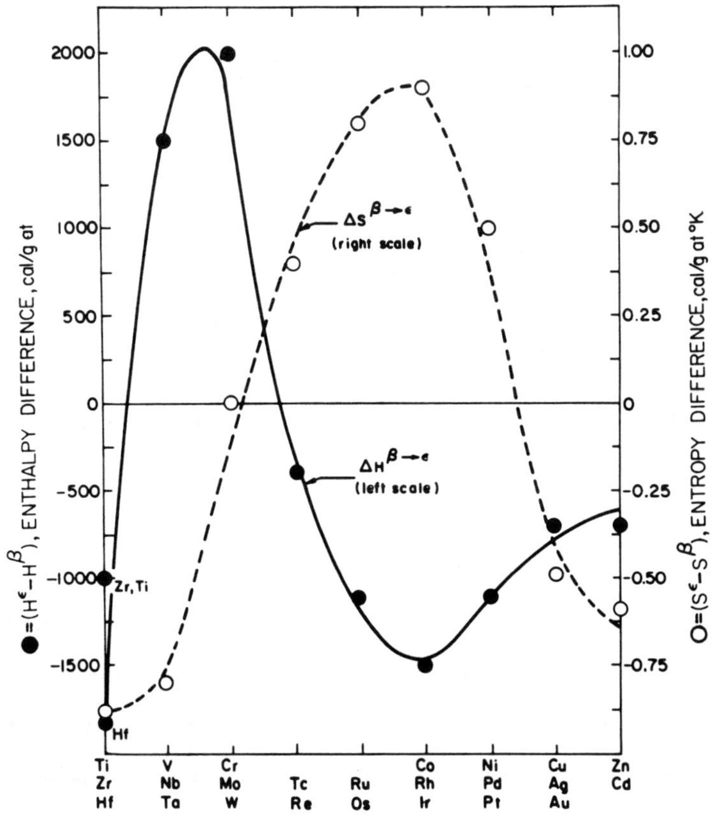

Fig. 6. Enthalpy and entropy differences between the h.c.p. (ϵ) and b.c.c. (β) forms of the transition metals (31).

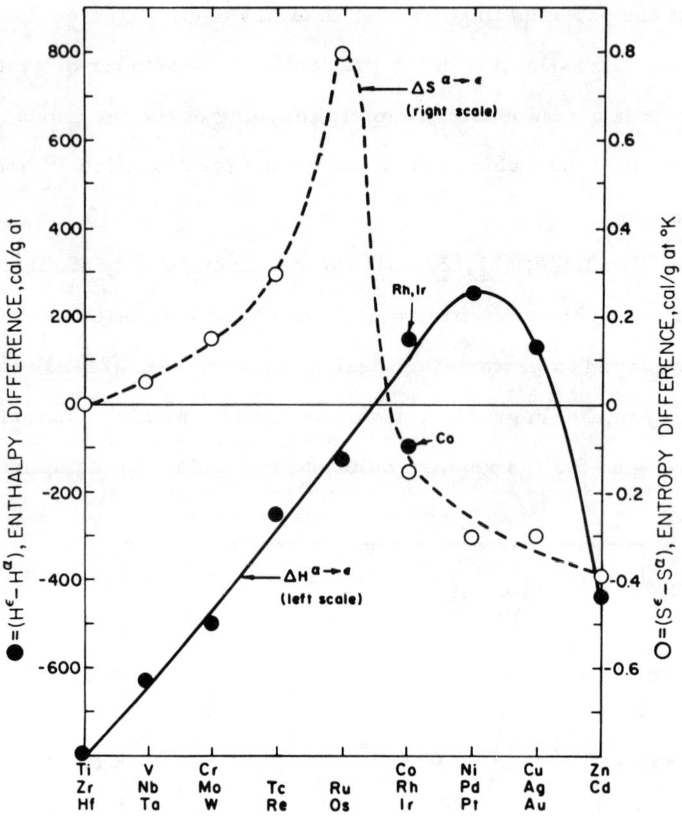

Fig. 7. Enthalpy and entropy differences between the h.c.p. (ε) and (α) forms of the transition metals (31).

and have recently been extended to calculate ternary phase equilibria (23). The calculations have been also extended to consider Laves type, λ, and $AuCu_3$ type, ξ, compound phases which form in transition metal alloy systems. Phase competition between the bcc, hcp, fcc, liquid, Laves and $AuCu_3$ type phases for each of the seventy-two binary phase diagrams between second and third row transition element is examined. The program chooses the most stable phases and then computes the equilibria for temperatures up to 4000°K. The same thermodynamic description predicts vaporization characteristics for the above mentioned phases. Although the present description is imperfect due to the simple models employed for describing the competing phases, comparison of the seventy-two computed phase diagrams with sixty-one cases where observed diagrams are available yields relatively good agreement (22). Figure 8 shows the computed Zr-Nb, Zr-Ta, Hf-Nb and Hf-Ta regular solution phase diagrams (heavy lines) along with the observed diagrams (thin lines). Figure 9 shows the computed Zr-Mo, Zr-W, Hf-Mo and Hf-W regular solution phase diagrams (with Laves phase, λ, intrusion) compared with the observed diagrams. The computed vapor pressure-composition relations for the Hf-Mo system at 2000°K indicating congruent vaporization in the $HfMo_2$-λ phase is shown in Figure 10. While Figure 11 shows a computed regular solution ternary section in the Zr-Ta-W system at 2600°K.

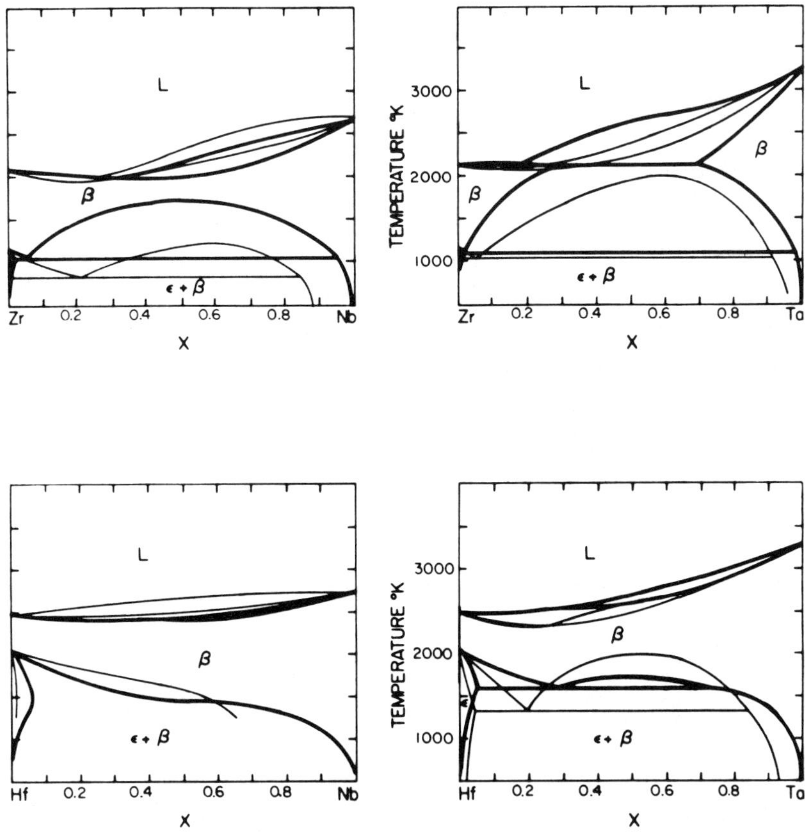

Fig. 8. Comparison of observed (thin lines) and regular solution computed (thick lines) phase diagrams (22).

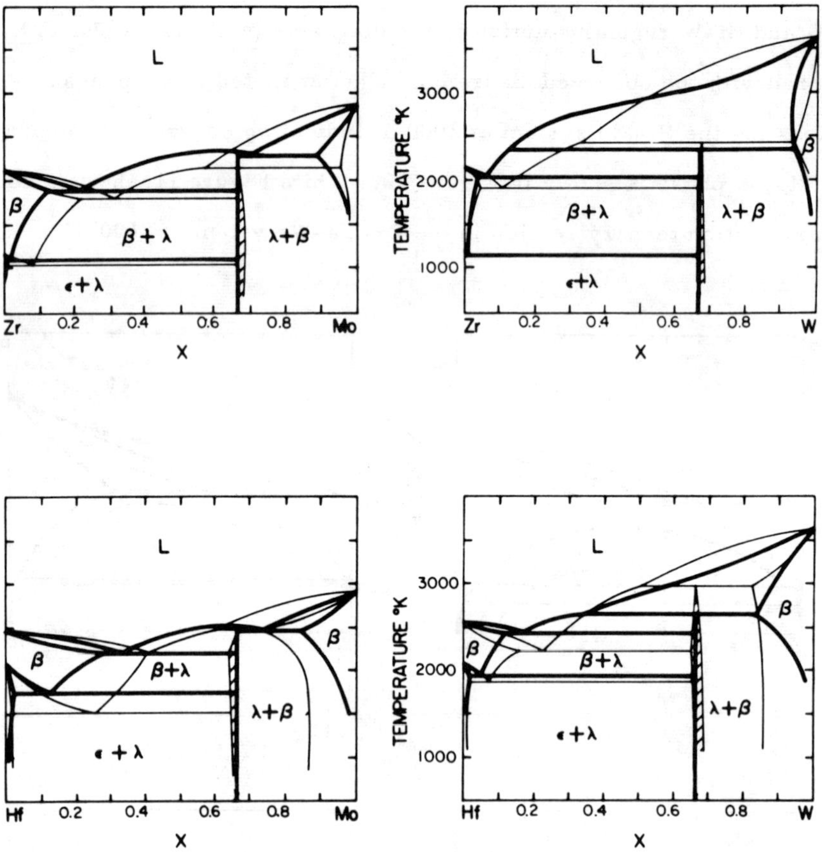

Fig. 9. Comparison of observed (thin lines) and regular solution computed (thick lines) phase diagrams (22).

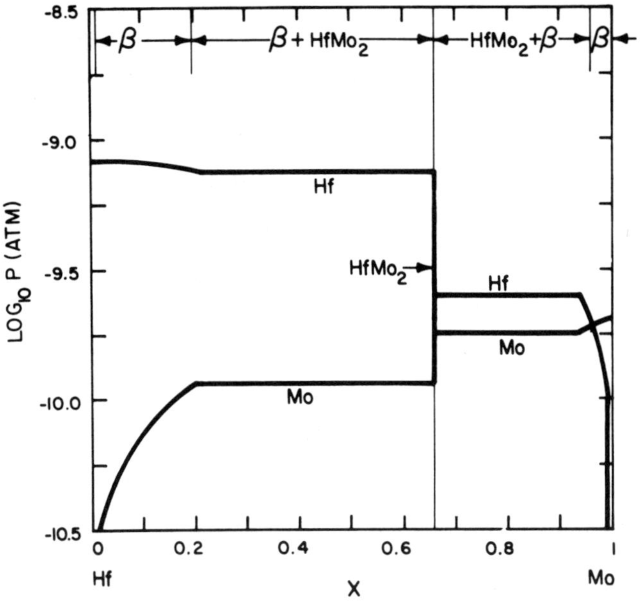

Fig. 10. Computed vapor pressure–composition relations in the Hf–Mo system at 2000°K (22).

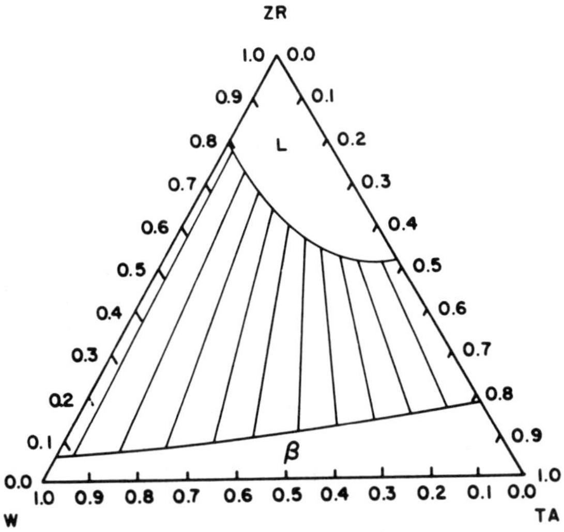

Fig. 11. Computed isothermal section of the ternary system Zr–Ta–W regular solution computed at 2600°K (23).

VI. SUMMARY

Current needs for composite material systems capable of functioning for extended periods of time at elevated temperatures over a wide range of environmental conditions have expanded the requirement for phase equilibria and vaporization information. In spite of the rapid acceleration of experimental studies, complete reliance on observed phenomena is impractical especially if one considers multicomponent systems. An auxiliary approach to this problem is generation and utilization of computational schemes for predicting high temperature phenomena. As indicated above, computer methods based on simple models are currently being employed for this purpose. In spite of the simplicity of the models employed and the crudeness of the current description of unstable phases, some successes have been achieved by using these methods. Thus, present methods have been employed to predict binary and ternary phase diagrams, compositions for minimum and congruent vaporization, and compositions of congruent melting points. Rudy (page 141 of Reference 19) has pointed out that prediction of tie lines and free energy gradients in ternary systems provides a unique advantage in designing the composition of ternary or higher order composites containing two or more phases. Under these circumstances it is likely that utilization of computational methods will expand. As a consequence, the models employed currently, as well as the present description of unstable phases, will be refined and improved leading to greater reliability and usefulness. In this regard, the cycle of review and refinement should parallel that employed in generating of classical thermodynamic information.

VII. ACKNOWLEDGEMENTS

The authors would like to acknowledge the technical contributions of Mr. F.W. Vahldiek, Project Manager on Contract AF 33(657)-9826 and Mr. H. Marcus, Project Manager on Contract AF 33(615)-2352 in performing this research.

REFERENCES

1. Kaufman, L. and Clougherty, E. V., "Investigation of Boride Compounds for Very High Temperature Applications", AFML-TDR-63-4096 Part 1 (December 1963), Part 2 (February 1965), Part 3 (March 1966).

2. Kaufman, L., "Stability Characterization of Refractory Materials under High Velocity Atmospheric Flight Conditions", AF-33-(615)-3859, Progress Report No. 1 (August 1966), Progress Report No. 2 (December 1966).

3. Clougherty, E. V., "Research and Development of Refractory Oxidation Resistant Diborides", AF33(615)-3671, Progress Report No. 1 (Jul6 1966) Progress Report No. 2 (October 1966).

4. Kaufman, L. and Clougherty, E. V., Metallurgy at High Temperatures and High Pressures, K. Gschneider, N. Parlee and M. Hepworth, Eds. Gordon and Breach Science Publishers, New York, N.Y. (1964) p 322.

5. Kaufman, L. and Clougherty, E. V., Metals for the Space Age, Plansee Proceedings (1964), F. Benesovsky Ed. Metallwerk Plansee, Reutte, Tyrol, Austria p. 722.

6. Kaufman, L., Compounds of Interest in Nuclear Reactor Technology, J. T. Waber, P. Chiotti and W. N. Miner, Eds., AIME, New York, N.Y. (1964) p. 193.

7. Kaufman, L. and Sarney, A. Ibid,p. 267.

8. Bernstein, H. Ibid, p. 609.

9. Kaufman, L. and Stepakoff, G. Third Conference on the Performance of High Temperature Systems, (1964) G. Bahn Ed., Gordon and Breach Science Publishers, New York, N.Y. (1967).

10. Hultgren, R., Orr, R. L., Anderson, P. D. and Kelley, K. K., Selected Values of Thermodynamic Properties of Metals and Alloys, John Wiley and Sons, New York, N.Y., 1963, Supplementary Data Sheets May 1965-December 1966.

11. Chang, Y. A.,"Ternary Phase Equilibria in Transition Metal-Boron-Carbon-Silicon Systems" AFML-TR-65-2, Part IV Volume 1, "Thermodynamic Properties of Group IV, V and VI Binary Transition Metal Carbides, September 1965.

12. Schick, H. L. et al., Thermodynamics of Certain Refractory Compounds, Academic Press, New York, N.Y., 1966.

13. Trulson, O. C. and Goldstein, H. W., Research on "Physical and Chemical Principles Affecting High Temperature Materials for Rocket Nozzles", Union Carbide Research Institute, Tarrytown, N.Y., Report 30 September 1963, p III-49, Project PRINCIPIA.

14. Trulson, O. C. and Goldstein, H. W., Ibid Final Report, August 1965, Volume I, p XI-1.

15. Huber, E. J., "The Heat of Formation of Titanium Diboride" University of California, Los Alamos Scientific Laboratory (January 1966) Submitted for Publication to the Journal of Physical Chemistry.

16. Bernstein, H. and Kaufman, L., "Thermodynamics of Interstitial Solid Solutions and Refractory Compounds", ASD-TR-61-445 Part IV October 1964.

17. Bernstein, H. and Kaufman, L., Ibid Part V, December 1965.

18. Rudy, E. and Chang, Y. A., Metals for the Space Age, Plansee Proceedings (1964), F. Benesovsky Ed., Metallwerk Plansee, Reutte, Tyral, Austria p. 786.

19. Rudy, E., "Ternary Phase Equilibria in Transition-Metal Boron-Carbon-Silicon Systems", AFML-TR-65-2, Part IV Volume II, Thermodynamic Interpretation of Ternary Phase Diagrams, January 1966.

20. Rudy, E., Ibid Part II Volume VIII, "The Ta-W-C System", September 1966.

21. Bernstein, H. and Kaufman, L., "Development and Application of Methods for Predicting the Temperature-Composition Stability of Refractory Compounds", AFML-TR-66-193, June 1966.

22. Kaufman, L. and Bernstein, H., Ibid Part II, "Regular Solution Phase Diagrams of Refractory Transition Metals", AFML-TR-67-108, April 1967.

23. Kaufman, L. and Bernstein, H., Ibid Progress Report No. 4, March 1967.

24. Hoch, M., Blackburn, P. E., Dingledy, D. P. and Johnston, H. L., J. Phys. Chem. (1955) 59, 97.

25. Worrell, W. L. and Chipman, J., J. Phys. Chem. (1964) 68, 860.

26. Eliot, J. F. and Gleiser, M., Thermochemistry of Steelmaking, Addison-Wesley Pub. Co., Reading, Mass. (1960).

27. Fries, R. J., J. Chem. Phys. (1961) 37, 3210.

28. Kaufman, L., Acta Metallurgica (1959) 7, 575.

29. Rudy, E., Z. Metallkunde (1963) 54, 112.

30. Kaufman, L., Energetics in Metallurgical Phenomena III (1967), W. Mueller Ed., Gordon and Breach Science Publishers, New York, p 55.

31. Kaufman, L. Phase Stability of Metals and Alloys, R. Jaffee, P. S. Rudman and J. Stunger, Eds. McGraw-Hill Book Co., New York, N. Y. (1967).

32. Willens, R. H. and Buehler, E., Tr. AIME (1966) 236, 171.

33. Bublik, A. I. and Pines, B. IA., Dokl. Akad Nauk SSR (1952) 87, 215.

34. Denbigh, P. and Marcus, R. B., J. Appl. Phys. (1966) 37, 4325.

35. Chopra, K. L., Randlett, M. R. and Duff, R. H., Applied Physics Letters (1966) 9, 402.

36. Chopra, K. L., Randlett, M. R. and Duff, R. H., "Face Centered Cubic Modification in Sputtered Thin Films of Tantalum, Molybdenum, Tungsten, Rhenium, Hafnium and Zirconium. Phil. Mag (1967) 16, 261.

37. Harmon, D. P. and Brukl, C. E., "Ternary Phase Equilibria in Transition Metal-Boron-Carbon-Silicon-Systems", AFML-TR-65-2 Part II, Volume III The Zr-Ta-C System, January 1966.

38. Chang, Y. A., Ibid Part IV, Volume III, "Computational Approach to the Calculation of Ternary Phase Diagrams, October 1966.

SOME PROPERTIES OF EUTECTIC "CRYSTALS" OF MnO—MnS

John W. Moore

Experiment Station
E. I. du Pont de Nemours
Textile Fibers Pioneering Research
Wilmington, Delaware

Lawrence H. Van Vlack

Department of Chemical and Metallurgical Engineering
The University of Michigan
Ann Arbor, Michigan

ABSTRACT

The lamellar eutectic structure of MnO and MnS has a [112] growth direction and a (111) interface plane for both phases. Thus the microstructure presents a two-phase "crystal" with one orientation. Slip bands were not noticeably influenced by phase boundaries. Hardness values vary with microstructural alignment. Wider lamellae introduced greater hardness. The thermal expansion of the eutectic is slightly less than either MnS or MnO alone.

INTRODUCTION

A number of eutectic microstructures between ionic compounds form "crystals" in which both phases of the eutectic pair have

This paper is a part of a dissertation submitted by Dr. Moore in partial fulfillment of his Ph.D. requirements at The University of Michigan.

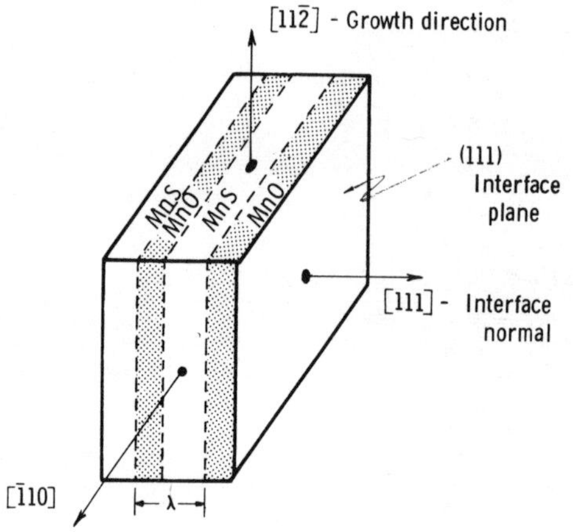

Fig. 1. Orientation of eutectic lamellae. Both MnO and MnS grew in the [11$\bar{2}$] direction and possessed (111) interfaces to produce a two-phase "crystal" with a single orientation[1].

identical orientations.[1] Manganous sulfide (MnS) and manganous oxide (MnO) provide such a two-phase crystal. Both phases have the NaCl-type structure and form a lamellar eutectic with a common [112] growth direction and a $(11\bar{1})$ interface plane (Fig. 1). Eutectic pairs involving other NaCl-type compounds behave similarly in that both phases of the pair have identical orientations to give two-phase "crystals." These orientations vary, however, from eutectic pair to eutectic pair (Table I).

TABLE I

Eutectic	Structure Types	Interface Planes	Growth Directions
MnO-MnS	NaCl-NaCl	(111)-(111)	$[11\bar{2}]$-$[11\bar{2}]$
NaCl-NaF	NaCl-NaCl	(110)-(110)	[001]-[001]
NaBr-NaF	NaCl-NaCl	(110)-(110)	$[1\bar{1}1]$-$[1\bar{1}1]$
LiF-NaF	NaCl-NaCl	$(1\bar{1}1)$-$(1\bar{1}1)$	[110]-[110]
FeO-FeS	NaCl-NiAs	(100)-(0001)	$[011]$-$[10\bar{1}0]$

This paper examines several properties and characteristics of two-phase MnO-MnS "crystals." They were grown by directional solidification as an eutectic mixture of purified MnO and MnS. The MnO was prepared by reduction of reagent grade manganese dioxide (MnO_2) of 99.9% purity in hydrogen at 950°C. The MnS was prepared using a technique previously reported by Chao and Van Vlack.[2] This involved the deoxidation of reagent grade manganous sulfate powder ($MnSO_4$) of 98.8% purity by sulfur according to the reaction

$$MnSO_4 + S_2 \longrightarrow MnS + 2SO_2$$

The sulfur was vaporized at 410°C in the lower part of vitrified silica tube by a resistance furnace. The sulfur vapor then reacted with a charge of $MnSO_4$ powder in the upper part of the furnace tube held at 900°C by a second resistance furnace.

The resulting MnS powder was then melted into ingots in an induction furnace using a graphite crucible—susceptor. Neither optical examination nor X-ray powder patterns indicated that phases other than MnS were present. The lattice constant, a_o, for the MnS was checked by calculating a_o's for a series of powder pattern lines and extrapolating the curve obtained to $\theta = 90°$ versus a Nelson-Riley factor. The value obtained was 5.223 Å compared with 5.225 Å reported by Chao[2], 5.224 Å by ASTM[3], and 5.165 to 5.260 Å by Boniszewski and Baker[4].

The raw materials described above were mixed as powders in the proper weight ratios as indicated in the phase diagram determined by Chao, Smith and Van Vlack[5], and after mixing in a Spex Mixer-Mill the eutectic compositions were melted in a Ferrovac iron crucible, 1/4 in. i.d., 1/2 in. o.d., and 18 in. long. An ingot of 1/4 in. o.d. and about 6 in. in length was obtained. The general procedure involved completely melting the powder charge by passing the crucible vertically upward through a silicon carbide resistance furnace under an argon atmosphere. Subsequent unidirectional solidification occurred by withdrawing the crucible downward from the heating zone of the furnace (Fig. 2). The rate of solidification was controlled through the use of an Instron testing machine which permitted movement ranging from 7 to 50 x 10^{-5} cm/sec.

The resulting ingots were sectioned and polished both transversely and longitudinally and photographed in the areas selected for x-ray study. The details of the x-ray study are described elsewhere.[1]

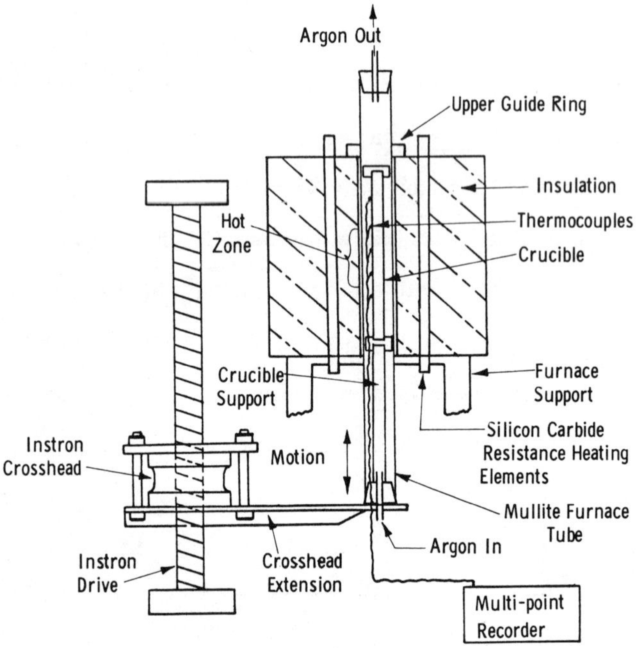

Fig. 2. Eutectic unidirectional solidification apparatus.

Thermal expansion coefficient of manganous sulfide (MnS) was determined from room temperature to 1200°C using a Rigaku-Denki high temperature x-ray camera with provision for recording powder patterns for several temperatures on a single film. The MnS powder was ground to -325 mesh in an agate mortar and inserted in a 0.2 mm vitrified silica tube. The silica tube was then sealed and cemented to an alumina rod 1.5 mm in diameter and 3.5 cm long with a Sauereisen cement. A 28 gauge platinum wire was cemented beside the silica tube to serve as a standard for calibrating the temperature of the specimen. The furnace was evacuated by a mechanical pump to reduce air scatter and oxidation. Unfiltered chromium radiation was used at 35 kv and 10 ma to obtain the maximum number of back-reflection lines for measurement. Standard methods for precision lattice parameter determination by extrapolation of a_o versus the Nelson-Riley function to $0 = 90°$ were utilized. The specimen temperatures indicated by the platinum-10% rhodium thermocouple adjacent to the specimen were corrected to the actual specimen temperature by calculating the lattice parameter of the platinum standard at each temperature and determining the true temperature of the specimen from the linear expansion versus temperature curve for platinum reported by Campbell.[6]

The thermal expansion coefficient of a polycrystalline MnS specimen was determined by the same dilatometric procedure used on the MnO-MnS eutectic specimen to verify the comparability of the x-ray and dilatometer results.

A Leitz HTV dilatometer was used which compared the linear expansion of a 50 mm long specimen with that of a 50 mm Chronin standard at a 200X optical magnification. The linear expansion of the Chronin standard was reported in the Leitz manual[7] from room temperature to 1000°C so that the Chronin could be used as a temperature calibration.

The deformation and fracture modes of single crystal MnO and lamellar MnO-MnS eutectic grains were determined from cleavage and Vickers indentation hardness tests. The fracture behavior was determined by cleaving the MnO and MnO-MnS samples along various crystallographic planes with a sharp blade and observing the predominant form of fragments caused by crushing. The fracture patterns observed around large Vickers indentations also contributed to this analysis.

The deformation behavior of MnO and the MnO-MnS eutectic was studied by observing the slip lines around Vickers indentations in oblique reflected light at 100 to 500X magnifications. The MnO sample used was a flame-fusion grown single crystal boule obtained from the Marubeni-Iida-America Co., Chicago, Illinois. The low index crystallographic planes on which the indentations were made were cleaved or ground and checked with Laue back-reflection x-ray photographs.

The ground or cleaved planes of the MnO crystal and the MnO-MnS eutectic ignots were used for the Knoop microindentation hardness tests. The Knoop indenter was a pyramid with a longitudinal angle of 172°30' and a transverse angle of 130°0'. It gave a diamond shape indentation with diagonals of approximately 7 to 1 length ratio. Very little elastic recovery occurred along the long diagonal of the Knoop indentation; therefore, the measured length of this diagonal and the dead weight applied load were used to determine the ratio of applied

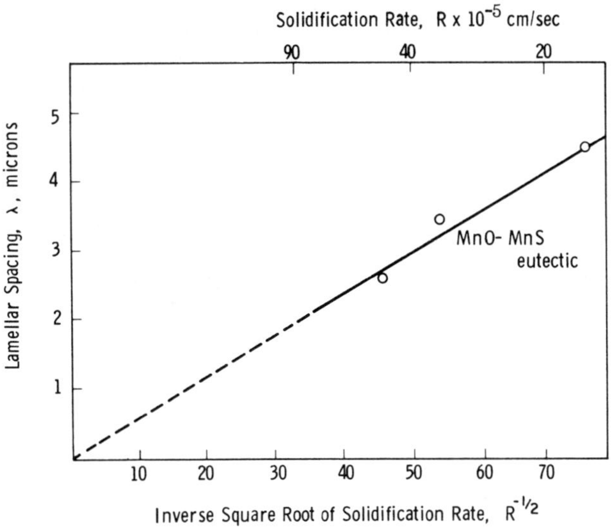

Fig. 3. Lamellar spacing, λ, versus solidification rate. Penfold and Hellawell[13] also observed a λ vs $R^{-1/2}$ relationship for a comparable LiF—NaF eutectic.

load to unrecovered projected indentation area in kg/mm^2, i.e. the Knoop Hardness Number (KHN).

RESULTS AND DISCUSSION

The crystal orientations described in Fig. 1 were found consistently for each phase of the MnO-MnS eutectic pairs.

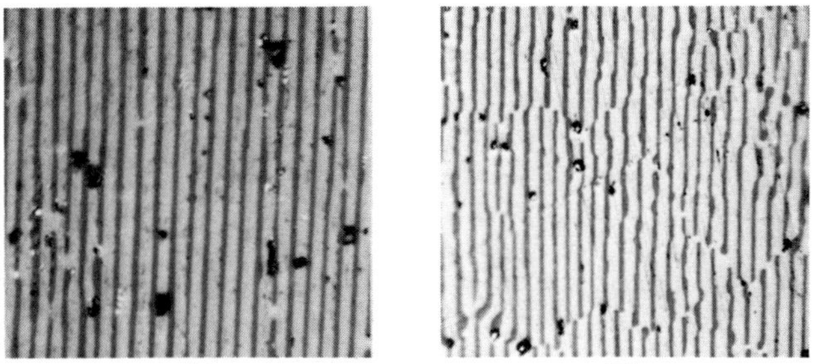

(a) (b)

Fig. 4. Photomicrographs of transverse (a) and longitudinal (b) cross-sections of a typical MnO—MnS unidirectionally solidified lamellar eutectic[1]. (500 ×. MnS — lighter; MnO — darker.)

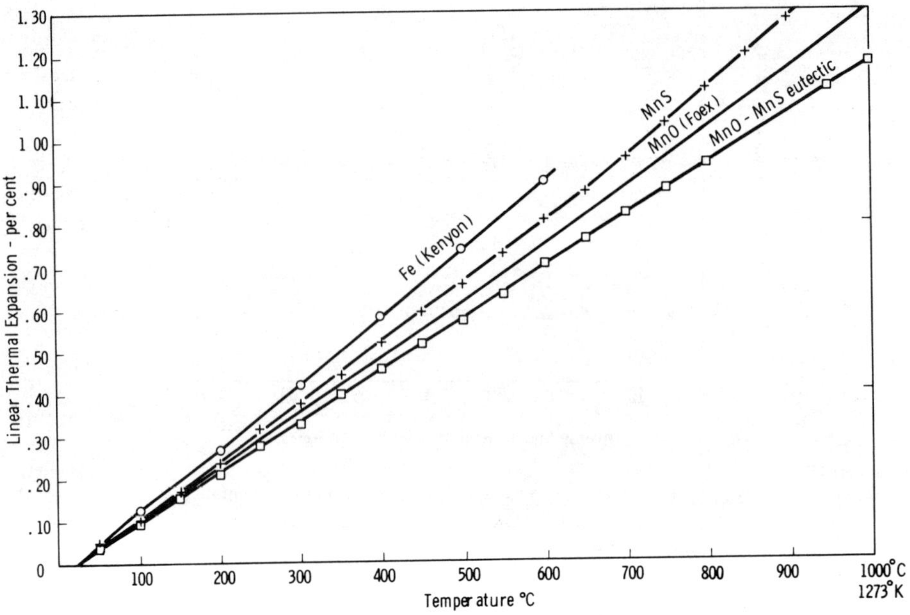

Fig. 5. Linear thermal expansions for MnO[8], MnS, MnO—MnS eutectic, and Fe[9] from room temperature to 1000°C. Dilatometric data.

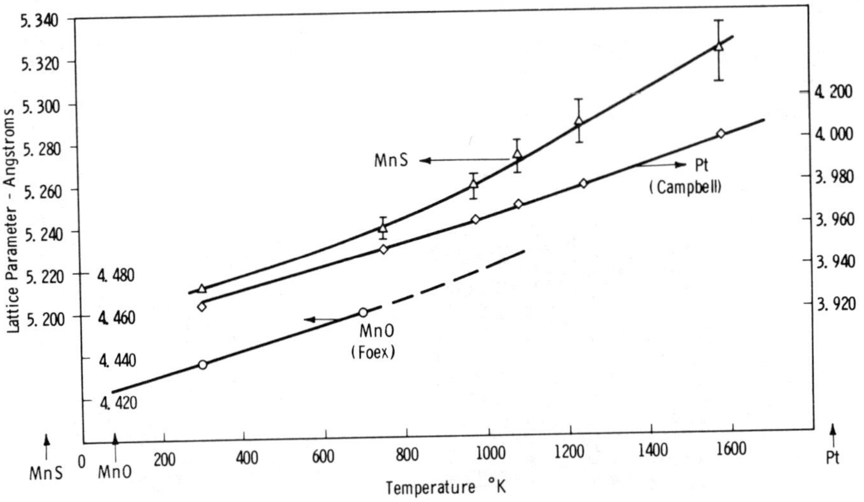

Fig. 6. Lattice parameter versus temperature for MnS, MnO[(8)], and Pt[(6)].

The lamellar spacing varied inversely with the square root of the solidification rate as shown in Fig. 3. Typical eutectic microstructures are shown in Fig. 4.

Thermal Expansion

The linear thermal expansion for MnO[8], MnS, MnO-MnS and iron[9] are shown in Fig. 5. The change in lattice parameter of MnS with temperature determined by a back-reflection x-ray technique is shown in Fig. 6, along with data for the platinum standard to calibrate the temperature.[6] and the data of Foex[8] for MnO converted to a lattice parameter basis. The x-ray data indicated a value of $15 \pm 1 \times 10^{-6}/°C$ while the dilatometer data indicated a value of $14.5 \pm 0.5 \times 10^{-6}/°C$ for the thermal expansion coefficient of MnS over the range from 25 to 1000°C. This is slightly higher than the value of $13.5 \times 10^{-6}/°C$ over the same range reported by Foex[8] for MnO.

The mean thermal coefficient of the MnO-MnS eutectic between 25 and 700°C was found to be $12 \pm 0.5 \times 10^{-6}/°C$ by the dilatometric method. Thus the MnO-MnS eutectic exhibits not only a slightly lower thermal coefficient than MnS as would be expected, but also a lower value than that of MnO. The apparent anomaly is greater than the indicated experimental error; however, the exact reason for the lower expansion value is not know.

Deformation and Fracture

The deformation and fracture of MnS single crystals during micro-indention on various crystallographic planes have been studied by Chao, Thomassen and Van Vlack.[2] They found that the primary glide system was $\langle 110 \rangle \{110\}$, as in most NaCl-type crystals. They also observed an unusual $\langle 110\rangle\{111\}$ system subsequent to $\{110\}$ fracturing but no other secondary glide systems. The primary fracture mode for unconstrained MnS crystals was on $\{100\}$ as found in other NaCl types, but $\{110\}$ fracture occurred around indentations on $\{100\}$ surfaces and was attributed to the interaction of (011) and $(\bar{1}01)$ slip near the surface.

The deformation and fracture of MnO was determined in this study by similar indentation procedures and was found to be identical to that described above for MnS, specifically, a $\langle 110\rangle\{110\}$ slip system with a strong tendency for $\{100\}$ cleavage fractures and a lesser tendency for $\{110\}$ fracture. Thus any differences to be encountered in the two-phase eutectic "crystal" are a consequence of the interfaces between the phases rather than the behavior of either phases individually.

The observations of deformation and fracture of the MnO-MnS eutectic were made on the transverse $(11\bar{2})$ and longitudinal (110) and (111) sections as sketched in Fig. 1. The slip lines on the transverse (112) surface were very straight and not altered by crossing the lamellar phase boundary. They were at 66° to each other corresponding to the angular variation of the lamellae to $(10\bar{1})$ and $(0\bar{1}1)$ slip planes. There were no noticeable slip lines around indentations on the (110) or (111) surfaces.

Fracture was predominantly $\{100\}$ but also propagated along the (111) interface boundary when indentation was performed near the free edge of the specimen or when cleaved with a sharp blade. In all MnO-MnS samples checked, this cleavage path was entirely in the MnO phase, adjacent to the MnS boundary. Indentations exhibited fewer (111) and more (100) fractures under light loads and in the interior of the transverse (112) cross-sections where a more symmetric strain was applied.

The above results indicate that the plastic deformation of the MnO-MnS eutectic arises primarily from slip on $\{110\}$ planes. The necessity of operating through two different phases and their phase boundary does not introduce other slip planes. The fracture behavior is more significantly altered by the two-phase composite since a fracture mode along the phase boundaries is observed in addition to normal $\{100\}$ fracture in the MnO and MnS single crystals. No well defined $\{110\}$ fracture was observed in the eutectic microstructure. This contrast in fracture behavior may arise from the slight difference in thermal contraction during cooling from the solidification temperature. It may be suggested that since the MnO had a slightly lower thermal expansion than the MnS, differential contraction would introduce compressive stresses parallel to the interface in the MnO phase with a Poisson's expansion in a [111] direction. The tensile stresses of the expansion could be relieved by a cleavage fracture along the interface and through the MnO.

<u>Hardness</u>

The variation in hardness with orientation of the Knoop indenter on the (111), (110), and (100) planes of single crystal MnS has been reported by Chao, <u>et al.</u>[10]. Their data, with the data on MnO, are plotted in Fig. 7, based on the assumption that there are six-fold, two-fold, and four-fold symmetries respectively for the planes just cited. These symmetries

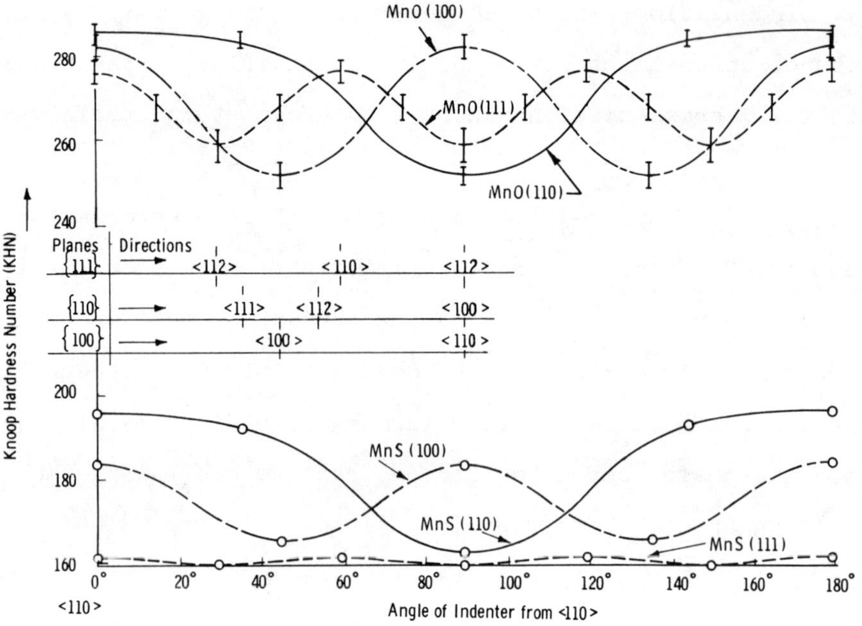

Fig. 7. Crystallographic variations in Knoop Hardness Number (KHN) for MnO and MnS[10] single crystals.

have been observed in previous hardness studies on NaCl-type crystals[11] and are consistent with the predominant (110) [1$\bar{1}$0] slip system. The hardness of MnO is significantly higher than MnS for all planes and orientations. This is as expected since Wooster[12] has shown that in an isomorphous structural series of compounds, the hardness generally increases with melting point (1830°C and 1620°C for MnO and MnS respectively). The hardnesses on the (111) plane of MnO vary more than that of MnS ($\pm$ 8 KHN for MnO, $\pm$ 1 KHN for MnS) and exceed the hardnesses of the (110) and (100) planes more than it does in MnS.

The hardness of the two-phase "crystals" of MnO and MnS were determined (cf. Fig. 1):

Plane	Knoop Direction
(1$\bar{1}$0)	[11$\bar{2}$], [111]
(111)	[110], [11$\bar{2}$], [110] + 45°
(11$\bar{2}$)	[110], [111].

These data are connected in Fig. 8 with a solid line. From a crystallographic standpoint, Knoop hardnesses on these planes with a long axis of the indenter oriented in these directions are all that can be compared between the single crystals and the two-phase "crystal". For this reason, only the hardness on the (110) and the (111) planes of single

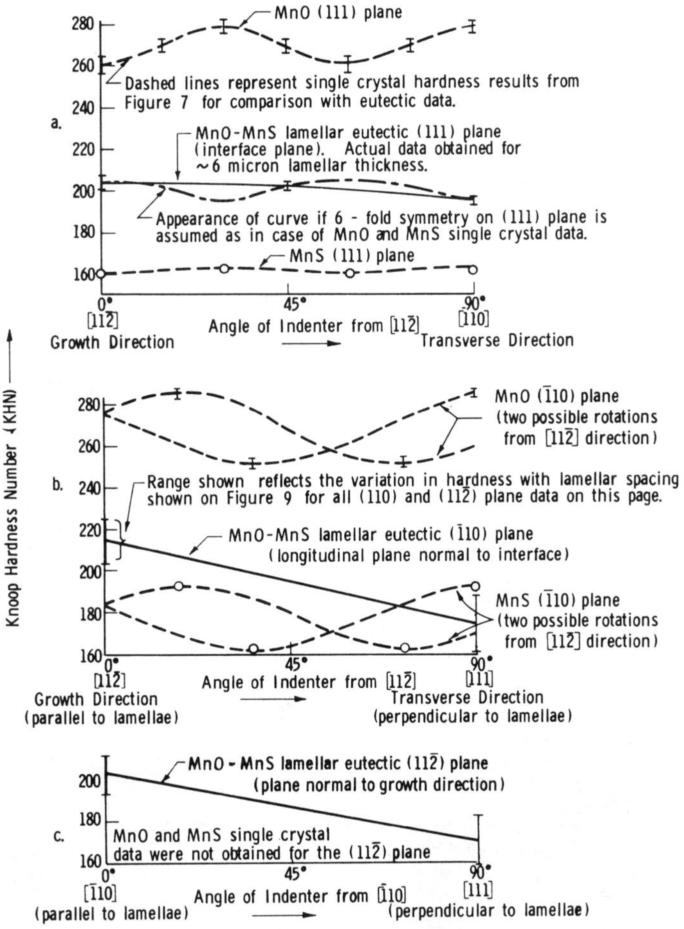

Fig. 8. Knoop hardness versus indentor-direction on the (111), (110) and (112) planes of the MnO–MnS lamellar eutectic and comparison with MnO and MnS single crystal hardnesses.

crystal MnO and MnS are plotted for comparison with the two-phase data in Fig. 8.

One fact stands out in Fig. 8. The hardness of the two-phase "crystal" is higher when the indenter length is parallel to the lamellae than when it is perpendicular to the lamellae. This occurs on both the (110) and $(11\bar{2})$ plane, and does not coincide with the hardness variations found in the single crystal of either MnO or MnS (Fig. 8b). Therefore, it must be concluded that the lamellar microstructure is more important in determining directional hardness in the two-phase "crystal" than the individual hardness characteristics of either crystal. This poses an interesting situation since we noted earlier that the deformation pattern of the two-phase "crystal," like the individual phases, depends heavily

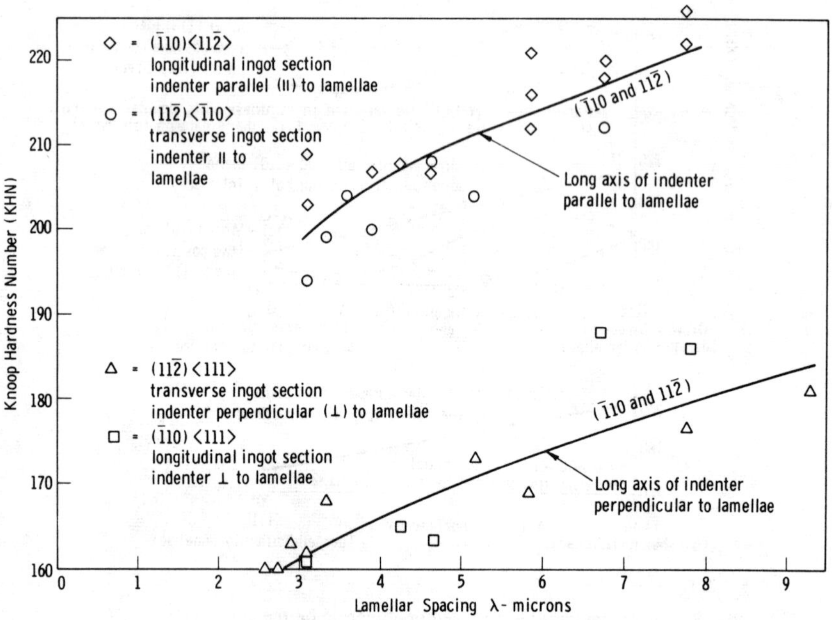

Fig. 9. Microstructural variations in the Knoop Hardness Number (KHN) for the MnO–MnS eutectic. Each point represents the average of five hardness readings.

on the (110)[110] slip system. In brief, it must be concluded that while the same slip system holds, the boundary offers resistance to deformation, particularly when the indenter axis is **parallel** to the boundary, when most of the displacement is lateral and occurs **perpendicular** to the interface.

In Fig. 3, it was revealed that the lamellar thickness of two-phase "crystals" could be varied. The effect of lamellar spacing on the hardness is indicated in Fig. 9. As previously noted, a higher KHN was noted when the long axis of the indenter was parallel to the lamellae than when the long axis of the indenter was perpendicular to the lamellae. In addition, the hardness decreased about 3kg/mm^2 per micron decrease in lamellar width. Thus while it can be interpreted that there is a resistance to deformation <u>across</u> the lamellae, the lamellae themselves do not represent a source of strengthening of the two-phase "crystal." In fact, the opposite is true. A final answer is not possible, but two suggestions are proposed for subsequent consideration.

First, as discussed earlier, finer spacing may give more possible fractures which combine with the actual plastic deformation to reduce the hardness. Secondly, the lattice disregistry at the boundary within the two-phase "crystal" involves numerous dislocations which in turn may generate slip at lower stresses. Thus a boundary could be a source of weakness. Other considerations which will require further study are the effects of residual impurities and thermostrains.

CONCLUSIONS

Two-phase "crystals" produced by eutectic solidification of NaCl-type compounds provide an interesting means of determining the role of phase boundaries without the concurrent variation of crystal orientation. The thermal expansion of the two-phase "crystal" is close to that of the individual phases but with a possible negative deviation. The boundary between phases does not alter the slip mechanism. It does, however, introduce additional fracturing parallel to the boundaries on an otherwise non-cleavable crystal plane, (111). It is suggested that this may be a result of residual stresses. The hardness of the two-phase "crystal" is more sensitive to the phase boundaries than to the primary slip direction. Specifically, displacement occurred more readily parallel to the phase boundaries than across the phase boundaries even though the former may involve a more difficult slip direction. Although the boundary offers interference to deformation, it is also a source of microstructural softening because the hardness decreases about $3 kg/mm^2$ for each micron decrease in lamellae thickness. Two possible explanations are suggested for future examination.

ACKNOWLEDGEMENT

This study was sponsored by the Office of Naval Research, Metals Branch, under Contract Nonr-1224(47). This support is gratefully acknowledged.

REFERENCES

1. Moore, J.W., and L.H. Van Vlack, "Preferred Orientations in Microstructures of Eutectics Between Compounds," Fall Meeting, Basic Science Division, American Ceramic Society, (1966).

2. Chao, H.C., L. Thomassen and L.H. Van Vlack, "Deformation and Fracture of MnS Crystals," ASM Trans. Quart., 57, p. 386 (June 1964).

3. Swanson, H.E., R.K. Fuyat and G.M. Ugrinie, "Standard X-ray Diffraction Powder Patterns," NBS Circular 539, 4, p. 11 (1955).

4. Bonizewski, T. and R.G. Baker, "Dislocations in Manganese Sulfide Inclusions in Steel," Acta Met., 11, pp. 990-992 (1963).

5. Chao, H.C., Y.E. Smith and L.H. Van Vlack, "The MnO-MnS Phase Diagram," Trans. AIME, 227, p. 796 (1963).

6. Campbell, W.J., "Platinum Expansion Values for Thermal Calibration of High Temperature X-ray Diffraction Cameras and Deffractometers," U.S. Bureau of Mines Information Circular IC8107 (1962).

7. "Instructions for Assembly and Operation of the Dilatometer Model HTV," Ernst Leitz Wetzlar, p. 15. (Data from Phys. Techn. Reichsanstalt).

8. Foex, Marc, "A Type of Transformation Common in the Lower Oxides of Mn, Fe, Co and Ni," Compt. rend., 227, pp. 193-194 (France) (1948).

9. Kenyon, R.L., "Structure and Properties of Iron," Metals Handbook, T. Lyman, Ed., ASM, Metals Park, Novelty, Ohio, p. 427 (1948). (Data after Hidnert).

10. Chao, H.C., L.H. Van Vlack, F. Oberin and L. Thomassen, "Hardness of Inclusion Sulfides," ASM Trans. Quart., 57, pp. 885-891 (1964).

11. Kuznetsov, V.D., Surface Energy of Solids, Trans. from Russian, Her Majesty's Stationery Office, London (1957).

12. Wooster, W.A., A Textbook on Crystal Physics, Cambridge University Press, London (1938).

13. Penfold, D. and A. Hellawell, "Microstructures of Alkali Halide Eutectics LiF-NaF and NaCl-NaF," J. Am. Cer. Soc., 48, p. 133 (1965).

SELECTED PROBLEMS IN THE ANALYSIS OF REFRACTORY MATERIALS

C. T. Lynch, F. W. Vahldiek, and S. A. Mersol

Air Force Materials Laboratory
Wright-Patterson Air Force Base, Ohio

ABSTRACT

The development of refractory materials for structural and protective applications at very high temperatures has received increased emphasis in the last decade. Research on refractory compounds such as the transition metal carbides, borides, and nitrides has brought into focus attendant problems with analytical methods and chemical standards. As research efforts have become more sophisticated and the available materials have improved in quality, the role of impurities in determining properties has been found of increasing importance. The limitations on methods through lack of standards, problems of non-stoichiometry, localization or segregation of impurities, and inherent deficiencies of methods are discussed. The potential of some of the newer methods of analysis is considered in the context of these specific problems with refractory materials analysis.

Introduction

Adequate characterization of materials demands that suitable standards of comparison be available and the methods of comparison be well understood. Unfortunately, much research on refractory materials in the last two decades has been wasted because the analytical characterization has been inadequate, or indeed, even non-existent. In this paper only the chemical compositional characterization will be considered, as a basis which must be established before other characterization unique to a given material becomes unequivocal in validity. Because new methods are developed so rapidly today, and available standards are in continual change, whatever can be said now will be outdated before it is printed. Nonetheless, the slowness with which adequate refractory chemical standards have been developed suggests that a review of selected analytical problems is most appropriate.

The processing of refractory materials is closely related to analytical considerations. The introduction of impurities accidentally or purposefully, purification itself, and the control of stoichiometry are inevitably associated with problems in analysis. In the instance of preparation of refractory standards these two facets of materials research merge to pose a severe challenge to the analyst. The need for systematic approach to strengthen the scientific basis of processing will not be treated here [(1)].

A considerable number of techniques are available for cation and elemental analysis of refractory solids which are sufficiently sensitive and precise to meet current demands. Analyses in the parts per million (PPM) range are standard and analyses in the parts per billion (PPB) range are often possible with the limitation that there is a paucity of suitable reference standards. Methods for anion analysis are not nearly so well developed. Commercial analyses often include detailed results for metals and omit consideration of common anion impurities such as sulfate, carbonate, nitrate, chloride and hydroxyl groups.

Even when determination of gross impurities of cations is routine, the determination of location can be extremely difficult in the low concentration ranges. The microstructure of ceramic materials is generally such that the segregation of impurities at a grain boundary is found to play a significant role in bulk material properties. The determination of a given impurity at a grain boundary is a critical problem, which is accentuated with fine-grained ceramic bodies.

Crystalline intermetallic compounds have been found to fail intergranularly. This failure has been associated with a solute induced grain boundary hardening [(2)]. There are a number of studies of

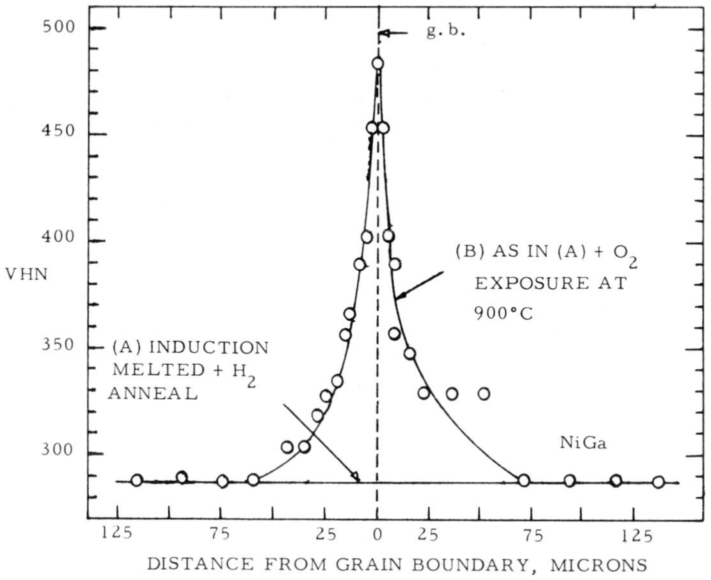

Fig. 1. Grain boundary hardness profile in NiGa before (A) and after (B) exposure to oxygen.

this phenomenon which have been carried out by Westbrook et al. Intermetallics investigated include AgMg [(3)], NiGa[(4)], and NiAl[(5)]. In Figure 1, for example, the grain boundary hardness profile of NiGa before and after exposure to oxygen, is seen to change markedly[(4)]. The determination of the oxygen in the total sample of NiGa is much simpler than determining the segregation of oxygen at the grain boundary. There is a significant change in hardness at 25 microns from the grain boundary. However, if one considers the case where the grain boundary may be only a micron or less in cross-section (a large grain boundary for many conventional ceramic materials) the determination of segregated impurities at this boundary by instrumental techniques such as the electron microprobe is virtually impossible. Nitrogen as well as oxygen may play a role in this grain boundary hardening, but there are extreme difficulties in analysis to determine nitrogen and oxygen simultaneously in this case. Therefore, the problem is not so much detecting the impurity, but once the impurity is detected, locating the impurity.

TABLE I.

3.16. Ceramic Materials

This group of standards is supplied in the form of powders, usually 100 mesh or finer. They are intended to provide materials for checking the accuracy of methods used in the analysis of similar materials, primarily in the glass and steel industries. Note that Silica brick No. 102 is a density sample with density of 2.33 g/cm³ at 25 °C.

Sample Nos.	Kind	Approximate weight	Price	Sample Nos.	Kind	Approximate weight	Price
76	Burned refractory (40% Al_2O_3)	60 g	$10.00	92	Glass, low boron	45 g	$10.00
77	Burned refractory (60% Al_2O_3)	60 g	10.00	93	Glass, high boron	45 g	10.00
78	Burned refractory (70% Al_2O_3)	60 g	10.00	165	Glass sand (low iron)	60 g	10.00
103a	Chrome refractory	60 g	10.00	1a	Limestone, argillaceous	50 g	10.00
198	Silica refractory (0.2% Al_2O_3)	45 g	10.00	102	Silica brick	60 g	10.00
199	Silca refractory (0.5% Al_2O_3)	45 g	10.00	104	Burned magnesite	60 g	10.00
89	Glass, lead-barium	45 g	10.00	112	Silicon carbide	85 g	10.00
91	Glass, opal	45 g	10.00	154a	Titanium dioxide	40 g	10.00

ANALYSES

Sample Nos.	Kind	SiO_2	Al_2O_3	Fe_2O_3	FeO	TiO_2	ZrO_2	MnO	P_2O_5
76	Alumina refractory	54.7	37.7	2.4	-------	2.2	0.07	-------	0.07
77	Alumina refractory	32.4	59.4	0.90	-------	2.9	.09	-------	.45
78	Alumina refractory	20.7	70.0	.79	-------	3.4	.12	-------	.62
103a	Chrome refractory	4.6	29.96	-------	12.43	0.22	.01	0.11	.01
198	Silica refractory	-------	0.16	.66	-------	.02	<.01	<.01	.02
199	Silica refractory	-------	.48	.74	-------	.06	.01	<.01	.01

Sample Nos.	Kind	V_2O_5	Cr_2O_3	CaO	MgO	Li_2O	Na_2O	K_2O	Loss on ignition
76	Alumina refractory	0.02	-------	0.27	0.58	0.11	0.15	1.54	0.22
77	Alumina refractory	.03	-------	.26	.50	.35	.06	2.11	.21
78	Alumina refractory	.05	-------	.38	.51	.20	.06	2.83	.26
103a	Chrome refractory	-------	32.06	.69	18.54	-------	-------	-------	-------
198	Silica refractory	-------	-------	2.71	0.07	.001	.01	0.02	.21
199	Silica refractory	-------	-------	2.41	.13	.002	.01	.09	.17

The instrumentation boom in recent years, which includes the adaptation of the laser beam to emission spectrographic and mass spectrometric analysis, can be utilized to obtain location determination of impurities to the low micron region. The electron microprobe can sometimes be used on a one-micron cross-section, but as yet cannot be used at high sensitivity for soft x-ray elements. Thus, some of the elements of greatest interest such as carbon, boron, oxygen, and nitrogen cannot be determined routinely unless they constitute major concentrations in the matrix.

Refractory Standards

No analytical technique is going to be better than its standards. All materials scientists must be concerned with the lack of refractory materials standards for doing analysis in the low PPM and PPB ranges. Reproduced in Table I are the entire refractory ceramic materials standards for materials listed in the catalogue of standard materials issued by the National Bureau of Standards. Currently, the Bureau is working on a series of four glasses which will be doped with 50 elements at the levels of 50 down to less than 1 PPM [7]. This is important and should be followed up by serious concern for improving the standards for non-glassy oxides, borides, carbides, nitrides, and intermetallic compounds. Without such work the development of ultra-high purity refractory inorganic non-metallic materials is seriously hindered since there is no accurate basis for comparing results of analysis on a laboratory to laboratory basis.

There was a conference on chemical compounds of certified high purity held in June 1959 at the National Academy of Sciences in Washington, D. C. It was sponsored by the NSF and NAS-NRC. The recommendations made were typical: improving sources of standard materials, the characterization of these materials, and increasing the number of compounds covered, improving communications between individual researchers and materials sources [8]. A recent symposium found the same problems needing attention with little progress having been made since the previous report [9].

There are alloy base standards such as Cu-base, Ti-base, and Al-base alloys which in some instances can serve through the low PPM range for standards in which the matrix does not influence the results. Even methods which have been considered relatively free of matrix effects such as Atomic Absorption Spectroscopy have recently been found, on closer examination, to have matrix effects which definitely limit the precision of results when sensitivity limits are approached.

For most refractory non-metallic compounds there is no arbiter party method of determining analytical accuracy and thus results from laboratory to laboratory cannot be compared or collated with any degree of confidence.

Method Sensitivity and Precision

Interdisciplinary materials scientists need to be aware of the development of analytical techniques appropriate to their materials analysis requirements. Lack of communication and understanding between the research investigator and research analyst often mean routine classical methods are applied, appropriate or not, and that the problems attendant with sensitivity and precision are ignored.

Gas Chromatography is a simple, sensitive, and excellent quantitative tool for volatile materials analysis. Gas Chromatography also has great potential for analysis of adsorbed gases, metal chelates such as the β-diketonates, and sufficiently covalent salts such as many metal halides. There are an estimated 50,000 or more GC units in the world making it likely the most widely used instrumental tool save perhaps the simple colorimeter and pH meter. Quantitative techniques are well established but the 1964 ASTM Committee E-19 got some sobering results when a sample mixture was sent to 32 major analytical laboratories for GC analysis. These results are in Table II [10]. Table III on the same page gives some results in an instrument laboratory on the type of results that can be achieved using various quantitative methods. Since the standards in Table II were accurately known to the 0.1% level, the Relative Standard Deviation is also a measure of the relative accuracy of measurement. Now if we consider that the method is capable of 0.1 to 1% relative standard deviation as seen in Table III, a 25% deviation for a simple hydrocarbon like toluene or 19% on n-butanol is discouraging. When a method such as emission spectroscopy is utilized, relative errors of 10% or more are expected in the PPM range. Practical errors on a laboratory to laboratory basis often exceed 100%. This is seldom considered by the scientist when an analyst's report is in front of him. It is impossible to perform a routine quantitative analysis on 20 or 30 elements in a matrix where no standard samples are available and come even close to the 10% figure. In these cases a 100% relative deviation is often optimistic.

TABLE II.

Compounds	Actual %	Found %	Std. Dev.	Rel. Std. Dev.
Meth-eth ketone	12.0	11.96	0.907	7.6%
Meth-isobut ketone	17.2	17.87	3.288	18.3%
Toluene	16.9	17.84	4.516	25.3%
n-Butanol	31.5	30.04	5.666	18.8%
p-Xylene	22.4	22.29	1.885	8.5%

TABLE III.

COMPARISON OF INTEGRATION METHODS

	#1	#2	#3	Avg	σ abs	σ rel
Model 471 Digital Integrator						
n-Nonane	39.156	39.150	39.193	39.166	0.0716	0.184%
n-Decane	60.844	60.850	60.807	60.834	0.0227	0.037%
Disc Integrator						
n-Nonane	39.33	38.97	38.91	39.07	0.22	0.56 %
n-Decane	60.67	61.03	61.09	60.93	0.23	0.38 %
Triangulation						
n-Nonane	40.77	40.68	40.07	40.51	0.38	0.94 %
n-Decane	59.23	59.32	59.93	59.49	0.39	0.66 %
Weighing Paper						
n-Nonane	42.58	41.73	42.83	42.38	0.58	1.37 %
n-Decane	57.42	58.27	57.17	57.62	0.58	1.01 %

Electrical Methods. - Problems in Non-specificity

In addition to providing analytical data a useful method should have some universality, an adaptability of determination in a variety of matrices. The method should also be commercially available. Electrical Methods of Analysis are at once singularly precise and sensitive, often are nondestructive, and are rapid. A list of a number of these methods and estimated detectibility limits is given in Table IV [11]. Residual resistivity and the resistivity ratio ($R_{4.2°K}$ to $R_{273°K}$) are useful in determining purification of ultra-high purity materials. The trouble with electrical methods is that you often don't know what you are measuring. The Hall Effect and Conductance values do not tell the investigator what carriers are there in the material. The results are non-selective. High purity germanium as originally developed had 50 PPM oxygen in it, but no one knew it was there, because electrical measurements insensitive to oxygen constituted the test. The determination of the purity of water is a common example where the non-conductor organics must be analyzed separately, otherwise "conductivity" water isn't necessarily very pure. Electrical measurements for general determination of impurity distributions, purification, and general purity have great usefulness to refractory inorganic materials if the limitations are understood.

Ion-specific electrodes with treated membranes have considerable potential for future development. Electrodes are currently available for determinations of Cl^- to 10^{-5} M/l (M), Br^- to 10^{-6} M, I^- to 10^{-7} M, Ag^+ to 10^{-6} M. New electrodes are being developed for sulfate, phosphate, sulfide, fluoride, and a variety of common metals [12, 13]. The possibility for anion analysis makes this development particularly interesting. The problem of non-specificity remains although in subtle fashion here. "Specificity" varies with concentration of "neutral" species, the extent of this effect being a matter of current debate.

TABLE IV.

ELECTRICAL METHODS

Sensitivity Limit (M/l)	Method
10^{-4}, 10^{-5}	AC polarography, chronopotentiometry Thin layer coulometry
10^{-5}, 10^{-6}	Coulometry at Controlled V Classical Polarography, Specific Electrodes
10^{-6}, 10^{-7}	Linear Sweep, Derivative Polarography Eddy Current Decay
10^{-7}, 10^{-8}	Pulse polarography Amperometry, Coulometric Titrations
10^{-8}, 10^{-9}	Anodic Stripping with hanging Hg drop
10^{-9}, 10^{-10}	Anodic Stripping with thin film electrodes or solid electrodes

Phase Equilibria and Stoichiometry

The analysis of refractory hard metals is often complicated by variations in stoichiometry within a stable phase. The zirconium-carbon phase diagram shown in Figure 2 is particularly interesting because of the broad single phase field of zirconium carbide from 38.5 to approximately 50 atomic % carbon (14). This is an interstitial compound where carbon atoms occupy the interstitial sites of the FCC lattice. The significance of this broad homogeneity range is that approximately one-fourth of the interstitial sites can be empty without a change in phase. Extreme care must be exercised in preparing a zirconium carbide body with given stoichiometry. Zirconium carbide is presented here as an example of the specificity in the process step which may be demanded when the material of desired characteristics is close to a phase boundary (15). For zirconium carbide a material containing 39 atomic % carbon

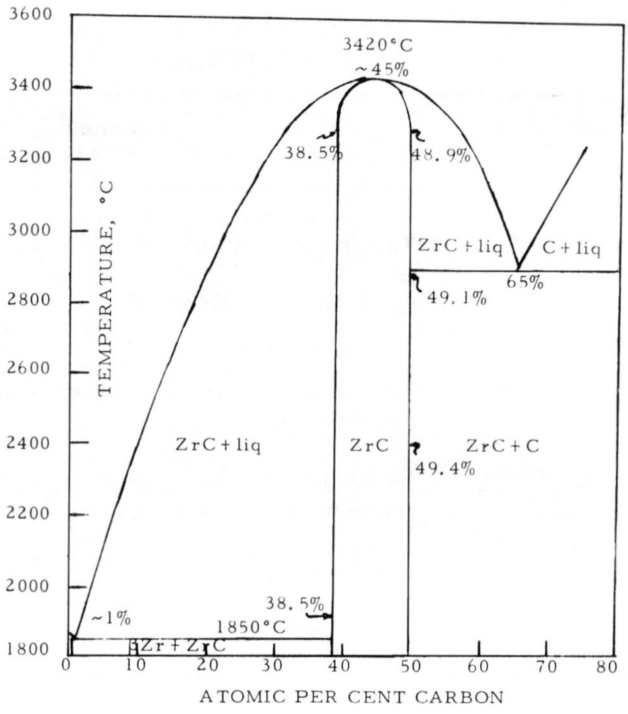

Fig. 2. Zirconium–carbon phase diagram.

would not melt below 3300°C, although a body containing 38 atomic % carbon would produce a liquid phase at 1850°C. In the zirconium carbide system this 1 atomic % difference in carbon is equivalent to approximately 3.5 parts per thousand by weight. This small weight change is extremely difficult to control in the complicated operation of high temperature processing. At the other side of the zirconium carbide homogeneity region, the boundary between zirconium carbide and zirconium carbide plus carbon changes with temperature from 49.4 at 2400°C to 48.9 at 3250°C. A starting composition of single phase zirconium carbide containing over 49.1 atomic % carbon would contain a liquid phase at 2850°C which is 500°C below the expected melting point for the material. Since impurities may effect the zirconium-carbon ratio, they become increasingly important whenever a material is desired with a composition close to the phase boundary.

The stoichiometry of compounds of AgMg and NiAl has a tremendous effect on the ductility of the material. A fractional excess of Mg or Al embrittles their respective intermetallics and increases the grain boundary hardness significantly over the grain bulk hardness. In Figures 3 and 4, the variation of hardness with stoichiometry for these intermetallics is shown [16]. Ternary solute additions have been found to modify the grain boundary hardening effect which further complicates consideration of the mechanism responsible for increased hardness. Regardless of mechanism, however, stoichiometry and the specific location of the impurities in these intermetallic compounds are often the most important factors determining strength and ductility. The demands placed on analysis for these compounds has been met by improvised alterations to existing wet analytical techniques.

The determination of nitrogen at the grain boundary was accomplished by a controlled dissolution of the individual grains at a measured rate followed by determination of nitrogen in the dissolved sample. It is difficult to determine oxygen and nitrogen simultaneously [17]. A method for this determination in refractory metals was recently reported using a DC carbon arc, gas chromatographic technique [18]. The refractory metal was fused with a platinum flux and the liberated oxygen and nitrogen analyzed by gas chromatography. This is an example of a recent approach to

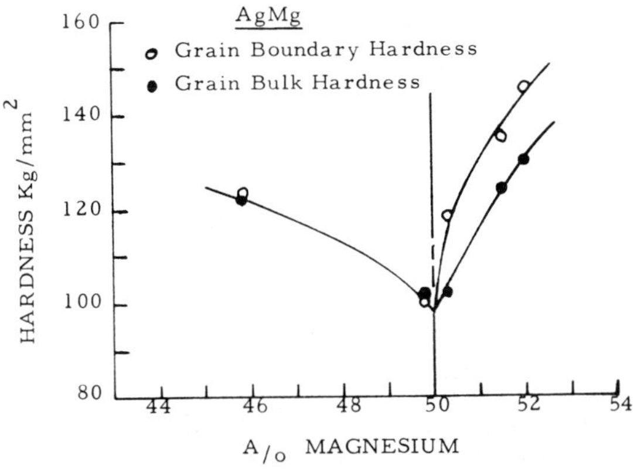

Fig. 3. Bulk and grain boundary hardness of AgMg as a function of composition.

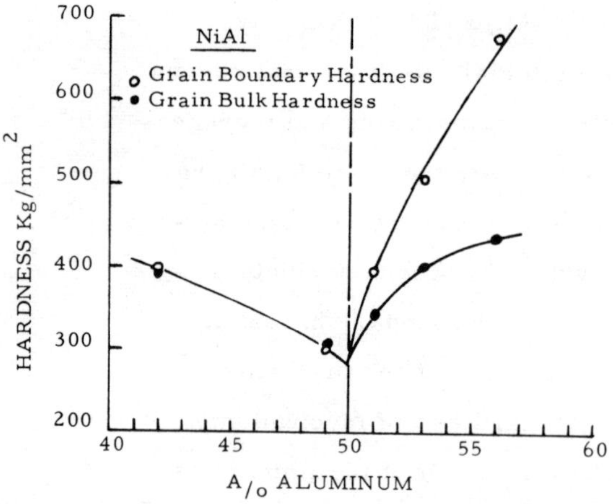

Fig. 4. Bulk and grain boundary hardness of NiAl as a function of composition.

refractory metals which might have interesting ramifications in the analysis of inorganic materials and intermetallic compounds. The measurement of rate of evolution of the gas vs. time should give equally good analytical data for grain boundary vs. matrix impurities as is obtained by timed dissolution experiments. Since the analysis is by gas chromatography the results would be much easier to obtain.

Wet-Chemical Methods

Wet Chemical methods are the backbone for standarization of instrumental techniques, a factor often overlooked by instrumentation analysts. Many wet chemical methods have applicability to refractory materials analysis through techniques of microanalysis. The field of catalytic analysis has been generally neglected but offers an excellent tool for low PPM range analysis. When the catalytic reaction is first order with respect to the catalyst and the uncatalyzed path reaction is of negligible rate, the rate of reaction is generally found proportional to the amount of catalyst present. Some examples are the determination of 0.1 to 10 PPM iodide in solid sodium chloride utilizing iodide catalysis of the Ce(IV) oxidation of As (III) [19]. The decomposition of H_2O_2 in alkaline medium is catalyzed by Mn, Pd, and Cu, providing a method for analysis of these metals by measurement of oxygen evolution with time [20]. Fe can be determined by its catalysis of the oxidation of p-phenylene diamine by H_2O_2 [21]. The reaction of H_2O_2 with iodide in acid medium is catalyzed by many metals and has been studied for the determination of Mo, W, Fe, and V [22]. A prior separation is necessary in such a nonspecific case as this, but the sensitivity for Mo in terms of minimum detectability has been reported as 2.6×10^{-9} M which is extremely low, i.e. in the PPB range.

The problems of determining free carbon in carbides and free boron in borides, and the general methods of refractory hard metal analysis are in large measure those of wet chemistry [23]. The Kjeldahl method, sometimes reported in reviews as inadequate for nitrogen determination because nitride nitrogen is determined, is a good illustration of inadequate understanding of methods. Nitrogen present in nitro- or nitrate forms can be determined by prior sample treatment with a salicylic acid-sulfuric acid mixture followed by a reduction of the nitro compounds to amines with thiosulfate [24].

One of the major problems in analysis is the determination of free carbon in the carbides and free boron in the borides. The insolubility of free graphitic carbon in acid mixtures which can be used to dissolve these compounds is widely used as a method of analysis [23]. Critical evaluation of results indicates that there is a small error due to incomplete dissolution of the sample and/or partial oxidation of the graphite. The method of determining free carbon is, however, superior to the methods that have been developed for determining free boron in the borides. New wet chemical methods have been developed in recent years, however, for some of these compounds. Free boron has been determined in boron carbide by a method based on the different rates of oxidation in acid solution of free boron and carbide-bonded boron [25]. The oxidation of amorphous boron into boric acid has been used to determine free boron in zirconium diboride with a relative error of less than 1% [26]. In this particular study a hydrogen peroxide-nitric acid solution was employed. This solution and other oxidizing solutions such as potassium iodate, cerium sulfate, and sulfuric acid partially dissolved refractory diborides of titanium, niobium, tantalum, etc. so that an accurate free boron analysis was not achieved [26]. More work should be done on differential dissolution techniques to develop a satisfactory free boron determination for all these compounds. A simplification in metal determination in refractory borides has been reported wherein the metal is complexed and the boron titrated directly [27]. This eliminates a precipitation step in the determination. Another new technique is pyrohydrolysis to determine total boron as the acid. The latter method eliminates the solubilizing step [28].

Difficulties in the analysis of ZrB_2-$MoSi_2$ composites are apparent from the results of wet chemical analyses recently obtained [29]. This is an example of refractory material on which data generation was being accomplished where the chemical techniques of analysis were inadequate. The analytical results were so erratic that attempts were made to use emission spectrographic and electron microprobe techniques for quantitative determinations. Neither of these methods are comparable in precision to wet analytical methods.

Instrumental Methods

The work-horse method for instrumental analysis is emission spectroscopy which serves as an excellent qualitative tool, but has definite quantitative analytical limitations. Inherent in this analysis is the problem of electric spark/arc stability. The laser beam microprobe offers an additional tool for microanalysis of ceramic materials. Used in conjunction with an emission spectrograph it gives localized impurity analyses of a volume of material 50 microns in diameter by 25 microns in depth. With further refinement, this area/depth limit can currently be reduced to half, and potentially a 1 to 2 micron cross-section must be considered possible. The sensitivity limit of emission spectroscopy is close to 1 PPM for most elements. The demands for low PPM into PPB range analyses have led to increased use of more sensitive techniques for trace analysis. Foremost of the more sensitive methods is the mass spectrograph. New sources for utilizing the mass spectrograph for refractory materials are the spark and the ion-beam [30]. PPB sensitivities are easily obtained for all elements except adsorbed gases (O_2, N_2, trace hydrocarbons, etc.) and water, where the background interference is severe. For other elements a 1 PPB sensitivity limit is the experimental limit when good technique is used to enhance signal/noise ratio. An example of this is the split-plate technique to reduce fogging of analytical plate by the high intensity of the primary beam of the matrix element(s). At a routine nominal 20 KV excitation potential consideration must be given to the variable energy distribution of ions of various materials. Ions may be present at 18-22 KV for an element such as Si and sharply close to 20 KVA for Ge. This gives rise to Relative Sensitivity Coefficients and results must be corrected accordingly. Difficulties in trace analysis from laboratory to laboratory become apparent in the large scatter in results from laboratory to laboratory using spark source double-focusing mass spectrographs. Knowing a trace element to a factor of 10 has often been the case under <u>favorable</u> conditions. Some problems encountered are due to the small amount of material sparked and analyzed which is not representative of the whole sample. Most "standards" in practice are not homogeneous under such mass-spec analysis. Results for matrix elements are always considerably better, and have been reported within 5% standard deviations for "ideal" examples. The results of Nicholls et al [30] and experience at the National Bureau of Standards [7] indicates that excellent results can be achieved in mass spectral analysis for trace elements with careful consideration of improved procedures. The spark source mass spectrometry has been employed to determine impurities in non-conducting ceramics [31]. A high purity gold probe was placed close to MgO in conducting mount. With a minimum projection of the MgO beyond the sample holder, a 50 KV, 1 Mc rf spark was struck between the gold probe and the MgO producing a low yield spark. As the sample temperature increased a high yield spark was obtained.

Neutron activation analysis [32] is an extremely sensitive tool for trace analysis and should be widely employed whenever standard methods can be established. Advantages include the lack of matrix effects which make standards simple and obtainable, nondestructive testing, the use of large or small samples, speed, high precision, and accuracy (better than ±10% relative precision routinely), and no need for sample preconcentration or pre-treatment (except radiochemical separation). Difficulties include the nature of equipment required with the associated radiation hazards, self shielding of specimens, variable isotope lives, and interfering nuclear reactions. The specimen often contains long-lived isotopes, after activation, that make subsequent routine handling impossible. Some isotopes have such short half-lives that short-irradiation times are necessary which create experimental problems with other isotopes requiring longer irradiation before obtaining the gamma-ray spectra.

Many light elements are not adequately activated by thermal neutrons (C, N, O, e.g.) and need high energy gamma-ray activation (using a linear accelerator). This produces short-lived isotopes which are hard to discriminate spectrally. Sensitivity for most elements is also much lower using this technique. Spark source mass spectrometry and the emergence of atomic absorption spectroscopy coupled with the classic flame emission technique for lighter elements seem to circumscribe the future applications of neutron activation analysis. Atomic absorption is almost as sensitive as neutron activation. Matrix effects are fairly small and development of standards is not difficult. The method is very good quantitatively, utilizes simple apparatus, and is rapid and

TABLE V.

DETECTION LIMITS IN ATOMIC ABSORPTION

Detection Limit (μg/ml)	Element
0.5	Aluminum *
0.2	Antimony
1.0	Arsenic
1.0	Barium
0.05	Beryllium *
0.2	Bismuth
0.01	Cadmium
0.01	Calcium
0.05	Cesium
0.01	Chromium
0.15	Cobalt
0.005	Copper
1.0	Gallium
0.1	Gold
0.5	Indium
0.05	Iron
0.15	Lead
0.005	Lithium
0.003	Magnesium
0.01	Manganese
0.5	Mercury
0.2	Molybdenum
0.05	Nickel
1.0	Palladium
0.5	Platinum
0.005	Potassium
0.3	Rhodium
0.02	Rubidium
1.0	Selenium
0.02	Silver
0.005	Sodium
0.02	Strontium
0.5	Tellurium
1.0	Titanium *
0.2	Thallium
2.0	Tin
0.5	Vanadium *
0.005	Zinc

*Requires organic solvent and oxyacetylene burner.

reproducible. For refractory materials analysis it should become a principal quantitative tool following qualitative/quantitative estimate work by emission spectroscopy [33].

The application of atomic absorption is extremely broad after only a few years of commercial development. Hollow cathode lamps are now available for some 70 elements and multiple-element lamps are made. One is available, for example, for Cr, Co, Cu, Fe, Mn, and Ni. Another is available for Ba, Sr, Ca, and Mg. Over 25 such combinations are produced. The detection limits for a number of elements are given in Table V [33]. A recent comprehensive study of igneous rock standards by Atomic Absorption and X-ray Fluorescence analysis for principle elements shows the former method equally as good on a routine basis for principle elements. For trace elements in igneous materials atomic absorption results were favorable to those obtained colorimetrically. It is the specificity of the spectral determination which

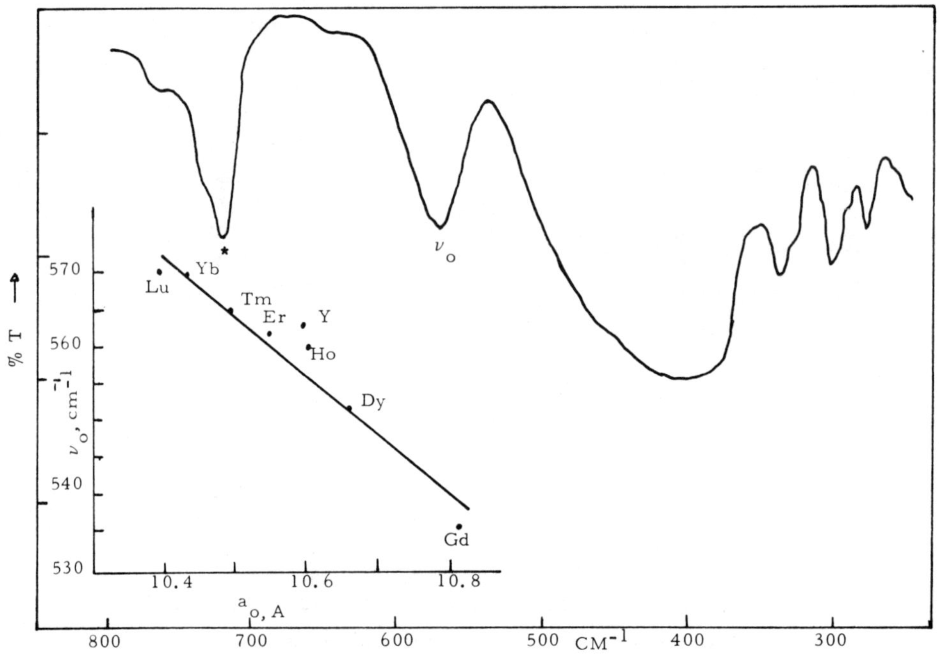

Fig. 5. Spectrum of a normal type C rare-earth oxide, ytterbia. (Graph shows a plot of frequency against unit cell dimension of type C oxides.) * Nujol

makes it superior to colorimetric methods. Very few color complexes can be found specific to one element, making prior sample treatment necessary to remove interferences.

Infrared spectroscopy is usually thought of in terms of organic analysis. The development of instruments for analysis in the medium and far-infrared regions of the spectrum place many inorganic compounds within analytical possibility. The refractory oxides have significant spectral changes in the medium infrared region for phase changes, such as anatase and rutile TiO_2, monoclinic and cubic ZrO_2. Figure 5 shows the change in frequency of a principle absorption peak obtained for various rare earth oxides in the 500 to 600 cm^{-1} region [34]. As yet the infrared spectra of refractory compounds has received limited attention.

Nuclear magnetic resonance has generally proven disappointing in studying inorganic solids. As magnets get larger, however, the potential for useful analyses should increase. Nuclear quadrupole resonance is insufficiently developed at this stage for any conclusions on its usefulness for these materials. It is an extremely complex method to understand, though, which places limits on its impact. The Mossbauer effect is of tremendous importance because the supremely sharp gamma ray lines make possible measurements of the Mossbauer effect with great precision. Very small influences on the nuclear electric and magnetic properties can be detected. The method has definite applicability to inorganic solids, commercial instruments are now available, and it should develop into an important tool. Light elements are unsatisfactory, however, because gamma ray energies E tend to be high, increasing the recoil energy ($\sim E^2$). In this case light elements include all the first three principal series elements and some of the transition elements which makes it an important limitation.

X-ray diffraction is an applicable tool when the cross-section for analysis is fairly large. On a practical basis, this limit is often as large as 40 microns, and it is difficult to realize effective analyses at less than 20 microns. The development of electron microscope equipment with electron diffraction devices, on the other hand, has provided a tool for analyzing below 1 micron in cross-section for a skilled operator using thin film transmission, or, in some cases, reflection techniques. This method must be considered potentially as the most exciting of the instrumental methods for locational analysis.

The electron microprobe can routinely analyze to cross-sections of 1 micron with a limitation that elements of low atomic number (Na^{11} and below) cannot be satisfactorily analyzed. Soft x-ray methods are available which can be used to analyze for the light elements provided reasonable concentrations (for example, 1% C in Fe) are present. When one considers trace analysis, however, light element determinations by the electron microprobe are very unsatisfactory. An indirect method of measuring the beam current from the specimen is used to estimate light element concentrations qualitatively, since the back-scatter electron effect in the specimen is proportional to the average atomic number of the material on which the beam impinges. The laser beam and the ion beam can be used as source exciters for a number of instrumental methods, such as the use of the laser beam with a mass spectrograph. This has been used with good results on graphite analysis [35]. A laser beam could also be coupled with atomic absorption equipment thus eliminating problems associated with the flame and offering locational determination. Practical limits of cross-sections with a laser beam are thus far unsatisfactory, but it should be within the state-of-the-art to develop a 1 micron cross-section capability.

In conclusion, the purpose of this report is to give some dimension to selected problems associated with developing satisfactory techniques for chemical analysis of refractory materials. Without attempting to cover every possible method or potentially critical problem, it is hoped that a sufficient breadth and depth has been given to this treatment to lead the materials scientist to a greater understanding of analytical problems associated with the characterization of refractory materials.

References

1. Materials Advisory Report on Ceramic Processing, MAB-197-M, November, 1963. Available from Office of Technical Services, Dept. of Commerce.

2. J. H. Westbrook and D. L. Wood, Nature, 192, 1280 (1961).

3. J. H. Westbrook and D. L. Wood, J. Inst. of Metals, 91, 174 (1963).

4. A. U. Seybolt and J. H. Westbrook, Acta Met., 12, 449 (1964).

5. A. U. Seybolt and J. H. Westbrook, "Oxygen-Induced Grain Boundary Hardening in the Intermetallic Compounds AgMg, NiGa, and NiAl," Plansee Proceedings, 1964.

6. NBS Publication 260. Office of Standard Reference Materials, NBS, Washington, D. C. Available from U. S. Govt. Printing Office, Washington, D. C., 20402.

7. John Hague, Office of Standard Reference Materials, NBS, Washington, D. C., private communication, 1967.

8. Report on Conference on Chemical Compounds of Certified High Purity, NAS, 22-23 June 1959, Washington, D. C. NSF, NAS-NRC, 1964.

9. Symposium on Standard Reference Materials and Symposium on National and International Standards, American Chemical Society, 148th National Meeting, Chicago, Illinois, Sept. 3, 1964. See also Anal. Chem., 36 [12], 23A (1964).

10. J. M. Gill and H. M. McNair, Aerograph Research Notes, pg. 1, Fall (1965).

11. H. A. Laitinen, "Trace Characterization by Electrochemical Methods," pp. 75-107 in Trace Characterization Chemical and Physical, NBS Monograph 100, Edited by W. W. Meinke and B. F. Scribner, U. S. Govt. Printing Office, Washington, D. C., 1967.

12. E. Pungor, J. Havas, and K. Toth, Z. Chem., 5, 9 (1965).

13. G. A. Rechnitz, M. R. Kresz, and S. B. Zamochnick, Anal. Chem., 38, 973 (1966).

14. R. V. Sarah, C. E. Lowell, R. T. Dolloff, "Research Study to Determine the Phase Equilibrium Relations of Selected Metal Carbides at High Temperatures," WADD TR 60-143, Part IV, February 1963.

15. C. T. Lynch and J. J. Krochmal, RTD Technology Briefs, 1 [10], 1 (1963).

16. J. H. Westbrook and D. L. Wood, "Effect of Basic Physical Parameters on Engineering Properties of Intermetallic Compounds," WADD TR 60-184, Part III, July 1962.

17. E. Booth, F. S. Bryant, A. Parker, Analyst, 82, 50 (1957).

18. R. K. Winge and V. A. Fassel, Anal. Chem., 37, 67 (1965).

19. E. B. Sandell and I. M. Kolthoff, Microchim. Acta, 1, 9 (1937).

20. T. Shiokarva and S. Suzuki, Sci. Rep. RITU, 3, 419 (1951).

21. H. Goto and S. Suzuki, Sci. Rep. RITU, 3, 429 (1951).

22. K. B. Yatsimirskii and L. P. Afanaseva, J. Anal. Chem. (USSR), 11 [3], 327 (1956). [English Translation]

23. O. W. Kriege, "The Analysis of Refractory Borides, Carbides, Nitrides, and Silicides," Los Alamos Scientific Laboratory, LA-2306, 1959.

24. W. C. Pierce and E. L. Haenisch, p. 140, Quantitative Analysis, 3rd edition, John Wiley and Sons, Inc., New York, 1948.

25. Ye. Ye. Kotlyar and T. N. Nazarchuk, J. Anal. Chem. (USSR), 15, 207 (1960).

26. L. N. Kugay, "Chemical Properties and Methods of Analyzing Borides of Transition and Rare Earth Metals," Trans. Acad. Sci. Ukranian SSR, Institute of Metal Ceramics and Special Alloys. Seminar on Heat-Resistant Materials, No. 6, Kiev, 1961.

27. V. G. Shcherbakov, R. M. Veytsman, and Z. K. Stegendo, "Analysis of Titanium, Chromium, and Zirconium Borides," ibid.

28. L. M. Litz and R. A. Mercuri, J. Electrochem. Soc., 110, 921 (1963).

29. F. M. Anthony and W. H. Dukes, "Selection Techniques for Brittle Materials," Bell Aerosystems Company Report No. 2178-900204, June 1964.

30. G. D. Nicholls, A. L. Graham, E. Williams, and M. Wood, Anal. Chem., 39, 584 (1967).

31. A. J. Socha and M. H. Leipold, J. Am. Ceram. Soc., 48, 463 (1965).

32. A. A. Smales, "Radioactivity Techniques in Trace Characterization," pp. 307-336, in Trace Characterization Chemical and Physical, op. cit.

33. Data on Atomic Absorption taken primarily from several recent issues of the Atomic Absorption Newsletter.

34. N. T. McDevitt and W. L. Baun, "A Study of the Absorption Spectra of Simple Metal Oxides in the Infrared Region 700 to 240 cm^{-1}," RTD-TDR 63-4172, January 1964.

35. P. D. Zavitsanos and L. E. Brewer, "Chemi-Ionization and Chemi-Excitation in the Oxidation of Gaseous Carbon," G. E. Rpt. R 66SD64, December 1966.

ON THE ROLE OF DIFFUSION IN THE PLASTIC DEFORMATION OF TRANSITION METAL CARBIDES

D. L. Harrod and L. R. Fleischer

Materials Department
Westinghouse Astronuclear Laboratory
Pittsburgh, Pennsylvania

1. INTRODUCTION

Diffusional processes are generally rate-controlling in high temperature plastic deformation. Emphasis upon the lattice resistance to dislocation glide motion in the transition metal monocarbides, however, has tended to temper interest in diffusion controlled mechanisms of deformation in these materials. If a simple Peierls mechanism, for example, is rate-controlling up to the melting point, then diffusion processes might be reasonably neglected altogether. However, it is the contention here that the high temperature mechanism(s) of deformation in the carbides has not yet been established, and that due consideration must be given to the possible role of diffusional processes. To this end it is necessary to know the relevant diffusion parameters, particularly the activation energies, for both metal and carbon diffusion.

The primary purpose of this paper is to summarize the available diffusion data on the transition metal monocarbides of the Group IVB and Group VB metals (TiC, ZrC, HfC and VC, NbC, TaC). In order to develop a case for diffusion controlled deformation, the first section considers the general deformation behavior of the carbides. The following section summarizes the diffusion data.

2. PLASTIC DEFORMATION

On the basis of the limited amount of testing done to date (Ref. 1-8), the following features have been established as probable characteristics of all of the carbides:

- The carbides are extremely brittle at low temperatures. Extensive deformation at normal strain rates is observed only at temperatures generally well above 1000°K.

- The yield stress is strongly temperature and strain rate dependent. No yield points have been observed.

- An extremely high Peierls stress is believed to dominate the plastic deformation behavior of the carbides, at least at the lower temperatures.

- Composition has a significant effect on the yield strength of the carbides. This effect, however, appears to go in opposite directions in the Group IV and Group V metal carbides. Thus, as the stoichiometric composition is approached, strength increases in the Group IV carbides (TiC) and decreases in the Group V carbides (TaC).

- The slip system has been identified as {111} <110>. The slip system, therefore, is like that observed in the f. c. c. metals, as opposed to the {110} <110> system usually observed in crystals with the NaCl-type structure, which is the structure of the monocarbides.

- The features of the dislocations observed in the carbides appear to be similar to those observed in f. c. c. metals of high stacking fault energy.

It is with the first three features listed above, and perhaps the last one also, that the following discussion is concerned.

2.1 DISLOCATION MOBILITY

The onset of plasticity at high temperatures, which in a practical sense gives rise to a ductile-brittle transition, is interpreted in terms of the intrinsic temperature and stress dependence of dislocation velocity, as opposed to some sort of physical or metallurgical change which gives rise to embrittlement. Although no direct measurements of dislocation velocities in carbides have been reported, Williams (Ref. 1-4) deduced, on the basis of the strain rate dependence of the yield stress, that the stress dependence is similar to that in the covalent crystals such as Ge. This low stress dependence of dislocation velocity leads to a strong strain rate dependence of the yield stress. A high Peierls stress in connection with thermally activated overcoming of the Peierls barrier leads to a strong temperature dependence of the yield stress even at high temperatures. These features are outlined below.

The plastic shear strain rate $\dot{\gamma}$ is related to the mobile dislocation density ρ, the displacement per dislocation or Burgers vector b, and the average dislocation velocity v by

(1) $\dot{\gamma} = \rho b v$

The dislocation velocity has been related to the applied stress at a given temperature by various empirical relationships (Ref. 9), for example

(2a) $v = \left(\frac{\tau}{\tau_o}\right)^m$ or (2b) $v = v_o e^{-D/\tau}$

where τ is the applied shear stress, τ_o is the stress required to produce unit velocity, v_o is the limiting velocity, and D is a constant which Gilman (Ref. 9) calls the characteristic drag stress. Although Equation 2b may be a better description of the data, Equation 2a is quite satisfactory, and more convenient for the present qualitative considerations. Assuming that the temperature dependence of v is proportional to the Boltzmann exponential exp $(-E/kt)$, where E is the activation energy for dislocation motion, Equation 1 becomes

(3) $\dot{\gamma} = \rho b v = \rho b \left(\frac{\tau}{\tau_o}\right)^m B e^{-E/kt}$

Thus the strain rate is determined by the number of moving dislocations and their velocity, and the dislocation velocity is determined by the applied shear stress raised to the power m and by the test temperature for thermally activated motion through the Boltzmann factor.

Examples of dislocation velocity-stress curves for Ge and LiF are shown in Figure 1, where the slope $\Delta\tau/\Delta v$ is equal to 1/m. The curves for Ge are from the work of Chaudhuri, Patel, and Rubin (Ref. 10), while the curve for LiF is from the work of Johnston and Gilman (Ref. 11). Representative values of m for different materials are as follows:

Material	m	Temperature Range	Ref.
Covalent Crystals (Ge, Si)	1 - 2	> .3 Tm	10
Carbides (TiC)	1 - 10	> .3 Tm	2
b. c. c. Metals (W, Mo)	5 - 10	< .25 Tm	12(W), 13(Mo)
LiF	~ 25	< .25 Tm	11
Fe-3%Si	~40	< .25 Tm	14
f. c. c. Metals	>100		

The value of m for TiC was obtained by Williams from the strain rate dependence of the yield stress, i. e., from Equation 3, $(\partial \ln\dot{\gamma}/\partial \ln\tau)_{T,\rho} = m$.

The hatched band in Figure 1 is simply an estimate of a possible dislocation velocity-stress curve for the carbides in relation to the other materials shown, as deduced from Williams' data on TiC. The significant point from the above considerations is the low value of m for TiC, and presumably for the other carbides as well. This leads to the conclusion that the stress dependence of dislocation velocity in the carbides is low, similar to that in the covalent solids and the b. c. c. transition metals at low temperatures.

The low stress dependence of dislocation velocity, i. e., low m, gives rise to a strong strain rate dependence of yield stress. This is immediately evident from Figure 1. Thus, if $\rho b = 1$, then from Equation 1, $\dot{\gamma}$ is numerically equal to v. For LiF a large change in γ, hence v, produces only a small change in τ, while for Ge and the carbides a very large change in τ is required to produce the change in dislocation velocity necessary to accomodate a change in the imposed strain rate $\dot{\gamma}$. This is also apparent from Equation 3, which can be rewritten explicitly in terms of stress as follows

$$(4) \quad \tau = \frac{\tau_0}{(\rho b B)^{1/m}} (\dot{\gamma})^{1/m} e^{+E/mkT}$$

Thus, for constant T, ρ, etc., the change in yield strength τ for a given change in strain rate $\dot{\gamma}$ is larger the smaller the value of m. For m = 1, τ is linearly proportional to $\dot{\gamma}$, while for large m, say 100, τ is quite insensitive to strain rate.

The temperature and strain rate dependence of the yield stress is made even more apparent if the data in Figure 1 are replotted in terms of the stress required to produce a given dislocation velocity as a function of temperature. This is done for Ge in Figure 2 (see the numbered points at v = 10^{-4} cm/sec in Figure 1). Also shown in Figure 2 is a curve for W (Ref. 12) and a plot of Williams' data on the yield stress of single crystal TiC at a strain rate of 10^{-4} sec^{-1}. The two curves shown for Ge again illustrate the strong strain

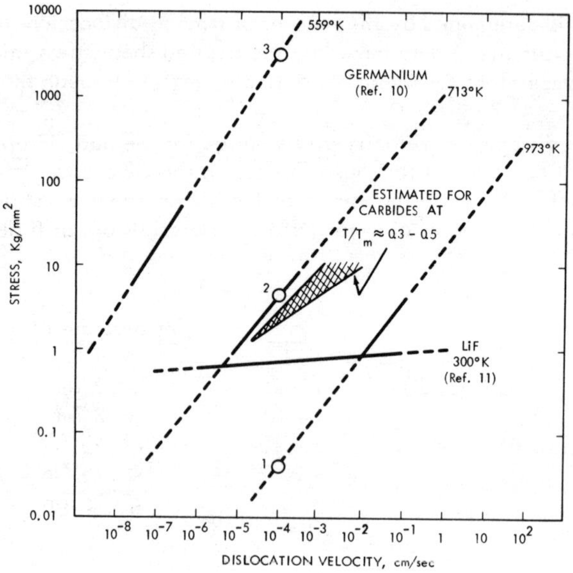

Fig. 1. Stress dependence of dislocation velocity.

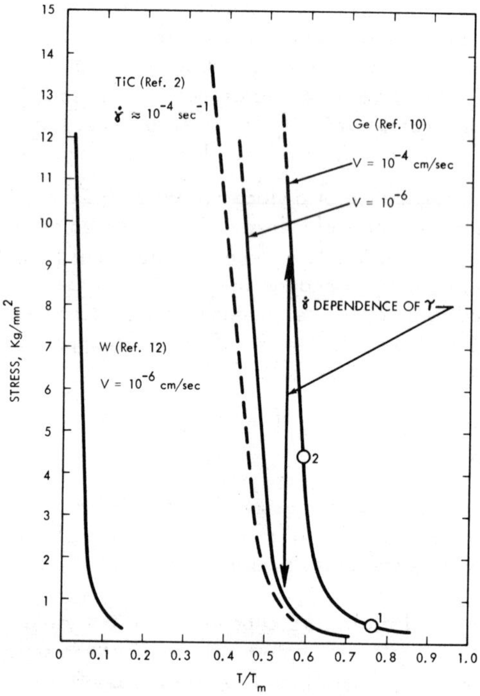

Fig. 2. Temperature and strain rate dependence of yield stress.

rate dependence of deformation in these materials. Since constant strain rate tests are generally run at rates from 10^{-5} to 10^{-3} sec^{-1}, it is clear that at low temperatures, e. g., room temperature, a very high stress is required to produce a sufficiently high dislocation velocity in the carbides such that the product ρbv, even for high ρ, equals the imposed strain rate. It follows that at the lower temperatures the competing process of fracture intervenes, and brittle fracture occurs before the applied stress ever gets high enough to produce extensive dislocation motion, hence plastic deformation.

The strong temperature dependence of each of the materials shown in Figure 2 is generally interpreted in terms of the Peierls mechansim. The strength of the temperature dependence, per se, is governed by the activation energy, which at low stresses is about 0.55 ev for V, 1.8 ev for W (Ref. 15) and about 1.6 ev for Ge (Ref. 10). A significant point about the temperature dependence of deformation in the carbides and covalent crystals is the high temperature at which the dependence is still strong. This implies that the Peierls stress at $0^{\circ}K$ must be extremely high in the carbides and covalent solids, as compared with values of about 122 Kg/mm^2 (175,000 psi) for W and about 62 Kg/mm^2 (88,000 psi) for V (Ref. 15).

2.2 MECHANISMS

The fact that dislocations in the carbides are relatively immobile, i. e., require a rather high stress to move, at low temperatures does not, in itself, specify the dislocation mechanism of deformation. However, as indicated above, the general opinion seems to be that the intrinsic resistance of the otherwise perfect lattice to the motion of dislocations, i. e., the Peierls barrier, is quite high in the carbides due to the strong metal-carbon bonding. No attempt will be made here to show that the available test data are consistant with the Peierls mechanism; rather, the point of view adopted is to assume that overcoming of the Peierls barrier is the rate-controlling dislocation mechanism at low temperatures, but that this is still consistent with the deformation at higher temperatures being diffusion controlled.

For purposes of perspective, Figure 3 shows a plot of yield stress vs. homologous temperature for three classes of solids, b. c. c. metals such as W, Ta, Nb, V, and Mo, the transition metal carbides, and pure covalent solids such as Ge, Si and diamond. This plot is intended to be primarily qualitative, but it is not believed that quantitative errors will alter the conclusions.

The upper curve, Curve 1, depicts the theoretical shear stress of the perfect (dislocation) free) crystal. The lower curve, Curve 2, for the pure metals depicts the marked decrease in strength from that of τ_{max} due to the presence of a high density of mobile dislocations in these metals. The strong rise in yield and flow stress in these metals at temperatures below 0.1-0.2 Tm has been interpreted by Conrad (Ref. 15) and later by Dorn and Rajnak (Ref. 16) as being due to a strong Peierls lattice friction stress. The yield stress at $0^{\circ}K$ is equal to the Peierls stress at absolute zero, τ^{o}_{p}, and for these metals τ^{o}_{p} is less than about 150,000 psi.

The hatched portion of the curve for the carbides depicts the present range of experimental data. Assuming that the Peierls mechanism is operative in this temperature range (0.3-0.6 Tm), it follows that the Peierls mechanism ought also to be controlling at all lower temperatures. Thus the extrapolated critical resolved shear stress (extrapolated analytically according to Williams' data on TiC) at $0^{\circ}K$ gives the Peierls stress τ^{o}_{p}. Any

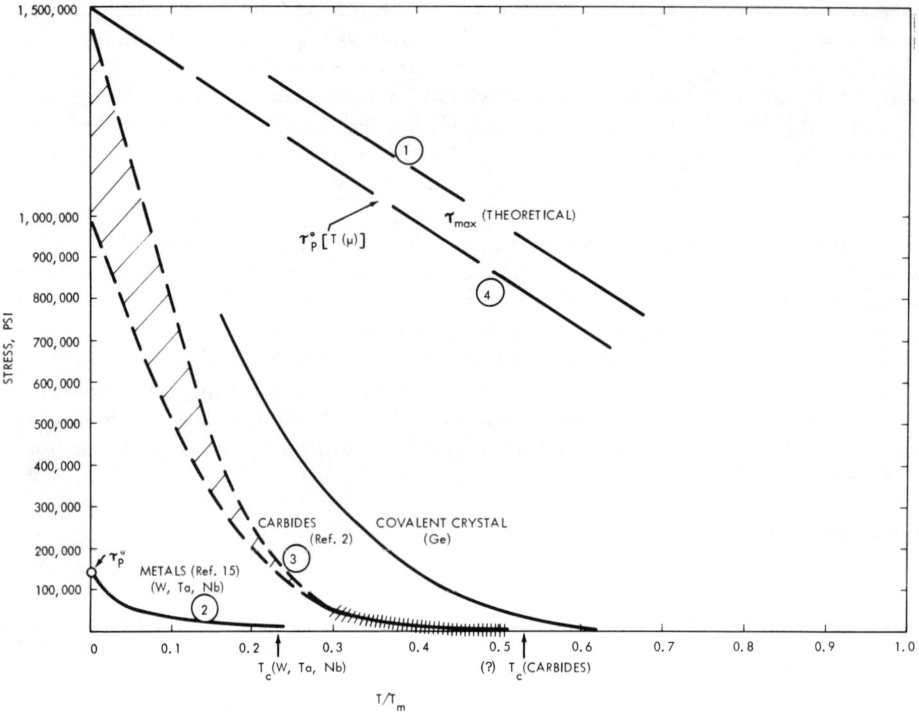

Fig. 3. Temperature dependence of yield stress.

reasonable extrapolation of the available high temperature data would therefore lead to a very high Peierls stress at 0°K. Curve 4 shows the assumed temperature dependence of τ^o_p in the absence of thermal fluctuations, i. e., the temperature dependence due to the shear modulus. The strong variation of yield stress with temperature is then due to thermal activation over the Peierls barrier by double kink nucleation and subsequent lateral spread of the kinks along the dislocation, thus moving the dislocation forward by one atomic distance.

It is the purpose now to show that it is quite possible that the Peierls mechanism in the carbides is giving-way to higher activation energy processes at temperatures in the range 0.4-0.5 Tm. For this purpose, the Dorn-Rajnak (Ref. 16) analysis of the Peierls mechanism is employed. Figure 4 shows the stress dependence of the activation energy for double kink nucleation, where U_n is the thermal activation energy, U_k is the kink energy, τ^* is the thermal component of the applied stress, τ_p is the Peierls stress, T is the test temperature and T_c is a critical temperature above which thermal fluctuations alone can overcome the Peierls barrier even in the absence of applied stress. This plot shows that when τ^* equals the Peierls stress ($\tau^*/\tau_p = 1$), the activation energy U_n is zero, and when $\tau^* = 0$ (at $T = T_c$) the activation energy is equal to $2U_k$. Thus, as the test temperature approaches T_c, the Peierls barrier is overcome with such ease by thermal fluctuations alone that higher activation energy processes become rate controlling, or else the athermal component of the stress τ_μ ($= \tau_{applied} - \tau^*$) dominates.

The cross-hatched area in Figure 4 depicts experimental data on the carbides assuming that $T_c = T_m$ and $\tau_p \sim 10^6$ psi. Thus, for yield stresses less than 10^5 psi at

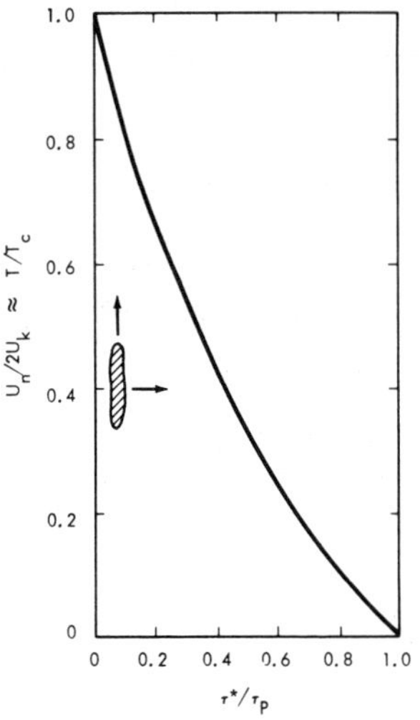

Fig. 4. Energy to nucleate a pair of kinks (after Dorn and Rajnak, Ref. 16). Hatched area is the estimated range of present experimental data assuming $T_c = T_m$ and $\tau_p \approx 10^6$ psi.

test temperatures less than $T_m/2$, $\tau^*/\tau_p < 0.1$ and $T/T_c < 0.5$. To move the hatched area onto the theoretical curve requires that either τ_p or T_c be considerably reduced. The former argues against a high Peierls stress, and the later argues against the Peierls mechanism being rate controlling above about $T_m/2$. Conrad (Ref. 15) has shown that the critical temperature T_c in the b. c. c. transition metals is in the range 0. 2-0. 25 T_m, as indicated in Figure 3, and it seems reasonable that the corresponding temperature in the carbides is no higher than about 0. 4-0. 6 T_m.

2. 3 CREEP

Following Weertman (Ref. 17), Figure 5 shows a normalized creep diagram in which the coordinates are the applied shear stress divided by the shear modulus and the test temperature divided by the melting point. The various regions of the diagram depict the types

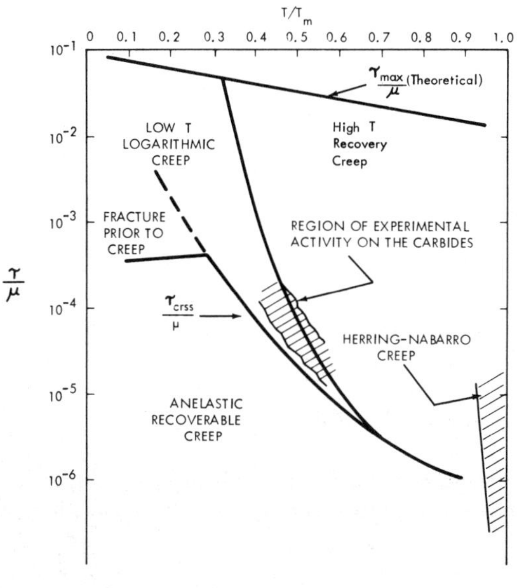

Fig. 5. Estimated creep diagram for carbides (after Weertman, Ref. 17).

of creep behavior usually observed, depending upon the stress and temperature. The diagram is drawn to be consistent with the assumption that the Peierls mechanism is giving way to higher activation energy processes at temperatures approaching $T_m/2$.

The hatched area in Figure 5 is an estimate of the region in which most of the experimental studies on the carbides have been conducted to date. The point to be made is that this may well be the region in which a change is occurring in the rate-controlling mechanism. If so, then this would be precisely the region in which it would be most difficult to experimentally isolate the deformation mechanism. Depending upon conditions, the Peierls mechanism might control, a higher activation energy process might control, or both might contribute. In this connection it is noted that any other mechanism, excepting Herring-Nabarro creep which is not a dislocation mechanism, must operate in series with the Peierls mechanism; that is, the Peierls barrier must always be overcome for dislocation glide motion.

Some experimental results from the literature on the activation energy and stress dependence of deformation in the carbides are summarized in Table 1. Because of possible differences in interpretation of the apparent activation energies and stress dependence of deformation for different types of tests, the expressions by which these quantities were calculated are listed in the table. From the results shown in Table 1 the following points are noted: (i) The test temperatures range from about $0.3\ T_m$ to $0.7\ T_m$, but for a given set of tests the temperatures were either in the low or the high part of the range. (ii) The range of activation energies is so broad as to encompass possible values for almost any thermally activated mechanism. Some of the results shown in Table 1 were interpreted by the indicated investigators in terms of a Peierls mechanism, while other of the results were explained in terms of diffusion controlled deformation. (iii) The stress dependence of deformation also covers a range, but the dependence in general seems to be low. (iv) The range of stresses employed in creep is not large. Compared with τ^o_p, and presumably τ_p at the test temperature, the ratio of applied stress to the Peierls stress is quite low, and according to Figure 4 the expected variation in activation energy due to this small range of stress variation is probably within experimental errors.

3. DIFFUSION

The previous section considered the qualitative aspects of the deformation of carbides, and it was recognized there that diffusional processes may play a role in deformation at high temperatures. Quantitative analyses of diffusion-controlled deformation mechanisms require a knowledge of the appropriate diffusivities, and in particular their temperature dependence as given by the activation energy. Therefore, this section summarizes the available diffusion data on the Groups IVB and VB transition metal carbides.

3.1 CARBON DIFFUSION IN CARBIDES

The field of carbon diffusion in the Group IVB and VB transition metal monocarbides was reviewed by Tobin, et al (Ref. 18) a year ago. Since that time some of the values reported have been changed by further work, and accounts of new determinations have been published. A summary of the revised and updated data is tabulated in Table 2.

Most of the values contributing to this summary were determined by the layer growth method, and therefore share the inherent inaccuracies arising from assumptions such

Table 1 - Literature Data on Deformation of Carbides

			Temp. Range, ^{o}K	T/TM (Absolute)	Apparent Activation Energy		Stress Dependence of Deformation	
					Q, ev	Expression for Q	m	Expression for m
Singlextl TiC	Williams (Ref. 1-4)	Compression τ_{crss}	1200-1380 1380-1750	.3 - .34 .34 - .43	.35 1.0	$Q = k\left(\frac{\partial \ln \tau}{\partial\ 1/T}\right)_{\dot{\varepsilon}}$	m = 1 to 10	$m = \left(\frac{\partial \ln \tau}{\partial \ln \dot{\varepsilon}}\right)^{-1}_{T=1400^{o}C}{}^{-1}$
		Creep $\tau_{rss} = 1.28$ ksi	1542-2021	.44 - .58	1.0	$Q = -k\left(\frac{\partial \ln \dot{\varepsilon}}{\partial\ 1/T}\right)_{t=1\ \&\ t=3\ hrs}$	$m \approx 2$	$m = \left(\frac{\partial \ln \dot{\varepsilon}}{\partial \ln \tau}\right)_{\dot{\varepsilon} \approx ss}$
							$m \approx 5.5$	$m = \left(\frac{\partial \ln \dot{\varepsilon}}{\partial \ln \tau}\right)_{\dot{\varepsilon}\ initial}$
Polyxtl TiC	Keihn & Kebler (Ref. 7)	Creep σ = 7-8 ksi	1911-2082	.47 - .52	5.6 - 7.6	$Q = -k\left(\frac{\partial \ln \dot{\varepsilon}}{\partial\ 1/T}\right)_{\dot{\varepsilon} = ss}$		
Polyxtl TaC	Steinitz (Ref. 6)	Creep σ = 6.6-9.8 ksi	2233-2373	.54 - .57	7.4	$Q = -k\left(\frac{\partial \ln \dot{\varepsilon}}{\partial\ 1/T}\right)_{\dot{\varepsilon} \approx ss}$	$m \approx 2$ to 4	$m = \left(\frac{\partial \ln \dot{\varepsilon}}{\partial \ln \sigma}\right)_{\dot{\varepsilon} = ss}$
Polyxtl ZrC	Leipold & Nielsen (Ref. 8)	Creep σ = 0.5-2 ksi	2073-2423 2423-2773	.55 - .64 .64 - .73	3.3 8.7		$m \approx 3$	$m = \left(\frac{\partial \ln \dot{\varepsilon}}{\partial \ln \sigma}\right)_{\dot{\varepsilon} = ss}$
Singlextl TiC	Hollox & Smallman (Ref. 4, 5)	Compression τ_{crss}	1173-1473	.33 - .42	.74	$Q = k\left(\frac{\partial \ln \tau}{\partial\ 1/T}\right)_{\dot{\varepsilon}}$		

Table 2 - Summary of Carbon Diffusion in Groups IV B and V B Metal Monocarbides

Carbide	Temp. Range, °C	Do, cm²/sec	Q, kcal/mole	D, cm²/sec	Method	Source	Ref.
TiC	1300-1500	0.1	62.0±9.0		Layer growth	Vansant & Phelps	25
	1800-2700	0.1	58.7		Layer growth	Adelsberg & Cadoff	26
	800-1400	2.44×10^3	17.5±5.7		Layer growth	Samsonov & Latysheva	27, 28
ZrC *	1800-2300		91.5		Vapor deposition	Blocher, et al	29
	1650-2000	10	79.0		Vapor deposition	Gert & Babad-Zakhryapin	30
	2000-2860	0.95	78.7		Layer growth	Adelsberg, et al	31
	1000-1600	4.52×10^3	17.9±5.6		Layer growth	Samsonov & Latysheva	27, 28
$ZrC_{0.96}$	1600-2100	3.3×10^2	114.0		C^{14} tracer-self diffusion	Andrievskii, et al	32
HfC	2300-3000	8.6	101.7		Layer growth	Adelsberg & Cadoff	26
VC	2160-2670	0.26	58.3		Layer growth	Adelsberg & Cadoff	33
NbC	1700-2200	7.6	88.2		Layer growth	Brizes, et al	23
	1900-2300	1.2	79.0		Layer growth	Resnick, et al	34
	1205-1816	0.102	72.7		Layer growth (in Nb-1Zr)	Bornstein, et al	35
$NbC_{0.98}$	1700-2200	1.0×10^{-6}	32.0		C^{14} tracer-self diffusion	Gel'd & Lyubimov	22
$NbC_{0.75}$	1700-2200	8.8×10^{-6}	35.3		C^{14} tracer-self diffusion	Gel'd & Lyubimov	22
$NbC_{0.94}$	2670			8×10^{-7}	Estimated from vaporization data	Fries	36
TaC	2100-2650	1.38	89.6		Layer growth	Brizes	37
	1800-2700	1.04	86.0		Layer growth	Resnick, et al	34
	1700-2700	0.18	85.0±3.0		Matano analysis in sub-stoichiometric carbide	Resnick & Seigle	19

*Note: Very recent tracer diffusion data of Sarian and Criscione (Ref. 38), $D^* = 1.32 \times 10^2 \exp(-113,200/RT)$, for C in ZrC agrees well with the data of Andrievskii, et al shown in Table 2.

Table 3 - Self Diffusion in Refractory Metals and Their Monocarbides

Carbide	Metal Diffusion in Carbide						Self Diffusion in Metal*		
	Estimated**	Experimental					Estimated**	Experimental***	
	Q, kcal/mole	Q, kcal/mole	D_o, cm^2/sec	Temp.	Method	Ref.	Q, kcal/mole	Q, kcal/mole	D_o, cm^2/sec
TiC	116.2(5.05 ev)						73.6(3.2 ev)	6.0 (2.61 ev)	1.09
$TiC_{0.99}$		121 (5.25 ev)		1200–1400	estimated from annealing rate of dislocation loops	4, 5			
ZrC	130.8(5.68 ev)						81.0(3.52 ev)	24–30(1.01–1.3ev)	4.25×10^{-5}–2.4×10^{-4}
HfC	149.3(6.49 ev)						94.8(4.12 ev)	38.7 (1.68 ev)	1.2×10^{-3}
VC	104.4(4.54 ev)						82.5(3.58 ev)	61 (2.65 ev)	1.1×10^{-2}
NbC	133(5.78 ev)						104.3(4.54 ev)	95 (4.12 ev)	1.3
$NbC_{0.98}$		55 (2.39 ev)		1700–2200	tracer diffusion	22			
TaC	151.5(6.58 ev)						124.2(5.4 ev)	98.7 (4.28ev)	1.24

*Values given for metals in high temperature allotropic form

**Estimated from LeClaire's Relation, Q = 38 Tm cal/mole ("Progress in Metal Physics", Vol. I, p. 306 (1949)

***J. Askill, ORNL-3795 (May 1965) and Supplement (June 1966).

as concentration independent diffusivities and linear concentration gradients. These values show good agreement between the different investigations.

All of the experimental methods except the C-tracer technique yield the chemical diffusivity D_c, whereas the tracer technique gives the self-diffusivity D_c^*. These two diffusivities are related by the Darken (Ref. 20) equation

$$(5) \quad D_c = D_c^* \left(1 + \frac{d\, \ln \gamma_c}{d\, \ln N_c}\right)$$

where γ_c is the activity coefficient of carbon and N_c is the carbon atom fraction. For negative deviations from ideal solution behavior, such as the carbides, the contribution of the thermodynamic term to the activation energy is usually of the same order of magnitude as the experimental error (Ref. 21) (about -6 kcal/mole or -0.26 ev at 3000°K). This suggests that the activation energy for chemical and tracer diffusion should not differ by more than about 0.25 ev for the same composition or for the case where the diffusivity is independent of composition. The activation energies for chemical and tracer diffusion are in reasonable agreement for ZrC but not for NbC. It is suggested that the low activation energy for the tracer data on NbC is due to high diffusivity paths and that the layer growth data are more appropriate for bulk diffusion. Similarly, the low activation energies of Samsonov and Latysheva (Ref. 27 and Ref. 28) for layer growth diffusion in TiC and ZrC is believed to be due to contributions of high diffusivity paths.

The layer growth diffusion values are considered an average diffusivity because of the variation of diffusivity across the composition range. Brizes (Ref. 23 and Ref. 37) and Resnick and Seigle (Ref. 19) have considered the variation of diffusivity with composition. They found that carbon diffusivity increases with decreasing carbon content, but the data cannot explicitly define the compositional dependence of activation energy.

It is interesting to note that Steinitz (Ref. 6) found increasing strength with increasing carbon vacancy concentration in TaC while Williams (Ref. 1 and Ref. 2) found the opposite dependence in TiC. This suggests that the nature of the bonding as influenced by carbon content may be different in the Group IVB and Group VB carbides. We speculate that this change in bonding might also influence diffusion; the result would be a smaller increase in diffusivity with increasing carbon vacancy concentration in the Group VB carbides than in the Group IVB carbides.

3.2 METAL DIFFUSION IN CARBIDES

Very little work has been done on the diffusion of the metallic species in the refractory metal carbides. Those few data that exist are presented in Table 3. Estimates of self-diffusion activation energies based on LeClaire's relation are also presented as are values of activation energies for self-diffusion in the corresponding metals. The tabulation for self-diffusion in the metals shows the disparity that can exist between experimental and estimated values of activation energy. In addition, since the activation energies for self-diffusion of metal in the carbides are expected to be higher than those of self-diffusion in the metals, the metal activation energies should provide a lower limit to the carbide activation energies.

Hollox and Smallman (Ref. 4 and Ref. 5) plastically deformed TiC single crystals and then studied the annealing kinetics of elongated dislocation loops produced by the deformation, presumably by cross slip of screw dislocations. The loops first "pinched-off" by pipe diffusion with an estimated activation energy of 3.4 ev (78.4 kcal/mole) and then annealed out with an estimated activation energy of 5.25 ev (121 kcal/mole). From their studies, they found that the activation energy for metal self-diffusion in TiC varied with carbon content. For example, they reported 4.88 ev (110 kcal/mole) for $TiC_{.88}$ and 5.25 ev (121 kcal/mole) for $TiC_{.97}$.

The activation energy for the self-diffusion of Nb in NbC was determined by Gel'd and Lyubimov (Ref. 22) using a tracer technique. Their value of 55 kcal/mole (2.39 ev) is much lower than expected as are their values for the activation energy of carbon diffusion. It has been suggested (Ref. 18 and Ref. 23) that the results were strongly influenced by the large surface area in their 10–15% porous powder metallurgy samples.

The layer growth method for carbon diffusivity discussed earlier utilizes as a simplifying assumption the negligibility of metal diffusion in the carbide compared to carbon diffusion. Brizes, et al (Ref. 23), and Resnick and Seigle (Ref. 19) performed inert marker experiments to justify the assumption. The markers used were empty microholes and "canals" respectively. Brizes, et al found the flux of Nb negligible with respect to that of C at 2100°C, but was unable to make a quantitative judgement because of cracks in the NbC layer. They found further confirmation to this conclusion in the absence of a central void when a 65 mil diameter Nb wire was completely carburized in lampblack at 2150°C. Resnick and Seigle's inert marker experiment in TaC at 2500°C gave similar results. They were able to estimate that the diffusivity of Ta was less than that of C by a factor of 80 or greater.

In view of possible interest in alloying carbides to improve their properties, the diffusivity of one metal in the carbide of another metal or alloy carbide is of interest. Baskin, Tret'yakow, and Chaporova (Ref. 24) measured the diffusion rate of radioactive Nb into TiC and equimolar NbC-TiC solid solution. Their results were of the form:

$$D_{Nb}^{TiC} = 2.4 \exp -\frac{84,000}{RT} \ cm^2/sec \text{ for Nb diffusion in TiC } (Q = 3.65 \text{ ev})$$

and $$D_{Nb}^{(Nb-Ti)C} = 4.7 \times 10^2 \exp -\frac{120,000}{RT} \ cm^2/sec \text{ for Nb diffusion in the solid}$$

solution (Q = 5.21 ev).

4. CONCLUSIONS

Some of the general aspects of the mechanical behavior of carbides have been established. For convenience, a distinction might be made between the behavior observed in three temperature ranges: (1) low temperatures, 0°K to $\sim 0.3\ T_m$, (2) intermediate temperature, ~ 0.3 to $\sim 0.5\ T_m$, and (3) high temperatures, $\gtrsim 0.5\ T_m$.

Extensive plastic deformation is observed in the intermediate temperature range. The evidence suggests a very high Peierls stress, and that overcoming the Peierls barrier is the rate controlling dislocation mechanism. However, the details of this lattice resistance to glide motion have not been established; for example, the role of double-kink nucleation versus kink motion, the role of pinning points such as jogs, point defects, precipitates, etc.

The brittleness observed in the carbides at low temperatures is not due to a discontinuity in the plastic deformation behavior. Rather, at temperatures below about 0.3 T_m the stress required to produce deformation is so high that the competing processes of fracture intervene. The technological utilization of the carbides at low temperatures is intimately related to the control and prevention of brittle fracture: the need for detailed studies of fracture mechanisms is evident.

The relative importance of conservative and nonconservative dislocation motion at temperatures near about 0.5 T_m is uncertain. However, it seems reasonable that diffusional processes will be rate controlling in the high temperature region. It is difficult to predict the appropriate diffusivity, and it might even be difficult to assess it experimentally. Simple metal or carbon self-diffusion may be controlling, or the appropriate diffusional process may involve some sort of coordinated atomic movement. The presently available diffusion data are inadequate for the purpose of analyzing diffusion-controlled deformation behavior. The disparity between self-diffusivities and chemical diffusivities casts doubt on the carbon diffusion data, and values for metal diffusion rates in the carbides are lacking. It follows that a more detailed knowledge of diffusional processes in the carbides is needed.

The details of the deformation and fracture behavior of the carbides have yet to be determined. These details must be established at all temperatures if the fundamentals of deformation and fracture are to be used to technological advantage.

ACKNOWLEDGEMENT

The authors would like to express their appreciation to William F. Brizes and J. Martin Tobin of Westinghouse Astronuclear Laboratory for helpful discussions of their current research in refractory metal carbides.

REFERENCES

1. W. S. Williams and R. D. Schaal, J. Applied Physics 33 No. 3, March, 1962, p. 955

2. W. S. Williams, J. Applied Physics, 35 No. 4, April 1964, p. 1329

3. W. S. Williams and R. G. Lye, Technical Documentary Report No. ML-TDR-64-25, Part I March 1964 and Part II April 1965.

4. G. E. Hollox and R. E. Smallman, Proc. British Ceramic Soc., July 1964, p. 211

5. G. E. Hollox and R. E. Smallman, J. Applied Physics, 37 No. 2, February 1966, p. 818

6. R. Steinitz, in Proc. of the Conf. on Nuclear Applications of Nonfissionable Ceramics, held in Washington, D.C., May 9-11, 1966, ed. by A. Boltax and J. H. Handwerk, pub. ANS, Hinsdale, Illinois (1966)

7. F. Keihn and R. Kegler, J. Less-Common Metals, 6 1964, p. 485

8. M. H. Leipold and T. H. Nielsen, J. American Ceramic Society, 47 No. 9, September 21, 1964, p. 419

9. J. J. Gilman, J. Applied Physics, 36 No. 10, October 1965, p. 3195

10. A. R. Chaudhuri, J. R. Patel and L. G. Rubin, J. Applied Physics, 33 No. 9, September 1962, p. 2736

11. W. G. Johnston and J. J. Gilman, J. Applied Physics, 30, 1959, p. 129

12. H. W. Schadler, Acta Metallurgica, 12, August 1964, p. 861

13. H. L. Prekel and H. Conrad, Acta Metallurgica, 15, 1967, p. 955

14. D. F. Stein and J. R. Low, Jr., J. Applied Physics, 31 No. 2, February 1961, P. 362

15. H. Conrad, in High Strength Materials, V. F. Zackay (Ed.), Proc. Second Berkeley International Materials Conference, June 15-18, 1964, University of California, John Wiley and Sons, 1965

16. J. E. Dorn and S. Rajnak, Trans. AIME, 230 No. 5, August 1964, p. 1052

17. J. Weertman and J. R. Weertman, pp. 793-819 in Physical Metallurgy, ed. R. W. Cahn, N. Holland Pub. Co., Amsterdam (1965)

18. J. M. Tobin, L. M. Adelsberg, L. H. Cadoff, and W. F. Brizes, in Proc. of the Conf. on Nuclear Applications of Non-fissionable Ceramics, held in Washington, D.C., May 9-11, 1966, ed. by A. Boltax and J. H. Handwerk, pub. ANS, Hinsdale, Illinois, (1966)

19. R. Resnick and L. Seigle, Trans. AIME 236, 1732-8 (1966)

20. L. S. Darken, Trans. AIME 175, 184 (1948)

21. J. E. Hilliard, Acta. Metallurgica 5, 38-40 (1957)

22. P. V. Gel'd and V. D. Lyubimov, Isv. Akad, Nauk. S.S.S.R., Otd. Tekhn, Met. i Toplivo, No. 6, 119 (1961). Trans. AEC-TR-5954

23. W. F. Brizes, L. H. Cadoff, and J. M. Tobin, J. Nuclear Materials 20, 57-67 (1966)

24. M. L. Baskin, V. I. Tret'yakov, and I. I. Chaporova, Fiz. Metal Metalloved 12, No. 6, 860-864 (1961). Trans. Phys. Metals and Metallography (USSR) 12, No. 6, 72-76 (1961).

25. C. A. Vansant and W. C. Phelps, Jr., Trans. ASM 59, 105-112 (1966)

26. L. M. Adelsberg and L. H. Cadoff, to be published in Trans. AIME (June 1967)

27. G. V. Samsonov and V. P. Lutysheva, Fiz. Metal i Metalloved 2, 309-19 (1956) and 14, 479-80 (1962)

28. G. V. Samsonov and V. P. Lutysheva, Doklady Akad Nauk S. S. S. R. 109, 582 (1956)

29. J. M. Blocher, C. J. Ish, D. P. Leiter, L. F. Plock, and I. E. Campbell, BMI-1200 (June 1957)

30. L. M. Gert and A. A. Bubad-Zakhryapin, Zavodskaya Laboratoriya 32, 970-3 (1966)

31. L. M. Adelsberg, L. H. Cadoff, and J. M. Tobin, Trans. AIME 236, 972-7 (1966)

32. R. A. Andrievskii, V. N. Zagryazkin, and G. Ya. Meshcheryakov, Symp. on Thermodynamics with Emphasis on Nuclear Materials and Atomic Transport in Solids, IAEA, Vienna, July 22-27, 1965, pub. IAEA, Vienna (1966).

33. L. M. Adelsberg and L. H. Cadoff, Personal communication (July 1966)

34. R. Resnick, R. Steinitz, and L. Seigle, Trans. AIME 233, 1915-17 (1965)

35. N. S. Bornstein, E. C. Hirakis, and L. A. Friedrich, TIM 927 (August 1965)

36. R. J. Fries, J. Chem. Phys. 37, 320-27 (1962)

37. W. F. Brizes, submitted to J. Nuclear Materials

38. S. Sarian and J. M. Criscione, J. Appl. Phys. 38, 1794-8 (1967)

ANISOTROPY IN EMISSIVITY OF SINGLE-CRYSTAL REFRACTORY MATERIALS

G. W. Autio* and E. Scala†

Cornell University
Ithaca, New York

ABSTRACT

The effect of anisotropy on normal spectral emissivity $\varepsilon_{n\lambda}$ has been investigated both experimentally and theoretically. Measurements include data on the two major crystallographic faces of single-crystal beryllium and zirconium, pyrolytic boron nitride, and pyrolytic graphite. The results were obtained at temperatures near 800° C over a wavelength range from visible to near infrared (0.60 - 11μ) in both purified argon and hydrogen atmosphere. Photomicrographs (up to 1000x), interference photographs, and x-ray orientation patterns were taken of all surfaces before and after the experiments.

The probing technique was employed to measure normal spectral emissivity which avoided complicated optics and separation of blackbody cavity and specimens from the same heating conditions. Thermocouple temperature measurements of specimen surfaces and blackbody cavity were made directly.

*Ph. D. Candidate †Professor

This research is supported by the Advanced Research Projects Agency.

A convenient correction for temperature difference between blackbody cavity and specimens was achieved by means of a macroscopic thermal relation involving the Planck Radiance.

Polarized light emissivities were measured for the A-face ("c" axes parallel to surface of the $(10\bar{1}0)$ faces) of the various anisotropic materials investigated. A comparison with theory was made to determine whether the polarized $\varepsilon_{n\lambda}$ ($\vec{E} \perp$ c-axis) of the A-face was equivalent to that of the unpolarized $\varepsilon_{n\lambda}$ of the corresponding C-face ("a" axes parallel to surface of the (0001) faces) .

Attempts to compare theory with experiment have been made based on optical constants measurements. Relating the normal spectral emissivity to other physical properties was achieved by coupling the electrical properties, conductivity and dielectric constant, with the optical properties, index of refraction and extinction coefficient. The comparison of theory with experiment depended on such practical considerations as surface preparation, degree of preferred orientation, and temperature. In the free-carrier intraband absorption region (Hagen-Rubens range, $\gtrsim 10\mu$) the anisotropy in $\varepsilon_{n\lambda}$ can usually be predicted from the observed anisotropy of dc electrical conductivity along the major crystallographic directions.

1. Introduction

Studies of the effects of anisotropy on normal spectral emissivity have become of interest only within the last few years[1-2]. The present investigation includes $\varepsilon_{n\lambda}$ data as a function of crystallographic orientation for single crystal beryllium and zirconium, pyrolytic boron nitride, and pyrolytic graphite. Effects of deposition temperature, surface roughness, and annealing on $\varepsilon_{n\lambda}$ for pyrolytic graphite and results for vitreous carbon and boron-doped pyrolytic carbon are reported elsewhere[3].

By definition, the normal spectral emissivity, $\varepsilon_{n\lambda}$, is the ratio of spectral radiant energy emitted by a real body to that emitted by an ideal blackbody, both under equivalent conditions. Equivalent conditions ideally imply that for both the blackbody and real body the spectral radiant energy must be emitted normally from the same amount of surface area into the same solid angle at the same temperature.

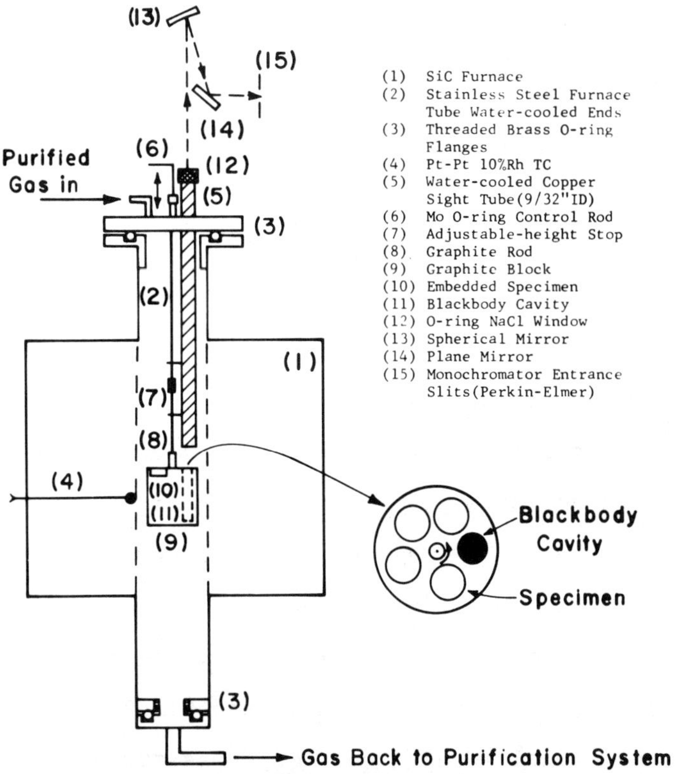

Fig. 1. Specimen furnace.

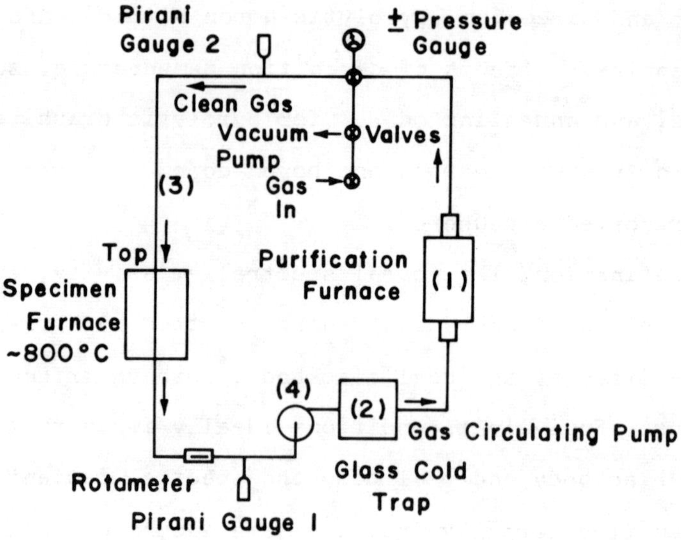

(1) (a) Resistance Tube Furnace(Max. T = 1070°C)
(b) Stainless Steel Tube Water-cooled ends
(c) Al_2O_3 boats filled with Grade R ZrH_2 powder to purify H_2 gas(at 810°C) or Titanium Sponge to purify Argon gas(at 925°C).

(2) Randolph Pump employing Tygon tubing

(3) Connections are made with O-ring Quick-connect Couplings and copper tubing

(4) Liquid N_2 for H_2 gas or dry ice and acetone for Ar gas

Fig. 2. Purification system.

2. Techniques

A schematic of the experimental apparatus employed is shown in Figures 1 and 2. The specimen furnace in Fig. 1 is similar to that described in previous work[1] but with certain modifications to incorporate the vacuum purification system. The gas purification system in Fig. 2 is similar to that used by D. F. Stein et. al.[4] to purify iron single crystals and the operating procedure followed has been described by them. The optical system, the experimental technique used to obtain the data, the blackbody cavity, and the analytical method used for temperature correction have all been adequately described in previous work[1].

3. Materials and Preparations

The polycrystalline beryllium investigated is a commercial vacuum cast and rolled material (99.0% Be, 0.50% BeO, balance each <0.15%) purchased from A. D. Mackay, Inc. and has a typical resistance ratio of about 2.[5] The single-crystal beryllium was grown at the Franklin Institute Research Laboratories using a starting material taken from a bar of Pechiney Co. SR grade beryllium (SR refers to a secondary electrolytic refining process). This bar was then vacuum melted and hot extruded. Six floating zone refining passes were then put through the material yielding an overall purity of 50 to 100 ppm (atomic basis) total impurities[5]. The resistance ratio $\rho 273/\rho 4.2^{\circ}$ K varies from 100 to 200[5].

The single crystal zirconium investigated was purchased from Materials Research Corp. This material is an electron beam zone refined (one pass) product (Grade I) grown in the solid state. Typical mass spectrometer analyses performed by Battelle Memorial Institute indicate about 130 ppm total impurities present in this single crystal zirconium.

The pyrolytic boron nitride investigated was prepared by High Temperature Materials, Inc.. This material, like pyrolytic graphite, is made by a vapor deposition process and is of high density (~2.2 g/cc) and purity (< 10 ppm total metallic impurities).[6]

The polycrystalline boron nitride investigated is a hot-pressed material (2.1 g/cc) fabricated by the Carborundum Co.. A typical analysis yields: 97.0% BN, 2.40% Methanol Soluble Borate, 0.10% Alkaline earth oxides, 0.20% Alumina and silica, and 0.008% Carbon[7].

The pyrolytic graphite investigated was vapor deposited ($\sim T_d = 2200^\circ$ C) by General Electric Co.. This material is of high density ($\sim$ 2.213 g/cc) and purity (< 15 ppm total impurities).

The grinding and polishing of the beryllium specimens were performed using the procedures given by R. D. Buchheit et. al.[8]. The final mechanical polish was performed using a 5% Oxalic Acid-water solution containing Linde B Al_2O_3 powder. The latter source was also adhered to in the mechanical polishing of the zirconium specimens.

The boron nitride and pyrolytic graphite specimens were mechanically polished using SiC 600 paper, diamond polish (3μ paste), and Linde Al_2O_3 powders C, A, and B.

Photomicrographs (up to 1000x), interference photographs, and x-ray diffraction analyses were taken before and after the experiments to check for any surface changes. None of the graphite or boron nitride specimens, which were run in purified hydrogen atmosphere, showed any surface changes due to oxidation or hydride formation. Only the beryllium single crystals, which were run in purified argon atmosphere showed very slight epitaxial oxidation. A weak (0002) BeO orientation on the (0001) Be face and a $(10\bar{1}0)$ BeO orientation on the $(10\bar{1}0)$ Be face were detected by x-ray diffraction after the experiment. No oxide lines were detected on the polycrystalline Be surface after the experiment and all three beryllium specimens exhibited their metallic appearances. A slight BeO transparent film will effect the $\varepsilon_{n\lambda}$ results only at the shorter wavelengths ($\lambda \lesssim 1\mu$) where interference becomes important.[9-10] The zirconium specimens were run simultaneously with titanium and diboride specimens which acted as getters keeping the Zr free from surface contamination. The lack of strong chemical oxidation causing complete breakdown and flakiness indicated that the partial pressure of oxygen was fairly low in the vacuum purified argon atmosphere (Fig. 2).

4. Theoretical Discussions

The normal spectral emissivity, $\varepsilon_{n\lambda}$, is related to the conductivity, $\sigma(w)$, and dielectric constant, $\varepsilon(w)$, which can be coupled with the optical constants, index of refraction, n, and extinction coefficient, k. This is achieved by using Maxwell theory of electromagnetism, the Fresnel relation of reflectivity for normal incidence, and Kirchhoff's Law for an opaque material. The important equations are given here for discussion purposes throughout the present paper:

$$\varepsilon_{n\lambda} = \frac{4n}{(n+1)^2 + k^2} \tag{1}$$

Complex index of refraction, N

$$N = \left[\varepsilon - i\,\frac{4\pi\sigma}{w}\right]^{\frac{1}{2}} \tag{2}$$

$$N = n - ik$$

$$\varepsilon(w) = n^2 - k^2 \tag{3}$$

$$\sigma(w) = nk\,\frac{w}{2\pi} = \frac{c}{2}\,\frac{2nk}{\lambda}$$

$\sigma(w)$ is proportional to absorption coefficient $\left(\eta \sim \frac{2nk}{\lambda}\right)$

$$\binom{n}{k} = \left[\frac{|\varepsilon|}{2}\right]^{\frac{1}{2}} \left[\pm\theta + \left\{1 + \left(\frac{4\pi\sigma}{w\varepsilon}\right)^2\right\}^{\frac{1}{2}}\right]^{\frac{1}{2}} \quad (4)$$

where choose n for $+ \theta$

k for $- \theta$

and

$\theta = +1, \quad \varepsilon_{>0}$

$\theta = -1, \quad \varepsilon_{<0}$.

For good conductors at long wavelengths ($\gtrsim 10\mu$), where $\left|\frac{\varepsilon}{2}\right| << \left(\frac{\sigma}{w}\right)$, equations (1) and (4) yield the celebrated Hagen-Rubens formula which to first order is:

$$\varepsilon_{n\lambda} = 36.50 \left(\frac{\rho_{dc}}{\lambda}\right)^{\frac{1}{2}} \quad (5)$$

where ρ_{dc} is the dc resistivity in Ωcm and λ is the wavelength in microns (μ). This equation is valid only in the free-carrier intraband absorption region ($\gtrsim 10\mu$) where the optical conductivity $\sigma(w)$ increases to the dc value σ_{dc}.

Considering the three relevant optical excitation processes in metals[13], the spectral variation of $\varepsilon_{n\lambda}$ for most metals is characterized by three regions. In the Hagen-Rubens range (intraband transitions, $\lambda \gtrsim 10\mu$), $\varepsilon_{n\lambda}$ varies inversely proportional to the optical conductivity $\sigma(w)$ which increases to the dc value σ_{dc} as the wavelength λ increases beyond about 10 microns. In the interband transition range ($0.40 \gtrsim \lambda \gtrsim 10$ microns), $\varepsilon_{n\lambda}$ varies directly proportional to $\sigma(w)$, which is proportional to absorption coefficient ($\eta \sim \frac{2nk}{\lambda}$, equation (3)), and usually exhibits peaks in the visible - near infrared region (0.40 - 4 microns). At the plasma resonance and shorter wavelengths ($\lambda \lesssim 0.40$ microns), $\varepsilon_{n\lambda}$ varies again inversely proportional to $\sigma(w)$ which decreases to small values ($\varepsilon_{n\lambda} \rightarrow 1.00$) as λ decreases still further. The latter is the ultra-violet transparent region (essentially no absorption, $\sigma(w) \sim \frac{2nk}{\lambda} \sim 0$) where the electrons behave free - like again.

The various metals exhibit different spectral dispersion of $\varepsilon_{n\lambda}$ over these three characteristic regions which is due mainly to differences in valency and band structure. The monovalent metals have half-filled s-bands overlapped by filled d-bands and they exhibit free-carrier intraband transitions ($\varepsilon_{n\lambda}$ remains fairly low $\lesssim 0.10$) right into the visible or ultra-violet regions before the onset of interband transitions occurs ($\varepsilon_{n\lambda}$ increases rapidly). Divalent metals like beryllium would be insulators were it not for the presumed overlap of higher bands with the nearly-filled valence s-bands. The small overlap means that the onset of interband transitions will occur at lower energies (longer wavelengths) than for the monovalent metals. The transition metals like zirconium have narrow partly-filled d-bands of high multiplicity which overlap wide s-bands and the Fermi level lies in the overlapped d-bands. Thus, a broad interband transition region sets in at even lower energies compared to a monovalent metal in which the d-band is still appreciably below the Fermi level[13]. Scattering of conduction s electrons into vacant d-band levels, where the density of states is high, accounts for the relatively low conductivity of the transition metals[14]. Hence,

$\varepsilon_{n\lambda}$ for a transition metal will be higher and exhibit an interband peak at longer wavelengths than for most other metals.

It can be shown from Fresnel's equations that the unpolarized normal spectral emissivity $\varepsilon_{n\lambda}$ is related to its polarized components by the following relation[15]:

$$\varepsilon_{n\lambda} \text{ (unpolarized)} = \frac{1}{2} (\varepsilon_{n\lambda}^{||} + \varepsilon_{n\lambda}^{\perp}) \tag{6}$$

where $\varepsilon_{n\lambda}^{||}$ is the polarized component when the emitted light wave has its $\vec{E}$ vector polarized parallel to the plane of incidence and $\varepsilon_{n\lambda}$ that when $\vec{E}$ is perpendicular to the same plane. For light waves emitted normal to a C-face (c-axis $\perp$ surface, (0001)), both $\vec{E}^{||}$ and $\vec{E}^{\perp}$ are perpendicular to the optic axis (c-axis, ordinary rays), so that no double refraction occurs and equation (6) becomes:

$$\text{(c-face)}\ \varepsilon_{n\lambda} \text{(unpolarized)} = \varepsilon_{n\lambda}^{||} (n_o,k_o) = \varepsilon_{n\lambda}^{\perp} (n_o,k_o)$$

$$= \frac{4n_o}{(n_o+1)^2 + k_o^2} \tag{7}$$

where n_o and k_o are the real index of refraction and extinction coefficient for the ordinary ray, respectively. Two cases are possible for the A-face (c-axis || surface) depending on whether the optic axis (c-axis) is oriented parallel or perpendicular to the plane of incidence. Both cases are equivalent for emitted unpolarized light and equation (6) becomes:

$$\text{(A-face)}\quad \varepsilon_{n\lambda}(\text{unpolarized}) = \frac{1}{2}\left[\frac{4n_o}{(n_o+1)^2 + k_o^2} + \frac{4n_e}{(n_e+1)^2 + k_e^2}\right] \quad (8)$$

where (n_o, k_o) and (n_e, k_e) are the real indices of refraction and extinction coefficients for the ordinary ($\vec{E}\perp c$) and extraordinary ($\vec{E}\parallel c$) rays, respectively. Nye[16] describes polarization due to double refraction in terms of a three-dimensional geometric figure called the indicatrix. Hexagonal layer type materials are described as positive uniaxial or negative uniaxial depending on whether $(n_e - n_o)$ is greater or less than zero. It is interesting to note from equations (7) and (8) that for an ideal single crystal the polarized $\varepsilon_{n\lambda}(n_o, k_o)$ of the A-face should be equivalent to the unpolarized $\varepsilon_{n\lambda}(n_o, k_o)$ of the corresponding c-face.

5. Results

5.1 Beryllium

The anisotropy exhibited by single crystal beryllium is shown in Fig. 3. Like pyrolytic graphite, single-crystal beryllium exhibits an $\varepsilon_{n\lambda}$ (C-face (0001)) < (A-face $(10\bar{1}0)$) for $\lambda \gtrsim 1\mu$ but the anisotropy is not as large. The single crystals have the typical metallic $\varepsilon_{n\lambda}$ dispersion at $\lambda \gtrsim 2.50\mu$ which is the free-carrier intraband absorption region for beryllium[11,12]. In this region, $\varepsilon_{n\lambda} \sim {}^1/\sigma(w)$ (inversely proportional to absorption, $\sigma(w) \sim \frac{2nk}{\lambda}$ where $\sigma(w) \simeq \sigma_{dc}$. Since, σ_{dc} (llc) < σ_{dc} ($\perp$c) at 0°C for single crystal Be[17,18], which may hold at ~790°C, then $\varepsilon_{n\lambda}$ C-face < $\varepsilon_{n\lambda}$ A-face Be ($\lambda \gtrsim 2.50\mu$) as is observed (Fig.3).

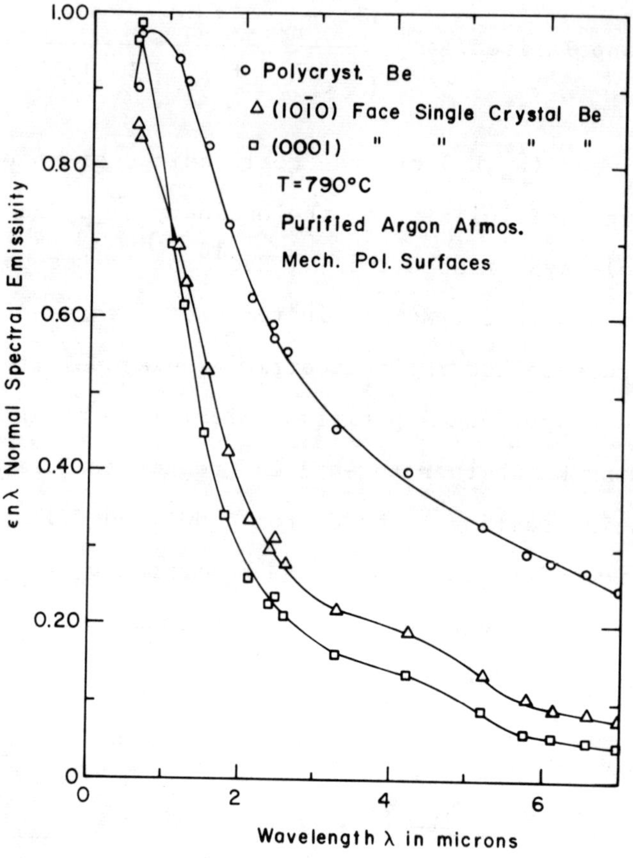

Fig. 3. Anisotropy in beryllium.

Beryllium being a divalent metal, the onset of its interband absorption region is expected to occur at longer wavelengths than for the monovalent metals. The present high temperature results (Fig. 3) indicate that the long wavelength edge for interband absorption ($\varepsilon_{n\lambda}$ ~ $\sigma(w)$ increases rapidly where $\sigma(w) << \sigma_{dc}$) occurs at about $\lambda = 2.50\mu$ which is in general agreement with room temperature data on polycrystalline Be[12]. This interband absorption region results from electron transitions between the valence (2s)-band and the overlapping (2p)-band as can be seen in J.H. Terrell's (E,k) curves for Be.

The long wavelength edge of interband absorption is steeper for C-face compared to A-face single crystal beryllium. This may be due to "parallel-band effects" in interband optical absorption as first discussed by W. A. Harrison[20]. These are absorption edges associated with nearly parallel bands arising from vertical transitions at a Brillouin-zone face. Harrison[20] finds, for various HCP metals (Be, Mg, Cd, Zn), as well as B.R. Watts[21] for Be, that the band gaps at $\{10\bar{1}0\}$ zone faces are smaller than those for {0002} zone faces. Thus, the onset of parallel-band absorption should occur at a lower energy (longer λ) and be steeper for light waves polarized with $\vec{E} \perp$ c-axis in these metals. This is actually observed in the room temperature absorption data ($2nk/_\lambda$ vs. λ) for single crystal Mg[22], Zn[23], Co[24], and Cd[25]. The steeper C-face edge may also follow from a discussion given by A. P. Lenham[26] and observation of J. H. Terrell's[19] (E,k) curves for beryllium. The former points out that if the surfaces of constant - energy - difference are parallel to the Fermi Surface, then the interband absorption edge is steep, as all the electrons on the Fermi Surface are excited at very similar photon energies. If, on the other hand, the constant - energy - difference surfaces cut across the Fermi Surface, the number of transitions increases more slowly and the interband absorption edge is more shallow. J. H. Terrell's[19] (E,k) curves indicate that the former situation of almost parallel bands occurs for C-face Be ($\vec{E}\perp$c, steeper edge).

The effect of impurity in beryllium on $\varepsilon_{n\lambda}$ is quite pronounced as can be seen by the higher $\varepsilon_{n\lambda}$ values for the commercial grade polycrystalline beryllium (Fig. 3). The latter is rather impure (99.0% Be, 0.50% BeO) compared to the single crystals (99.99% Be) which could result in a lower σ_{dc} (~790°C), hence, higher $\varepsilon_{n\lambda}$ using the Hagen-Rubens formula (equation 5) for the polycrystalline specimen. Further, the impure polycrystalline Be may be exhibiting an $\varepsilon_{n\lambda}$ curve which is characteristic of a metal having compounds BeO and MBe_5 (M is a metallic impurity)[27] which are known to exist in low purity Be. The higher $\varepsilon_{n\lambda}$ values for polycrystalline Be may also be due in part to a surface effect caused by small pits and inclusions compared to the relatively smoother surfaces of the single crystals. It was thought also that a difference in slight oxidation may have been the cause. However, no oxide lines were detected on the polycrystalline Be surface even though slight epitaxial oxidation was detected on both single crystal faces (transparent). Besides, the effect of any slight BeO film is important only at $\lambda \lesssim 2\mu$, whereas $\varepsilon_{n\lambda}$ for the polycrystalline Be is still much higher than for the single crystals at the longer wavelengths (Fig.3).

Polarization in $\varepsilon_{n\lambda}$ of A-face Be is small (Fig.4) but the measurements are in the wavelength range ($\lambda \lesssim 2\mu$) where the effect of a thin $(10\bar{1}0)$ BeO film on $(10\bar{1}0)$ Be is expected to be large. Like pyrolytic graphite, Be is negative uniaxial near $\lambda \sim 2\mu$ where $\varepsilon_{n\lambda}\ (\vec{E}\parallel c) > \varepsilon_{n\lambda}\ (\vec{E}\perp c)$.

5.2 Zirconium

The effect of anisotropy in single crystal zirconium is shown in Fig. 5. Unlike beryllium, zirconium exhibits

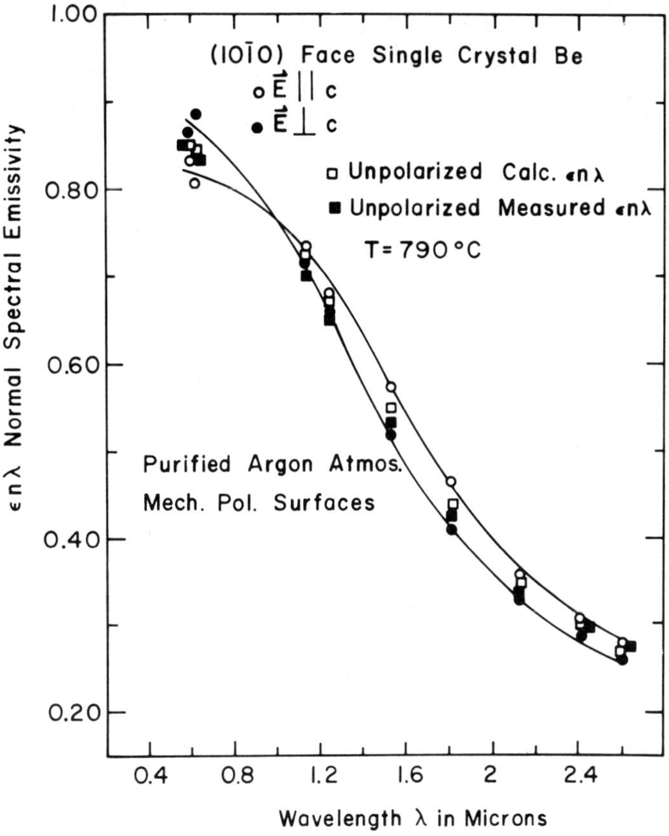

Fig. 4. Polarization in beryllium.

an $\varepsilon_{n\lambda}$ C-face > $\varepsilon_{n\lambda}$ A-face at the longer wavelengths which is of greater anisotropy (2.50 - 7μ) but still not as large as pyrolytic graphite. The greater anisotropy of Zr compared to Be may be partly due to the closeness of the test temperature (~790°C) to the HCP → BCC (860°C) transformation in Zr. The anisotropy ratio (C_{44}/C_{66}) in Zr exhibits a pronouced increase with temperature as the transformation is approached[28]. The possible effects of an increasing c/a ratio[29], just before a phase transformation, on band structure may be unique in Zr.

The onset of interband absorption is expected for Zr at a wavelength as long as $\lambda \lesssim 9\mu$ where $\varepsilon_{n\lambda} \sim \sigma(w)$ (directly proportional in λ variation) and $\sigma(w) << \sigma_{dc}$[24,26,30]. Since $\varepsilon_{n\lambda}$ C-face > $\varepsilon_{n\lambda}$ A-face (Fig. 5, 2.50 - 7μ), then absorption in the basal plane ($\vec{E} \perp$ c-axis) > absorption in the prismatic plane ($\vec{E} \parallel$ c-axis) of Zr which usually occurs for most other HCP metals[22-25,31] in the same wavelength range. The general dispersion of $\varepsilon_{n\lambda}$ vs λ exhibited by both single crystal Zr faces (Fig. 5) corresponds nicely to the absorption

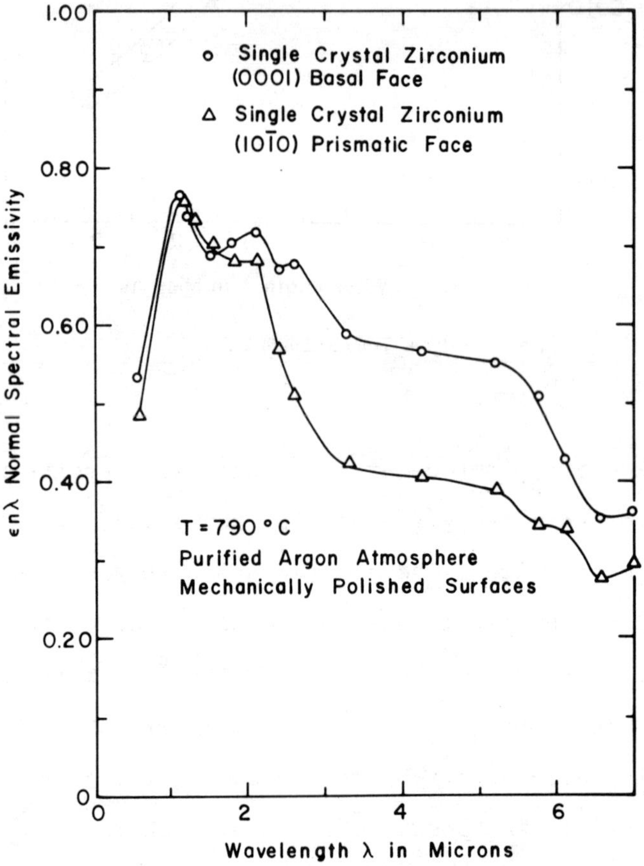

Fig. 5. Anisotropy in zirconium.

$\sigma(w)$, spectrum observed in bulk polycrystalline Zr at room T.[24] The $\sigma(w)$ spectrum of the latter exhibited absorption peaks at 0.654μ , 0.954μ , 2.07μ , and 8.85μ with a flat region (3 to 5μ) and a dip in between the 0.954 and 2.07μ peaks. After the 8.85μ peak, $\sigma(w)$ increases rapidly as λ increases into the intraband region ($\varepsilon_{n\lambda} \sim 1/\sigma(w)$ decreases). In the interband region (0.60 - 9μ), $\varepsilon_{n\lambda}$ varies as $\sigma(w)$ and $\sigma(w) << \sigma_{dc}$. The interband absorption peaks in Zr are a result of electron transitions from the narrow partly-filled (4d)-band to the wider overlapping (5s) and (5p) bands. The Fermi Surface of Zr is not a free-electron one but has a very complicated shape[32].

The broad interband absorption region of Zr observed in the present high temperature results (Fig. 5), from 0.60 - 3μ , may be explained from the very recent observations given by A. P. Lenham[26]. The latter has observed for most transition metals, that the long λ edges (~3μ for Zr, Fig. 5) of the interband absorption peaks tend to increase in steepness with increasingly filled d-bands. Thus, Zr, having a relatively empty d-band, compared to Cr, Fe, and W, exhibits a shallower long λ edge than the latter. On the other hand, Lenham[26] mentions that on the short λ edge of the interband absorption peak, several metals (Cr, Fe, W) show rather pronounced decreases of absorption ($\varepsilon_{n\lambda}$ drops off rapidly), which suggests that at these higher energies the transitions between the highly populated d-bands are weakening. Since, Zr has a relatively empty d-band to start with than (Cr, Fe, W), the drop off of $\varepsilon_{n\lambda}$ as λ decreases ($\lambda < 1\mu$, Fig. 5) is even steeper.

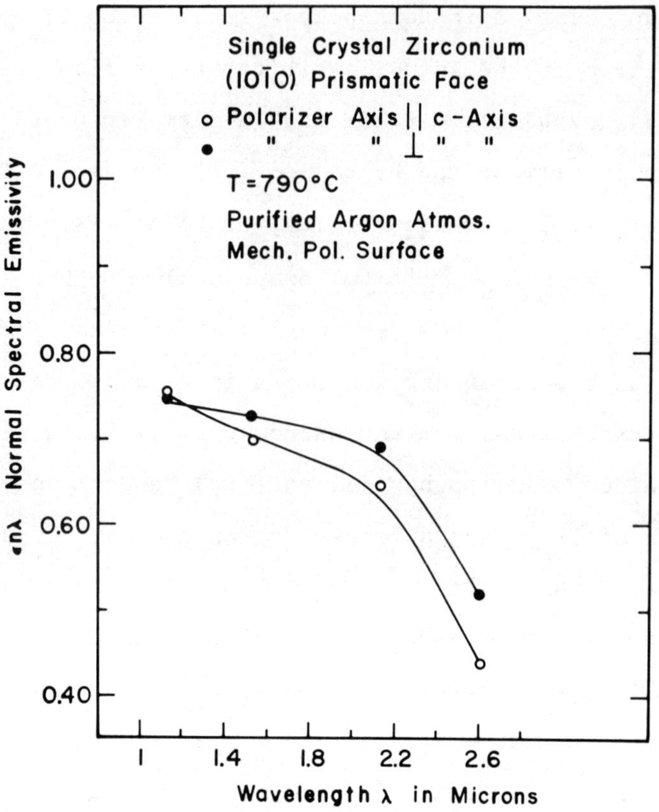

Fig. 6. Polarization in zirconium.

Since, $\lambda \lesssim 9\mu$ is near the start of the interband absorption region of Zr, agreement with a Hagen-Rubens formula (equation 5) is not expected until longer wavelengths. Further, Zr, like most transition metals, exhibits an optical conductivity $\sigma(w)$ which is much less than σ_{dc} ($\sigma(w)/\sigma_{dc} = 0.20$, room T) in the interband region[24]. Room T $\varepsilon_{n\lambda}$ for single crystal Co, calculated from optical constant data[24], like Zr (~790°C), exhibits $\varepsilon_{n\lambda}$ C-face > $\varepsilon_{n\lambda}$ A-face over a comparative wavelength range.

Polarization in $\varepsilon_{n\lambda}$ of A-face Zr is shown in Fig. 6 and in contrast to Be, Zr exhibits a polarized $\varepsilon_{n\lambda}$ ($\vec{E} \perp$ c-axis) > polarized $\varepsilon_{n\lambda}$ ($\vec{E} \parallel$ c-axis) (1.4 - 2.60μ) . It is positive uniaxial in this range, and like Co, the absorption in the basal plane ($\vec{E} \perp$ c-axis) > absorption in the prismatic plane ($\vec{E} \parallel$ c-axis) and both vary with λ.

5.3 Pyrolytic and Polycrystalline Boron Nitride

The anisotropy in pyrolytic BN and a comparison to hot-pressed polycrystalline BN is shown in Fig. 7. Like pyrolytic

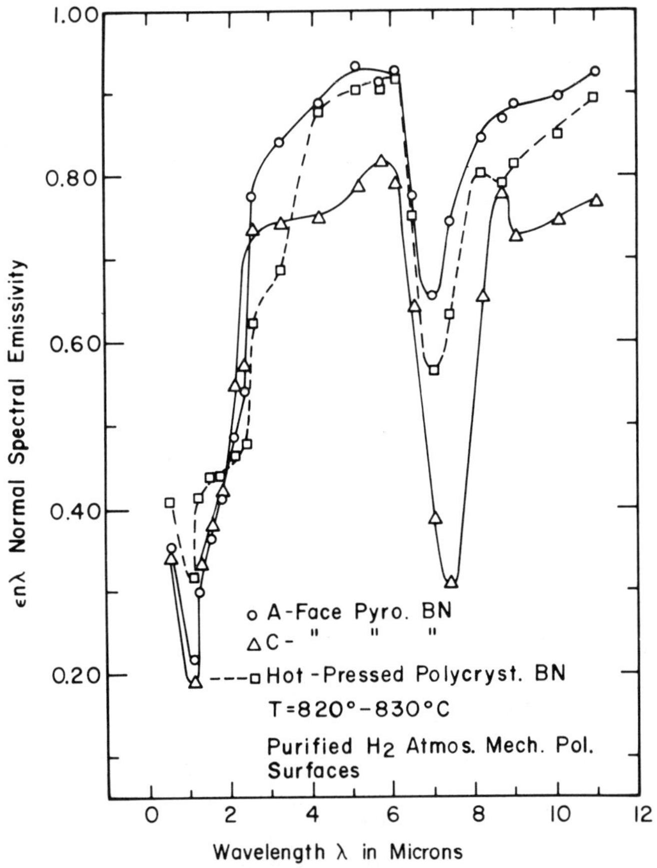

Fig. 7. Anisotropy in boron nitride.

graphite, pyrolytic BN exhibits maximum absorption $\sigma(w) \sim 2nk/\lambda$ when $\vec{E} \perp$ c-axis and minimum when $\vec{E} \parallel$ c-axis[33-34], hence, $\varepsilon_{n\lambda} \sim {}^{1}/\sigma(w)$ and $\varepsilon_{n\lambda}$ C-face < $\varepsilon_{n\lambda}$ A-face of pyrolytic BN. However, this behavior is a function of λ and for $\lambda \lesssim 2.50\mu$ the anisotropy between the BN specimens decreases (Fig. 7). Pyrolytic and polycrystalline graphite[1] also exhibit a decreasing anisotropy at the shorter wavelengths.

It is interesting to note that the polycrystalline BN exhibits a spectral dispersion of $\varepsilon_{n\lambda}$ similar to A-face pyrolytic BN (Fig. 7). A transmission Laue of the former is similar to the latter which may indicate that a majority of the BN crystallites are oriented with their c-axes almost parallel to the surface of the polycrystalline material. Previous work also indicated that the same situation occurs when comparison was made between polycrystalline graphite and A-face pyrolytic graphite[1].

Theoretically the spectral dispersion of $\varepsilon_{n\lambda}$ exhibited by BN is typical of a classical Lorentz insulator. Room temperature reflectance measurements (normal incidence) of A-face pyrolytic BN[33], converted by means of a Kramers-Kronig analysis, yield the real and imaginary parts of the complex dielectric constant, ε^*. Both of these exhibit the typical classical dispersion of an insulator.[13, 35] The imaginary part ($\text{Im}\ \varepsilon^* = \frac{4\pi\sigma(w)}{w}$) exhibits a pronounced peak at a classical (Hooke's law) spring frequency, w_o, associated with a lattice vibrational mode. Since, $\varepsilon_{n\lambda} \sim {}^{1}/\sigma(w)$, $\varepsilon_{n\lambda}$ exhibits a pronounced minimum at w_o. For pyrolytic BN the absorption is maximum when $\vec{E} \perp$ c-axis at $\lambda \sim 7.31\mu$ and maximum again when $\vec{E} \parallel$ c-axis at $\lambda \sim 12.78\mu$.[33] These two wavelengths correspond to the Reststrahlen bands for in-plane ($\vec{E} \perp$ c-axis) and out-of-plane ($\vec{E}$ || c-axis) lattice vibrations, respectively. They correspond to the classical spring frequency, w_o, for each polarization. The A-face Reststrahlen ($\vec{E}$ || c, 12.78μ) was not reached in the present measurements which ended at about λ ∿ 11μ due to lack of available energy. The less pronounced dip in $\epsilon_{n\lambda}$ near λ ∿ 7μ for A-face pyrolytic BN probably corresponds to the 2-phonon peak (superposition of in-plane ($\vec{E} \perp$ c) Reststrahlen and out-of-plane ($\vec{E}$ || c) 2-phonon peak).[33] The Reststrahlen, w_o, of the C-face pyrolytic BN ($\vec{E} \perp$ c) occurs at a higher frequency (shorter λ) than that for the A-face ($\vec{E}$ || c). This is due mainly to the stronger in-plane bonding and results in a higher w_o which depends on spring constants and masses.[36] The poly-crystalline BN exhibits a greater dip in $\epsilon_{n\lambda}$(∿ 7μ) than the A-face pyrolytic BN which is probably a superposition of the 2-phonon absorption peak (A-face) and the C-face Reststrahlen. The former seems to be dominant considering the closer comparison of $\epsilon_{n\lambda}$ vs. λ for the polycrystalline BN and A-face pyrolytic BN.

A-face pyrolytic BN exhibits little polarization in $\epsilon_{n\lambda}$ from 1.13 to 2.60μ as shown in Fig. 8. Like pyrolytic graphite, pyrolytic BN shows a polarized $\epsilon_{n\lambda}$ ($\vec{E}$||c) > polarized $\epsilon_{n\lambda}$ ($\vec{E} \perp$c) over all λ (1.13 - 2.60μ), hence, is negative uni-axial.

5.4 Pyrolytic Graphite

The anisotropy of $\epsilon_{n\lambda}$ in pyrolytic graphite along with polarization of the A-face is shown in Fig. 9. In comparing the semi-metallic C-face $\varepsilon_{n\lambda}$ with that of the insulator-like A-face (Fig.9), it is seen that the absorption, $\sigma(w) \sim 2nk/_{\lambda}$, is maximum when $\vec{E} \perp$ c-axis and minimum when $\vec{E} \parallel$ c-axis, thus, $\varepsilon_{n\lambda} \sim {}^{1}/\sigma(w)$ and $\varepsilon_{n\lambda}$ C-face $<$ $\varepsilon_{n\lambda}$ A-face pyrolytic graphite. Pyrolytic graphite exhibits an extreme anisotropy in $\varepsilon_{n\lambda}$ mainly due to the fact that the C-face is semi-metallic in properties while the A-face acts like an insulator. A less anisotropic metal is expected to exhibit essentially metallic-like properties for both of these major orientations.

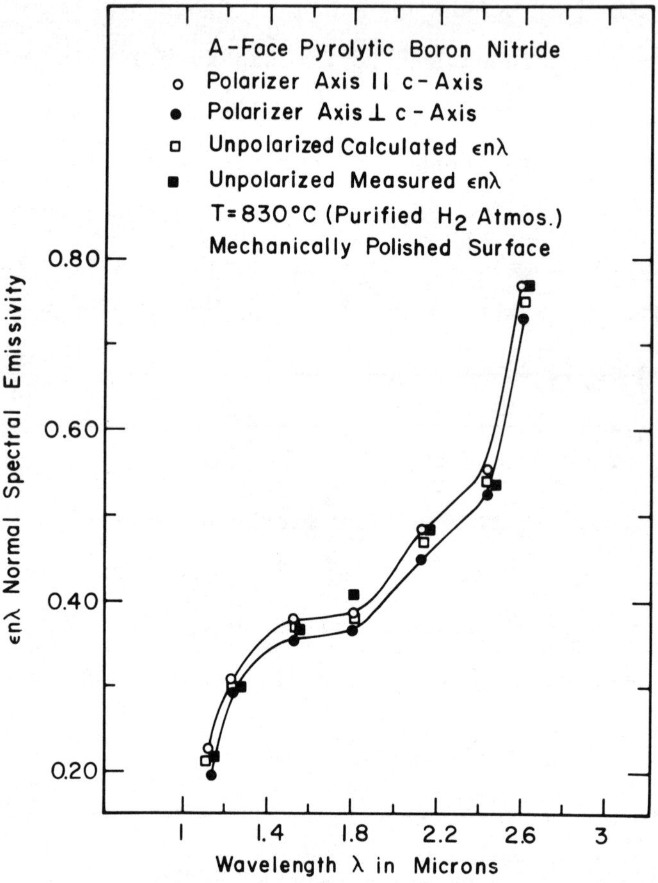

Fig. 8. Polarization in boron nitride.

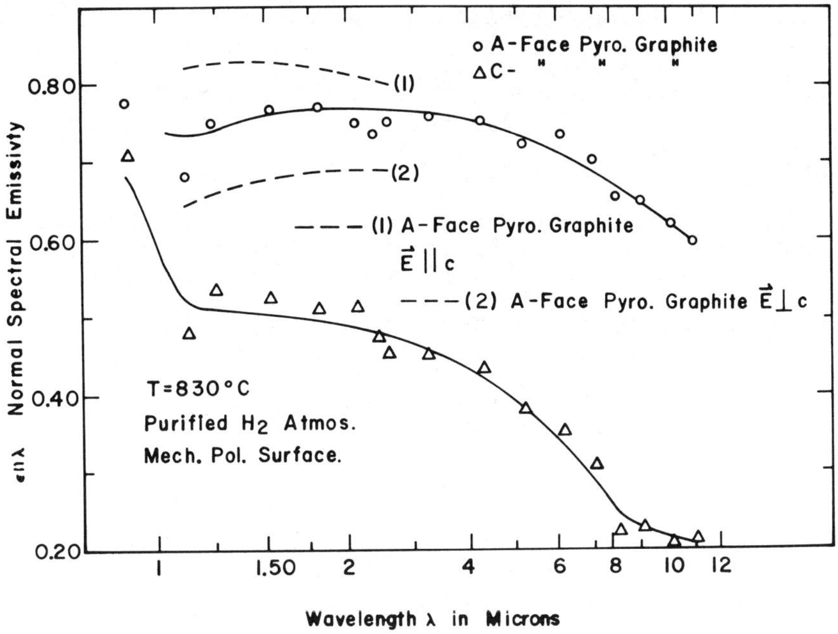

Fig. 9. Anisotropy in pyrolytic graphite.

The polarization in $\varepsilon_{n\lambda}$ of A-face pyrolytic graphite is fairly large (Fig.9). The polarized $\varepsilon_{n\lambda}$ ($\vec{E} \perp c$) < polarized $\varepsilon_{n\lambda}$ ($\vec{E} \,||\, c$) indicative of a negative uniaxial material. The A-face unpolarized $\varepsilon_{n\lambda}$ is a function of both the ordinary and extraordinary ray optical constants whereas the C-face depends only on the ordinary ray optical constants.

References

1. Autio, G. W. and Scala, E., Carbon 4, 13 (1966).
2. Dreshfield, R. L. and House, R. D., Spectral normal emittance of single crystals, AIAA Paper No. 65-134, 2nd Aerospace Sciences Meeting, N. Y. (January, 1965).
3. Autio, G. W. and Scala, E. (to be published in the journal Carbon).
4. Stein, D. F., Low, J. R., Jr., and Seybolt, A. V., Acta. Met. 11, 1253 (1963).

5. Rengstorff, G. W. P., High Purity Metals, DMIC Report 222, Jan. 3, 1966, pp. 7-10.

6. Boron Nitride Data Sheet, Feb. 1, 1965, High Temperature Materials, Inc., Lowell, Mass.

7. Carborundum, Boron Nitride, Latrobe, Penn.

8. Buchheit, R. D., Brady, C. H., Wheeler, G. A., "Procedures for the Metallographic Preparation of Beryllium, Titanium, and Refractory Metals," DMIC Memo 37, Oct. 26, 1959.

9. Vasicek, A., "Optics of Thin Films", North-Holland Publ. Co., Amsterdam, 1960, p. 309.

10. Malovetskaya, V. M., Vavilov, V. S., Galkin, G. N., Sov. Phys. Sol. St. 1, 1099 (1959).

11. Shklyarevskii, I. N. and Yarovaya, R. G., Opt. Spectry. 11, 355 (1961).

12. Givens, M. P., Sol. St. Phys. 6, 313 (1958).

13. Ehrenreich, H., IEEE Spectrum 2, 162 (1965).

14. NASA TN D-1523, "Optical Properties of Satellite Materials -- The Theory of Optical and Infrared Properties of Metals", March (1963).

15. Von Hippel, A. R., "Dielectrics and Waves", John Wiley and Sons, Inc., N. Y., 3rd ed. 1962, p. 52.

16. Nye, J. F., "Physical Properties of Crystals", Oxford at the Clarendon Press, 1960, Chpt. XIII, pp. 235-240.

17. Grüneisen, E. and Adenstedt, H., Ann. d. Phys. 31, 714 (1938).

18. Grüneisen, E. and Erfling, H. D., Ann. d. Phys. 38, 399 (1940).

19. Terrell, J. H., Phys. Rev. 149, 526 (1966).

20. Harrison, W. A., Phys. Rev. 147, 467 (1966).

21. Watts, B. R., Proc. Roy. Soc. (Lon.) 282A, 521 (1964).

22. Lenham, A. P., Treherne, D. M., Woodall, A. J., "Optical Properties of Anisotropic Metals", in Optical Properties and Electronic Structure of Metals and Alloys, ed. by F. Abeles, North-Holland Publ. Co., Amsterdam, 1966, p.40.

23. Lettington, A. H., "Optical Properties and Fermi Surface of Zinc", in Optical Properties and Electronic Structure of Metals and Alloys, ed. by F. Abeles, North-Holland Publ. Co., Amsterdam, 1966, p. 147.

24. Lenham, A. P. and Treherne, D. M., "The Optical Properties of the Transition Metals", in Optical Properties and Electronic Structure of Metals and Alloys, ed. by F. Abeles, North-Holland Publ. Co., Amsterdam, 1966, p. 196.

25. Lenham, A. P. and Treherne, D. M., Proc. Phys. Soc. 83, 1059 (1964).

26. Lenham, A. P., J. Opt. Soc. Amer. 57, 473 (1967).

27. Wolff, A. K., Gelles, S. H., Aronin, L. R., "Impurity Effects in Commercially Pure Beryllium Prepared from Powder", The Metallurgy of Beryllium, Chapman and Hall, Ltd., 1963, p. 150.

28. Fisher, E. S. and Dever, D., Trans. AIME Met. Soc. 239, 48 (1967).

29. Goldak, J., Lloyd, L. T., Barrett, C. S., Phys. Rev. 144, 478 (1966).

30. Kirillova, M. M. and Charikov, B. A., Opt. Spectry. 17, 134 (1964).

31. Lenham, A. P. and Treherne, D. M., J. Opt. Soc. Amer. 56, 752 (1966).

32. Altmann, S. L. and Bradley, C. J., Phys. Rev. 135, A1253 (1964).

33. Geick, R., Perry, C. H., Rupprecht, G., Phys. Rev. 146, 543 (1966).

34. Basche, M. and Schiff, D., "New Pyrolytic Boron Nitride", Materials in Design Eng., Feb., 1964.

35. Ziman, J. M., Principles of the Theory of Solids, Cambridge Univ. Press, Chpt. 8, p. 219 (1964).

36. Kittel, C., Introduction to Solid State Physics, John Wiley and Sons, Inc., N. Y., 1960, p.112.

ANISOTROPIC THERMAL EXPANSION CHARACTERISTICS OF SOME CRYSTAL STRUCTURES

L. Cartz
Metallurgy and Materials Science
College of Engineering
Marquette University
Milwaukee, Wisconsin

Abstract

Anomalous thermal expansion characteristics of some crystal structures can be related to the presence of distorted coordination polyhedra present in their atomic arrangements. The occurence of these distortions is discussed particularly in terms of the cationic radii, and examples include $PbTiO_3$ and R_2TiO_5 compounds. A related phenomena is the observed structural rearrangements under high pressure of structures containing distorted coordination polyhedra.

* * *

The thermal expansion behaviour of many compounds is known to be dependent on the nature of their atomic structure arrangements; Wooster 1938, Chebotarev 1962, Srinivasan and Krishnan 1958, Lonsdale 1959 & 1962, Shaw (1953), Candlin (1956), Zhdanov(1965). One particular example of this is afforded by lead titanate, $PbTiO_3$, which has a structure at room temperature which is a tetragonal distortion of the perovskite structure isomorphous with tetragonal $BaTiO_3$ (Shirane 1950). The lead titanate undergoes a thermal contraction upon heating from room temperature to 490°C where a transition to a cubic perovskite structure occurs. In the room temperature structure, the lead ions are surrounded by a distorted coordination

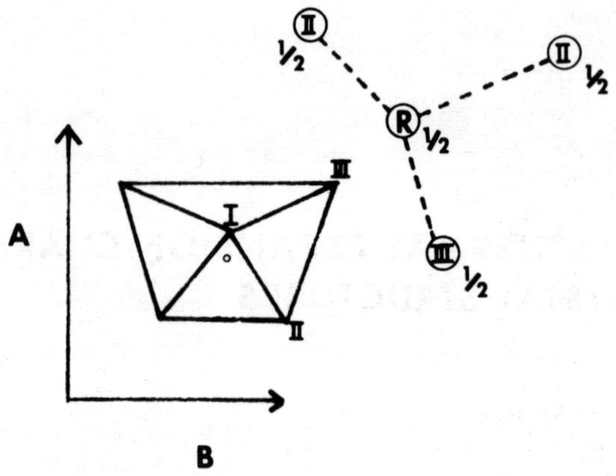

Fig. 1. Bond lengths about the Ti and R cations in Pseudobrookite. The Ti atom is indicated by a small circle in an oxygen octahedra outlined by straight lines.

Bond Lengths

	Fe_2TiO_5		Al_2TiO_5
Ti–01	1.91Å		1.91Å
Ti–011	1.95		1.90
Ti–0111	1.95		2.00
Fe–01	2.25Å	Al –01	2.05
Fe–011	1.90	Al–011	1.86, 1.96
Fe–0111	1.93, 2.25	Al–0111	1.84, 2.14

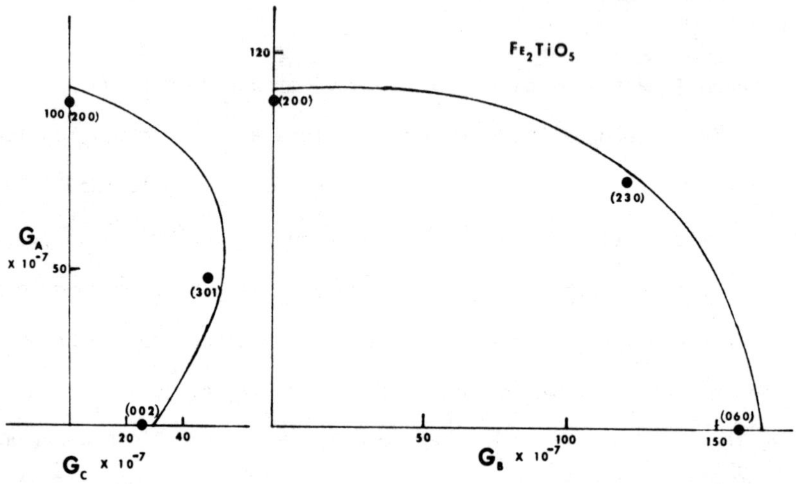

Fig. 2. Fe_2TiO_5 thermal expansion ellipsoid in the ac and ab planes. G is the thermal expansion coefficient. The experimental points are given for the various planes.

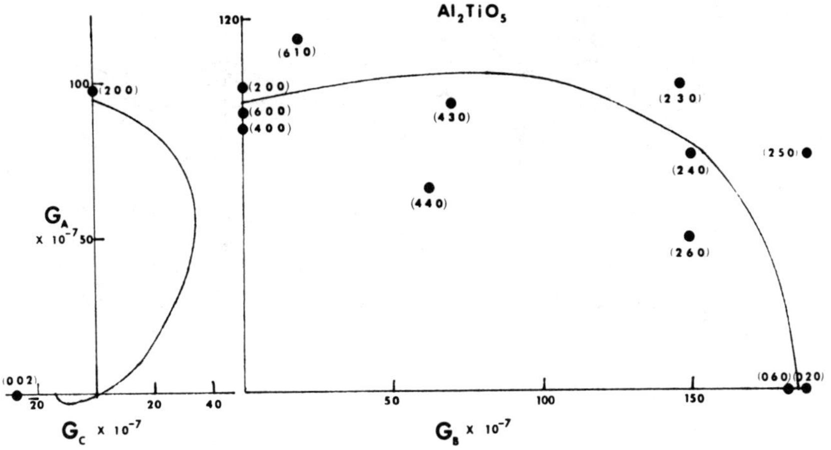

Fig. 3. Al_2TiO_5 thermal expansion ellipsoid.

polyhedra of twelve oxygens at distances varying from 2.53A to 3.20A and titanium ions are in a distorted octahedra of Ti-O distances varying from 1.78A to 2.38A. In the cubic, high-temperature form above 490°C, the Ti-O distances are 1.98A and Pb-O 2.80A. Thus the thermal con-

Table I

Thermal Expansion Coefficients for R_2TiO_5 pseudobrookite structures. The data for $MgTi_2O_5$ is taken from Bush & Hummel (1959).

Cation	Ionic Radius	Thermal Expansion Coefficient α a	α b	α c $\times 10^{-7}$/°C
Al	0.50A	95	190	−14
Fe	0.60A	109	163	30
(MgTi)	0.65A	120	120	20
Ti High Temperature Form	0.69A	-	-	-

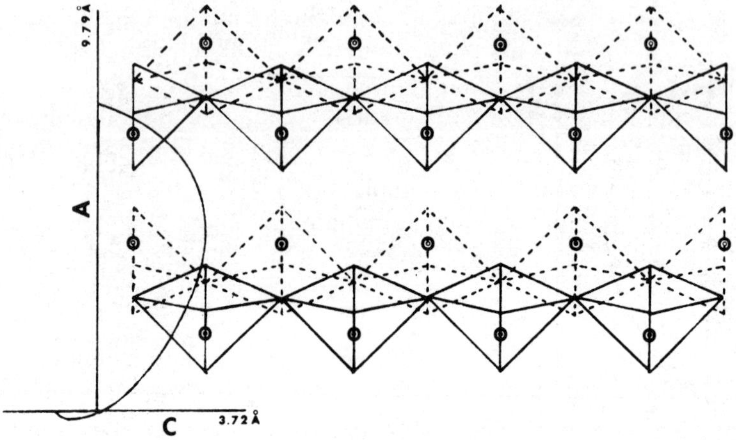

Fig. 4. Schematic representation of the R_2TiO_5 structure showing the ac projection. The trace of the thermal expansion ellipsoid is shown for Al_2TiO_5. The octahedral chains are at two levels with the Ti at 0.69 and 0.31. The (Al or Fe) R cations occur at levels of 0.44, 0.06, 0.56, 0.94.

traction with increase of temperature can be understood in terms of the ions being able to pack more closely and symmetrically about the cations so that an overall contraction in volume occurs.

One can anticipate anomalous thermal expansion behaviour for other crystal structures containing 'distorted' coordination polyhedra, that is where the bond lengths vary markedly about the cation. The structure of pseudobrookite (R_2TiO_5) contains distorted coordination polyhedra (Pauling 1930, Austin & Schwartz 1953) particularly about the R cation (Al or Fe); see Fig 1. The anisotropy of thermal expansion has been studied for these two materials by high temperature x-ray diffraction for several high angle x-ray reflexions, 17 values of (hkl) for Al_2TiO_5 and 12 for Fe_2TiO_5. Using a least square method of curve fitting, the principle coefficients of thermal expansion are obtained using the relationship $g_3=x_{11}C_{13}^2+x_{22}C_{23}^2+x_{33}C_{33}^2$ (Wooster 1938, Zhdanov 1965) where g_3 is the thermal expansion coefficient in a general direction, x_{ii} the principle thermal expansion coefficients and C_{i3} the direction cosines. The traces of the thermal expansion ellipsoids are shown in Figs. 2 & 3 together with the experimental determinations in those zones, and the data summarized in Table I. The data for $(MgTi)TiO_5$ is taken from Bush & Hummel 1959; high temperature Ti_2TiO_5 (anosovite) has the pseudobrookite structure though its thermal expansion behaviour is not yet known (Asbrink & Magneli 1959).

The pseudobrookite structure can be considered to consist of chains of Ti-O octahedra joined at an apex and running parallel to the c-axis. A diagram of the a-c plane is shown in Fig. 4 where the trace of the thermal expansion ellipsoid has been superimposed. For both pseudobrookite structures, the expansion is greatest in the ab plane, perpendicular to the Ti-O chains. Essentially, the separation of these chains increases with increase of temperature, particularly parallel to the b axis, and the smallest thermal expansion (contraction for Al_2TiO_5) is in the direction of the chains. This is in general agreement with observations for planar & linear structures where the main thermal expansion direction is perpendicular to the closely packed directions (Lonsdale 1962). Comparison with the direction of maximum and minimum bond lengths of Al-O and Fe-O in the ab plane shows that there is a relationship between those directions and those of minimum and maximum thermal expansion coefficients in the ab plane; see also α-U(Chebotarev 1962). The Ti-O chains separate, possibly 'twist', the more so for the structure containing the more highly distorted Al-O octahedra. The Al ion is small in size (see Table I) and presumably, where the cation size is near the limit of the ionic size for the atomic structure the degree of distortion in the coordination polyhedra is greatest and the anisotropy of thermal expansion behaviour most marked. It will be interesting to determine and compare the thermal expansion characteristics for the other compounds $(MgTi)TiO_5$ and Ti_2TiO_5. The thermal contraction along the c-axis may also be related to the observations of Siratori and Iida (1962) who observed a thermal contraction along the c-axis for CrO_2 and explained this by a Jahn-Teller effect.

Many tungstates, molybdates and uranates exist in distorted forms of the epalsoite structure (Wyckoff 1960). Steward & Rooksky (1951) have shown, using high temperature x-ray studies that the deviations from the ideal cubic structures decrease with increase of temperature until at high temperatures the cubic form is obtained. This represents an anomalous thermal expansion behaviour for the compounds in the temperature range leading to the transition. Studies on Ba_2SrWO_6 in particular have shown that from room temperature to ~500°C, where transition to the cubic form occurs, the Ba_2SrWO_6 structure can be considered as orthorhombic having along one axis a negative thermal expansion coefficient (Graf & Cartz, to be reported).

It is in general to be expected that anisotropic thermal expansion will be related to anisotropic compressibilities. It has been observed, for example, (Young & Schwartz 1962) that structures containing distorted coordination polyhedra, such as $CrVO_4$, undergo transformations at high pressure to a more symmetrical structure. Kabalkina et. al. (1962) has observed a large contraction under high pressure along the c-axis for $PbTiO_3$ which is the same direction along which $PbTiO_3$ contracts under high temperatures. Studies in progress on $CrVO_4$ have indicated that the linear compressibility is greatest parallel to the axis which possesses a negative thermal expansion coefficient.

Acknowledgements. Part of the x-ray studies on R_2TiO_5 compounds was carried out by Miss I. Covin of IITRI, Chicago, and W. Hammeter prepared a computor programme for the determination of the prinicpal thermal expansion coefficients. The high pressure studies on $CrVO_4$ have been carried out by Dr. J. Jamieson of the University of Chicago.

References

Asbrink, A. & Magneli, A. (1959) Acta Cryst., 12, 575

Austin, A. E. & Schwartz, C. A. (1953) Acta Cryst. 6, 812

Bush, E. A. and Hummel, F. A. (1959) J. Amer. Ceram. Soc., 42, 388

Candlin, R. Acta Cryst. (1956) 9, 545

Chebotarev, (1962) Reaction Sci. Technol. (GB) 16, 555

Kabalkina, S. S. et. al. (1962). Dokl. Akad. Nauk SSSR 143, 818, 144, 1019

Lonsdale, K. (1962). International Tables for X-ray Crystallography, 3, 125.
(1959). Z. Kristallography, 112, 188

Pauling, L. (1930) Z. Krist., 73, 97

Shaw, R. Acta Cryst., (1953) 6, 428

Shirane, G. and Hoshino, S. (1950), Phys. Rev., 80, 1105; (1951), J. Phys. Soc. Japan, 6, 265; (1950), J. Phys. Soc. Japan, 5, 453

Siratori, K. & Iida, S. (1962) J. Phys. Soc. Japan, 17, Supplement B-I, 208

Srinivasan, R. and Krishnan, R. S. (1958). Prog. Crystal Physics I, 10-54

Steward, E. G. and Rooksby, H. P. (1951) Acta Cryst., 4, 503; (1953) 6, 49

Wooster, W. A. (1938) "Crystal Physics" Cambridge University Press

Wyckoff, R. W. G. "Crystal Structures"; 3, 374 (1960).

Young, A. P. & Schwartz, C. M. (1962) Acta Cryst., 15, 1305.

Zhdanov, G.S. "Crystal Physics" Oliver and Boyd 1965.

ELECTRON PARAMAGNETIC RESONANCE ABSORPTION AND OTHER PHYSICAL PROPERTIES OF CHROMIUM OXIDE AND OTHER CRYSTALS CONTAINING CHROMIUM AS IMPURITY

S. C. Jain* and R. K. Jain

Physics Department
Indian Institute of Technology
Hauz Khas, New Delhi-29 INDIA

Abstract - The results of electron paramagnetic resonance absorption in chromium oxide in superparamagnetic and paramagnetic state are reported. The E.P.R. absorption of chromium doped in many non-metallic crystals has been measured recently. These results and the results on chromium oxide obtained in this laboratory and those by other workers are reviewed. Recent work on electrical properties of chromium oxide is also discussed. Some results of other 3d oxides are quoted and tentative reasons for the difference of behaviour between chromium oxide and other antiferromagnetic materials are suggested.

INTRODUCTION

The 3d metal oxides are of great practical and academic interest. Recently many different techniques for growing single crystals of these materials have

---

* Also at National Physical Laboratory, New Delhi-12 INDIA

been used. In this paper, the methods used for growing single crystals of the 3d metal oxides are quoted. Recent work on EPR absorption in chromia and in other related crystals is discussed.

METHODS USED FOR GROWING SINGLE CRYSTALS OF 3d OXIDES

1) High temperature methods. Single crystals of chromia are grown by Verneuil's[1] flame fusion method. In this method, finely ground powder of the material is dropped through an oxyhydrogen flame on to a ceramic pedestal. The tip of the flame is directed towards the pedestal and the powder becomes molten. The powder is fed regularly and the pedestal is slowly lowered as the crystal grows. The 3d oxide crystals have also been grown by a floating zone method using a carbon arc image furnace. Vernon and Lovell[2] grew single crystals from spectrographically standardized NiO powders used in the form of sintered rods at a rate of about 1 cm/hr. Recently Austin et al[3] have prepared single crystals of NiO by a new 'arc transfer' process. In this method, a d.c. arc is struck in air between two electrodes consisting of sintered oxide rods. Stable molten caps are formed on each electrodes and mass transport occurs across the arc gap, the anode growing at the expense of the cathode. By using a seed crystal for the anode, single crystal growth can be propagated. In these methods, very high temperatures are used and cooling rates are even faster than in the Verneuil method, resulting in a large concentration of quenched in defects and strain due to steep temperature gradients.

2) Halide decomposition method. The chemical reaction of the type

$$NiCl_2 + H_2O \rightleftharpoons NiO + 2HCl \qquad (1)$$

has been used by Cech and Alessandrini[4] and others to decompose the halides by water vapour and grow single crystals of 3d oxides. The metal halide decomposition occurs on the MgO crystal surface and the oxide is deposited in the form of an apitaxial single crystal film. The crystals upto 500 μ thick can be easily grown by this method. The oxide crystal can be removed from the substrate by dissolving MgO in a suitable solvent. The reaction temperature and the composition of atmosphere required for the growth of oxides of different 3d metals have been tabulated by Cech and Alessandrini[4].

Rambauske and Gruenzel[5] have reported that they could grow Fe_3O_4 whiskers by decomposing $FeCl_2$ in argon atmosphere with hydrogen contamination. However, Takei[6] found that water vapour is essentially needed to grow the whiskers. He could grow FeO whiskers on polycrystals of FeO by decomposing $FeBr_2$ in a mixed gas of nitrogen, hydrogen and water vapour. When the hydrogen was substituted by small quantities of oxygen, the rate of growth decreased and mixed whiskers of FeO and Fe_3O_4 were obtained. No whiskers could be grown if there was more than 10% oxygen in the atmosphere. Takei and Koide[7] have grown single crystals of vanadium oxide by decomposing $VoCl_3$ with water vapours on the inner surface of the crucible. Crystals of different oxides (V_2O_5, VO_2, V_3O_5 and V_2O_3) could be grown by changing the ratio of the concentrations of hydrogen and nitrogen mixed with the heated vapours.

The back reaction represented by eqn. (1) is endothermic and Stolpe has utilised[8] this fact to grow single crystals of NiO by chemical transport from a high temperature region towards a low temperature region in an evacuated sealed silica tube containing pure NiO powder and the transporting gas. Attempts to grow single crystals of NiO doped with Li by this technique were not successful.

ELECTRICAL CONDUCTIVITY

A search of literature revealed only a scanty amount of data on the physical properties of Rhombohedral Sesquioxide Cr_2O_3 with corrundum structure. The early work of Hauffe and Block[9] showed that the conductivity of chromia in the temperature range of $400^o - 800^oC$ is nearly independent of oxygen pressure. Fischer and Lorentz[10] extended the measurements on Cu and Ti doped samples upto 1750^oC and found no pressure dependence upto this temperature. Hagel and Seybolt[11] found a slight dependence on oxygen pressure below 1250^oC. More recently Crawford and Vest[12] measured the conductivity of single crystals of chromia and found that the conductivity is independent of the oxygen pressure in the high temperature ($>1200^oC$) intrinsic region and depends on the pressure in the low temperature defect controlled region ($< 1100^oC$). The slope of the log σ vs $\frac{1}{T}$ plot in the intrinsic region is 1.6 - 1.7 eV. Hagel[13] has measured the σ of sintered chromia, pure and doped with 2.5 and 10 moles per cent of Li_2O. He found a marked dependence of σ on oxygen pressure for pure samples but not for Li doped samples. However, 2.5 per cent Li_2O increased σ by a factor of $\sim$ 100 at

750°C. The increase of the concentration of Li_2O to 10% did not show much change.

J.F. Garcia De La Banda _et al_[14] have measured the electrical conductivity of sintered specimens of Cr_2O_3 in the temperature range of 350° to 500°C but they do not achieve the stabilisation of the values of resistance. They feel that it may be due to the diffusion of oxygen from the bulk of the chromia to the surface, or adsorption from the walls of the apparatus.

The effect of pressure of oxygen on σ is thus suppressed in Li doped samples and the fact that the earlier workers did not work with pure enough samples may have resulted in lack of dependence on oxygen pressure. Diffusion of radioactive Cr^{51} in single crystals does show the $pO_2^{3/16}$ expected dependence. Recently, conductivity[15] of Cr_2O_3 doped with BeO has also been measured. The conductivity increases upto 30 per cent of BeO and then starts decreasing again. Thermoelectric measurements show Cr_2O_3 is a p-type semiconductor. The available data on the 3d oxides points to the conclusion that there is a d band in scandium, titanium and vanadium oxides. The dependence of mobility on temperature in Cr_2O_3 has not been measured. The optical results show evidence that there may be a narrow d band in Cr_2O_3. The d orbitals are perhaps localized in the oxides starting with manganese and ending with nickel. However, the recent work of Austin _et al_[3] have shown the existence of a narrow d band polaron conduction in NiO. This supports the view that there is a d band in Cr_2O_3 also. The energy 1.6 eV obtained from the log σ vs $\frac{1}{T}$ plot in the intrinsic region must be that needed to free the holes in the d band.

EPR MEASUREMENTS

Very little work has been done on **EPR** absorption in chromium oxide. EPR of two different chromium oxide powders was examined in this laboratory. The first powder was from Albright and Wilson Mfg. Ltd. London and had a purity of 98-99 % and the second powder was from Chemical de Universe India and was of unspecified purity. Fig. 1 shows the EPR absorption of the two powders at room temperature. Residual EPR absorption below Neel temperature in some of the antiferromagnetic materials was first observed by Trounson et al[16], Okamura et al[17] and Maxwell et al[18]. However, the origin of this absorption was not understood at that time. It has been shown subsequently[19-20] that this absorption is due to the material being in the form of powder consisting of small particles. The powder behaves as superparamagnetic material. The EPR in superparamagnetic state at room temperature differed from sample to sample. This can be due to different particle size of the different powders or due to some magnetic impurity

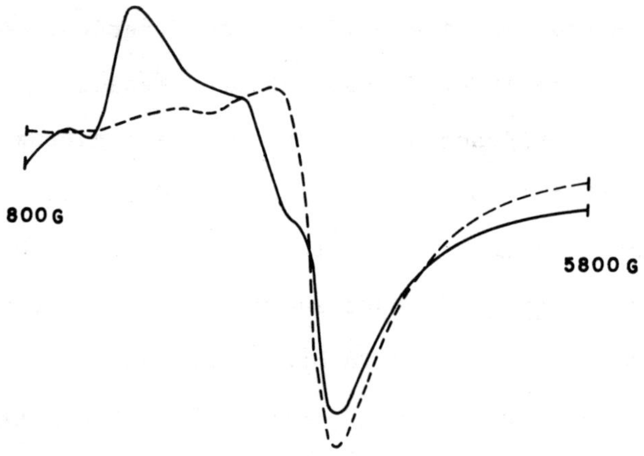

Fig. 1 E.P.R. absorption of two different Cr_2O_3 powders at R.T..

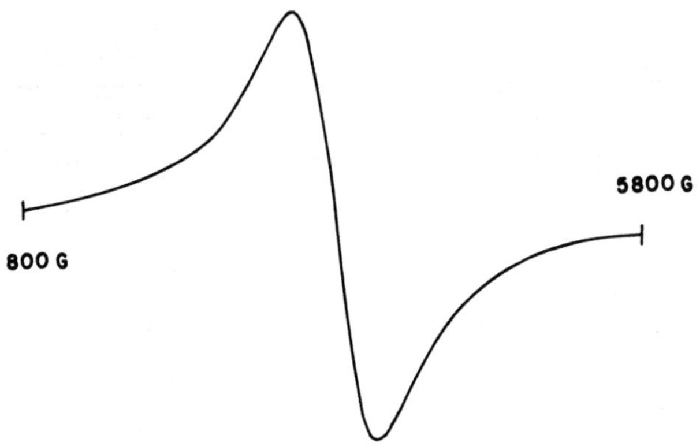

Fig. 2. E.P.R. absorption of Cr_2O_3 powder at R.T. (after heating it to 1250°C for ~4 hrs.).

present in the powder. The EPR of all powders examined at 40^oC or above gave single line with g value = 1.980±0.005 and half-width of 450 gauss. The shape of the line is Lorentzian. The change from superparamagnetic absorption to paramagnetic absorption was almost sudden at 35 ± 1^oC. This line is obviously due to paramagnetic state of chromium oxide. The EPR of chromium oxide powders was also examined at room temperature (R.T.) after heating them at different temperatures in air and in vacuum. On heating the same powder in air at low temperatures the spectrum changed considerably.

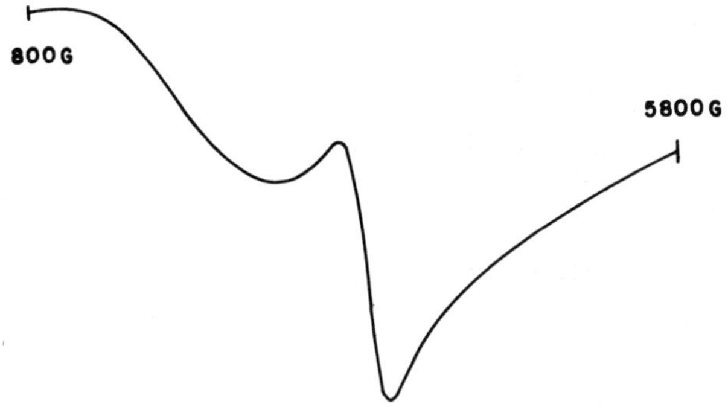

Fig. 3. E.P.R. absorption of Cr_2O_3 powder at R.T. (after heating it to 850°C in vacuum).

Table 1

Crystal	g value
MgO	1.978 [21]
CaO	1.973 [22]
SnO tetragonal axis	1.975 [23]
TiO (Rutile) Tetragonal axis	1.97 [24]
TiO (Anatese) tetragonal axis	1.973 [25]
Al_2SiO_5 (Kyanite)	$g_{\parallel}$ = 1.982, $g_{\perp}$ = 1.997 [26]
Li NbO_3	1.969 [27]
$ZnWO_4$	g_x = 1.954, g_y = 1.967, g_z = 1.966 [28]
$Be_3Al_2\ Si_6O_{18}$ (Emareld)	$g_{\parallel}$ = 1.973, $g_{\perp}$ = 1.97 [29]

On longer heatings at higher temperatures the spectrum shown in Fig. 2 was obtained. The measurements shown in this figure were made at room temperature after heating the powders to 1250°C for about 4 hours. The g value for this line is 1.985±0.005 and the half width is 600 gauss. The EPR absorption obtained after heating the powder in vacuum to 850°C is shown in Fig 3. It is now seen that this spectra consists of two lines. An attempt to resolve this into two separate lines gave g values of 1.97 and 3. Conductivity of the powders under identical conditions is being measured and a detailed discussion of the conductivity, optical absorption and E.P.R. will be published elsewhere.

Table 1 shows the g values of Cr^{3+} in different lattices. It is seen that the value of g in our case is close to observed values of g of Cr^{3+} in different crystals.

REFERENCES

1. M.A. Verneuil, Ann. Chim. et phys. 3, 20 (1904)
2. M.W. Vernon and M.C. Lovell, J. Phys. Chem. Solid, 27 1125 (1966)
3. I.G. Austin, A.J. Springthorpe, B. A. Smith and C.E. Turner, Proc. Phys. Soc. 90, 159 (1967)
4. R.E. Cech and E.I. Alessandrini, Tran. A.M.S. 51, 150 (1959)
5. W. Rambauske and P.R. Gruenzel, Japan J. Appl. Phys. 35, 408 (1964)
6. Humihiko Takei, Japan J. Appl. Phys. 4, 152 (1965)
7. Humihiko Takei and Shigenao Koide, J. Phys. Soc. Japan 21, 1010 (1966)
8. C. Vande Stolpe, J. Phys. Chem. Solids 27, 1952 (1966)
9. K. Hauffe and J. Block, Z. Physik, Chem. 198, 232 (1951)
10. W.A. Fisher and G. Lorenz, Z. Physik, Chem. 18, 308 (1958)
11. W.C. Hagel and A.U. Seybolt, J. Electrochem. Soc. 108, 1146 (1961)
12. J.A. Crawford and R.W. Vest, J. Appl. Phys. 35, 2413 (1964)
13. W.C. Hagel, J. Appl. Phys. 36, 2586 (1965)
14. J.F. Garcia De La Banda, J.A. Pajara Somoano and J. Soria An. Real Soc. Espain, Fis. Quim, (Spain) 61 (A), 311 (1965)
15. I. Ursu, O. Pop, L.S. Stanescu & Indiu Pop, Rev. Roum, Phys. 11, 751 (1966)
16. Trounson, Bleil, Wangsness and Maxwell, Phys. Rev. 79, 542 (1950)
17. Okamura, Torizuka and Kojima, Phys. Rev. 82, 285 (1951)
18. L.R. Maxwell and T.R. McGuire, Revs. Modern Phys. 25, 279 (1953)
19. M. D'Abbigne and K.G. Srivastava, Jl. De Phys. 24, 75 (1963)
20. K.G. Srivastava and R. Srivastava, Nuovo Cimento, 39, 71 (1965)

21. B. Henderson and T.P.P. Hall, Proc. Phys. Soc. 90, Pt. 2, 511 (1967)

22. W. Low and R.S. Rubins, Paramagnetic Resonance Edited by W. Low, Vol. 1, pp 79

23. S.L. Hou, R.W. Summitt and R.F. Tucher, Phys. Rev. 152, 258 (1967)

24. H.J. Gerritsen, S.E. Harrinson, H.R. Lewis and J.P. Wittke, Phys. Rev. Letters 2, 153 (1959)

25. T.I. Barry, Solid State Commun. 4 123 (1966)

26. D.R. Hutton and G.J. Troup, Brit. J. Appl. Phys. 15 275 (1964)

27. S. Burns, D.F. O'Kane and R.S. Title, Phys. Letters 23, 56 (1966)

28. J.W. Orton, A.S. Fruin and J.C. Walling, Proc. Phys. Soc. 87, 703 (1966)

29. P.T. Squire and J.W. Orton, Proc. Phys. Soc. 88, 649 (1966)

SUBJECT INDEX TO VOL. I

O

P

R

S

T